C0-AUY-483

Wheat Growth and Modelling

NATO ASI Series

Advanced Science Institutes Series

A series presenting the results of activities sponsored by the NATO Science Committee, which aims at the dissemination of advanced scientific and technological knowledge, with a view to strengthening links between scientific communities.

The series is published by an international board of publishers in conjunction with the NATO Scientific Affairs Division

A Life Sciences **B Physics**	Plenum Publishing Corporation New York and London
C Mathematical and Physical Sciences	D. Reidel Publishing Company Dordrecht, Boston, and Lancaster
D Behavioral and Social Sciences **E Engineering and Materials Sciences**	Martinus Nijhoff Publishers The Hague, Boston, and Lancaster
F Computer and Systems Sciences **G Ecological Sciences**	Springer-Verlag Berlin, Heidelberg, New York, and Tokyo

Recent Volumes in this Series

Volume 82—Receptor-Mediated Targeting of Drugs
Edited by G. Gregoriadis, G. Poste, J. Senior, and A. Trouet

Volume 83—Molecular Form and Function of the Plant Genome
Edited by Lous van Vloten-Doting, Gert S. P. Groot, and Timothy C. Hall

Volume 84—Behavioral Epidemiology and Disease Prevention
Edited by Robert M. Kaplan and Michael H. Criqui

Volume 85—Primary Photo-Processes in Biology and Medicine
Edited by R. V. Bensasson, G. Jori, E. J. Land, and T. G. Truscott

Volume 86—Wheat Growth and Modelling
Edited by W. Day and R. K. Atkin

Volume 87—Industrial Aspects of Biochemistry and Genetics
Edited by N. Gürdal Alaeddinoğlu, Arnold L. Demain, and Giancarlo Lancini

Volume 88—Radiolabeled Cellular Blood Elements
Edited by M. L. Thakur

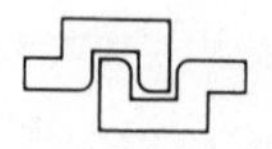

Series A: Life Sciences

Wheat Growth and Modelling

Edited by

W. Day

Rothamsted Experimental Station
Harpenden, United Kingdom

and

R. K. Atkin

Long Ashton Research Station
Bristol, United Kingdom

Plenum Press
New York and London
Published in cooperation with NATO Scientific Affairs Division

Proceedings of a NATO Advanced Research Workshop on
Wheat Growth and Modelling,
held April 9-12, 1984,
in Bristol, United Kingdom

Library of Congress Cataloging in Publication Data

NATO Advanced Research Workshop on Wheat Growth and Modelling (1984: Bristol, Avon)
Wheat growth and modelling.

(NATO ASI series. Series A, Life sciences; v. 86)
"Proceedings of a NATO Advanced Research Workshop on Wheat Growth and Modelling, held April 9-12, 1984, in Bristol, United Kingdom"—T.p. verso.
"Published in cooperation with NATO Scientific Affairs Division."
Bibliography: p.
Includes index.
1. Wheat—Growth—Mathematical models—Congresses. 2. Wheat—Growth—Data processing—Congresses. I. Day, W. II. Atkin, R. K. III. North Atlantic Treaty Organization. Scientific Affairs Division. IV. Title. V. Series.
SB191.W5N324 1984 633.1′1 85-3691
ISBN 0-306-41933-5

A Division of Plenum Publishing Corporation
233 Spring Street, New York, N.Y. 10013

Printed in the United States of America

. . . when they come to model Heaven and
calculate the stars, how they will wield the
mighty frame, how build, unbuild,
contrive to save appearances. . . .

—JOHN MILTON (1608–1674)

PREFACE

The concept of using mathematical models to investigate crop growth and productivity has attracted much attention in recent years. A major reason is that modelling can allow an expert in one area to assess the impact of his ideas in the light of other advances in our understanding of crop performance. Whether or not many of the claims made for or the demands made of models can ever be satisfied, this role as a focus for quantitative definition of crop growth is an important one. One consequence is that the development and appraisal of such models requires the efforts of scientists from a wide range of disciplines. This NATO Advanced Research Workshop was designed to bring together such a range of scientists to consider the wheat crop, and assess our understanding of the crop and our ability to model its growth and yield.

The ideas and organization behind the workshop involved many people. The U.K. interest in a computer model of wheat growth was instigated by Dr. Joe Landsberg (then of Long Ashton Research Station, but now Director of CSIRO Division of Forest Research), who in 1979 started a modelling exercise as part of a collaborative study of the causes of yield variation in winter wheat, involving four research institutes supported by the Agricultural and Food Research Council. Dr. Paul Bragg (then of Letcombe Laboratory) suggested the idea of an international meeting in 1981, and, between then and 1984, much thought and planning was put in by Drs Alan Weir and James Rayner, Rothamsted Experimental Station, Dr. John Porter, Long Ashton Research Station and Dr. Michael Kirby, Plant Breeding Institute. In the end, it was with financial support from the Scientific Affairs Division of NATO, and some assistance from the Underwood Fund, administered by the Agricultural and Food Research Council, that 39 scientists gathered at Burwalls in Bristol from 9-12 April, all agreed it was a most successful meeting.

The papers that follow in this publication speak for themselves. In addition many valuable contacts were made, and opportunities for scientific collaboration established. We hope there will be opportunities in future to follow up and extend some of the ideas and approaches considered here both through collaborative research and scientific meetings.

We would like to acknowledge the assistance given by Dr. Joe Ritchie (USA) and Dr. Herman van Keulen (Netherlands) who constituted an International Organizing Committee; all at the meetin were disappointed that Dr. van Keulen was prevented from attending because of illness. The smooth running of the meeting itself was du to the efforts of Dr. Roger Atkin and his staff from the Scientific Liaison Section at Long Ashton Research Station, and the staff of Burwalls, to all of whom the participants were most grateful. The preparation of these proceedings has been a substantial task, and a great education for the editors. We hope that the final product merits our efforts, and express our thanks to Janet Why for the production of the typescript, to Dick Chenoweth for artwork, to Virginia Day for advice on editing and indexing and to Anita Webb an Helen Clarke for keeping things on the strait and narrow.

W. Day
R.K. Atkin

December 1984

CONTENTS

ASSIMILATION AND CROP GROWTH

ENVIRONMENTAL RESPONSES OF PLANTS AND CROPS

WHEAT GROWTH AND MODELLING: AN INTRODUCTION

W. Day

Rothamsted Experimental Station
Harpenden, U.K.

Over the last twenty years there has been a tremendous growth of interest in crop models. This shows in the number of published models (for reference to some, see Legg, 1981; Baker, 1979) also in the increasing complexity of the models and the range of problems to which they are being applied. For the cotton crop, for example, Baker (1979) refers to four groups developing simulation models, and there are others. Many of these models contain routines defining upwards of eight plant, soil or atmospheric processes, and these models have been applied to such diverse problems as assessing the feasibility of various genetic changes to cotton (Landivar et al., 1983) and studying the effects of insect damage (Wallach, 1980). Elements of crop growth simulation have even been incorporated into a computer-based pest management scheme (Ives et al., 1984).

This NATO Advanced Research Workshop is concerned with progress and potential in the modelling of the wheat crop. Though simulation modelling of wheat has possibly not been as extensive as that of cotton, the crop has been the focus for many experimental and modelling studies. Some of the conclusions from wheat modelling have, as with cotton, now been included in a pest and disease control system (Rijsdijk et al., 1981) though at this meeting we will only consider the healthy crop. Underlying much of this work is the wish to establish quantitative relations between the environment, plant physiological processes and the growth and yield of crops, and then to put these relations together into general statements that define how wheat may be expected to perform in particular situations. In reviewing our progress in modelling wheat growth, we must first consider practical studies that show how well we understand the physiology of the crop, and then consider the performance of models of physiological processes and of crop growth. To assess potential for the future we must identify the problems to which models can appropriately be applied, and also the range of modelling techniques that is available. These various aspects of wheat growth modelling are the subject of our meeting.

PRINCIPLES OF MODELLING

General opinions of modelling and of comprehensive crop models in particular vary from an expectation of accuracy in relation to an problem, to scepticism as to whether there is any scientific merit i the technique. For the true value of modelling to be realised, it i essential to recognise the basic requirements of modelling and to ensure that the objectives of any particular study are matched by th modelling approach chosen.

Murray's (1981) guide to the art (*sic*) of good model building defines the following requirements:

1) A sound appreciation and understanding of the biological problem, though not necessarily including the most intricate details.

2) A realistic mathematical representation of the important phenomena.

3) Finding a solution, quantitative if possible, of the resulting mathematical problem.

4) A biological interpretation of the results, ideally giving biological insight and predictions.

These simple statements provide a good basis for anyone wishing to develop or use models of plant physiological processes and crop growth. They were formulated by a mathematician, but he recognised two important factors: that the mathematics is dictated by the biology, and that the mathematics may be trivial. The first two of his statements require that at this meeting we hear from plant and crop physiologists about their current appreciation of the important biological phenomena determining wheat growth and yield. The second and third require that we consider the mathematical analyses that ma be applied, and the form of the results they provide. It is the fourth statement that offers the greatest challenge; it requires us to define clearly our objectives when planning to model aspects of crop performance.

MODELLING IN PRACTICE

The objectives of any particular study must define the modelling methods, in particular the scale of the model employed. The range of interests of those present at this workshop cannot be satisfied by the same model, or even the same approach to modelling. The most wide-ranging approach is the comprehensive crop simulation model, which contains many descriptions of component processes and consequently many parameters. It is often considered as "a research tool developed for the enlightenment of research workers, and not directed to solving practical problems" (Legg, 1981), but such models can have specific applications in relation to agronomy (Chapter 23, 29 and 30). Despite their complexity, such crop models are never complete. They are based on empirical descriptions of processes at some lower level than whole crop performance. The proper definition of these descriptions, linked where possible to an understanding of their fundamental basis at a yet more detailed level of biological organization, is rightly the concern of separate exercises of experimentation and modelling. For some processes, such exercises can be undertaken in isolation from other aspects of crop performance: for example, the timing of phenological development stages (Chapter 2) seems largely independent of plant growth or nutrition. For other processes, the interactions are vitally important, and a good scientific basis for their description will require the definition of at least a subset of the elements in a comprehensive crop simulation model; e.g. in seeking to define how grain growth is limited (Chapter 12), the interaction between nitrogen demand of the grain and flag leaf photosynthesis (Chapter 14) may need to be taken specifically into account.

FINAL COMMENTS

Models of crop growth have had a mixed press. Passioura (1973) identified many dangers inherent in simulation modelling, largely resulting from the fact that "the gap between crop and yield biology and molecular and process biology is too wide, with the result that models are commonly too large and complex to be properly testable, or, if smaller, too simplistic to be at all realistic" (Lang and Thorpe, 1983). Yet we do need to gather quantitative information together clearly and concisely if we are to be able to make useful statements about the way wheat performs in different environments. It is sometimes claimed that crop simulation models will reduce the need for extensive, and expensive, field trials. If they do this it will not be by replacing those trials, but by aiding their planning and interpretation. The simulation model, like a good field trial, requires much effort to be put in if it is to be successful, and, like the results of many field trials, will be forgotten in ten years' time if there is not continuing effort to test it, refine it and apply it. Perhaps the major problem is construing the comprehensiveness of the crop simulation model to mean that such a model is the ultimate goal. Penning de Vries (1982) has identified not only the role for the comprehensive model, but also the need for summary models to make the model more accessible. Though comprehensive models can be used to study some agronomic factors, the summary model distilled from the more detailed approach is likely to be more appropriate to specific problems. In particular, for tactical decision making in crop management a flexible and informative model, using as much current information on crop status as is available, is required. Overall we must therefore not just build our quantitative descriptions of wheat growth into comprehensive models: we need also to develop models of component processes at a more detailed level, to ensure there are experimental programmes that test these models, and to identify summary models that may be more appropriate to particular applications.

Monteith and Elston (1971), in discussing how much easier it was to measure the environment than to predict how crops respond to it, identified the lack of contact between physicists and biologists as the root of the problem. Meetings such as this workshop will help to bridge that gap, and by doing so identify realistic approaches towards the goal of predicting crop responses.

REFERENCES

Baker, D. N., 1979, Simulation for research and crop management, in: "World Soybean Research Conference II: Proceedings". F. T. Corbin, ed., Westview Press, Boulder, Colorado.

Ives, P. M., Wilson, L. T., Cull, P. O., Palmer, W. A., Haywood, C., Thomson, N. J., Hearn, A. B., and Wilson, A. G. L., 1984, Field use of SIRATAC: an Australian computer-based pest management system for cotton, Prot. Ecol., 6:1.

Landivar, J. A., Baker, D. N., and Jenkins, J. N., 1983, Application of GOSSYM to genetic feasibility studies. II Analysis of increasing photosynthesis, specific leaf weight and longevity of leaves in cotton, Crop Sci., 23:504.

Lang, A., and Thorpe, M. R., 1983, Analysing partitioning in plants, Pl. Cell Environ., 6:267.

Legg, B. J., 1981, Aerial environment and crop growth, in: "Mathematics and Plant Physiology", D. A. Rose and D. A. Charles-Edwards, eds., Academic Press, London.

Monteith, J. L., and Elston, J. F., 1971, Microclimatology and crop production, in: "Potential Crop Production, A Case Study", P. F. Wareing and J. P. Cooper, eds., Heinemann, London.

Murray, J. D., 1981, Introductory remarks, Phil. Trans. R. Soc., Ser. B, 295:427.

Passioura, J. B., 1973, Sense and nonsense in crop simulation, J. Aust. Inst. agric. Sci., 39:181.

Penning de Vries, F. W. T., 1982, Phases of development of models, in: "Simulation of Plant Growth and Crop Production", F. W. T. Penning de Vries and H. H. van Laar, eds., Pudoc, Wageningen.

Rijsdijk, F. H., Rabbinge, R., and Zadoks, J. C., 1981, A system approach to supervised control of pests and diseases of wheat in the Netherlands, in: "Proceedings of the IXth International Congress of Plant Pathology", Washington.

Wallach, D., 1980, An empirical mathematical model of a cotton crop subjected to damage, Field Crop Res., 3:7.

SIGNIFICANT STAGES OF EAR DEVELOPMENT IN WINTER WHEAT

E. J. M. Kirby

Plant Breeding Institute
Cambridge, U.K.

INTRODUCTION

Wheat growth and yield simulation models often have an element which traces plant development, generally by taking a few selected development stages and estimating the time which elapses between them. The criteria which are used to select these stages, and thus the phases of development which they demarcate, are considered in this paper. Specific stages may be selected because they have some fundamental importance in the physiology of the plant and may signal a change from one phase of growth to another; for example, the onset of stem elongation, which may affect the partitioning of resources to other organs in the plant. Specifying stages that define the duration of a particular process allows numbers of parts formed to be estimated if a prediction of the rate of the process can also be made; for example, Rahman and Wilson (1978) have analysed numbers of spikelets per ear in terms of rate and duration between specified morphological stages (double ridge and terminal spikelet). It may also be necessary to identify developmental stages when there is a change in the response of development or growth to certain environmental factors; for example, after anthesis, development depends only on temperature and it does not respond to daylength, which affects the preceding phases of the life cycle.

Prediction of developmental stages may also have a direct relevance to crop husbandry where the purpose of the model is to predict stages which are important for a particular treatment; for example, application of growth regulator based herbicides which may cause damage if applied at the wrong stage. Similarly attainment of a certain stage may indicate a change in response to a stress factor;

for example, the greater susceptibility of the young apex to frost damage as the apex becomes florally initiated (George, 1982).

DESCRIPTION OF APEX AND SHOOT DEVELOPMENT

Central to any consideration of development are the changes which take place at the shoot apex or meristem. All the organs of the shoot are initiated at the shoot apex and the rate and duration of the formative processes determine the final number of organs. More detail than can be given in the present brief review may be found in Bonnet (1966), Williams (1975) or Kirby and Appleyard (1981). The initiation and development of the root system is dealt with in Chapter 8.

In the mature seed there is an embryo plant with an organized meristem which has already initiated the coleoptile and three or four foliage leaves. When the seed is sown and has imbibed water, growth and elongation of the coleoptile and the leaves resumes, and the meristem initiates further leaves. After a period of time, depending on temperature and depth of sowing, the seedling emerges. At this stage the apex is conical (Fig. 1a) and continues to initiate further leaves. Growth of the leaf initials takes place and leaves emerge in succession. The leaf initiation phase continues for a variable period which may be expressed in time, accumulated temperature or phyllochrons (a phyllochron is the period from the emergence of one leaf until the emergence of the next leaf). During this phase tiller buds are initiated and tillers may emerge from the ensheathing main shoot leaf. Each tiller develops in a more or less similar manner to the main shoot.

When a certain number of leaves has been initiated, the apex begins to produce spikelet primordia and it becomes elongated and cylindrical in form (Fig. 1b). As initiation continues the spikelets begin to differentiate and the double ridge stage is seen (Fig. 1c). Subsequently, each spikelet primordium initiates the glumes and a number of floret initials (Fig. 1d). Each of these floret initials produces primordia which in turn form the lemma, palea and other floral structures. The last formed spikelet, the terminal spikelet, is characteristic in that it forms at right angles to all other spikelets on the ear (Fig. 1d). In the phase between reaching total number of spikelets and anthesis, floret initiation continues for a period and the first initiated florets continue their development. Some of the more distal florets die in the period before anthesis. After differentiation and growth of the stamens and carpel, meiosis takes place, followed by differentiation of the pollen and the embryo sac. Stem growth starts during the spikelet initiation phase, at about the stage when the lemmas become visible. The main period of growth of the stem occurs during the period from terminal spikelet to anthesis.

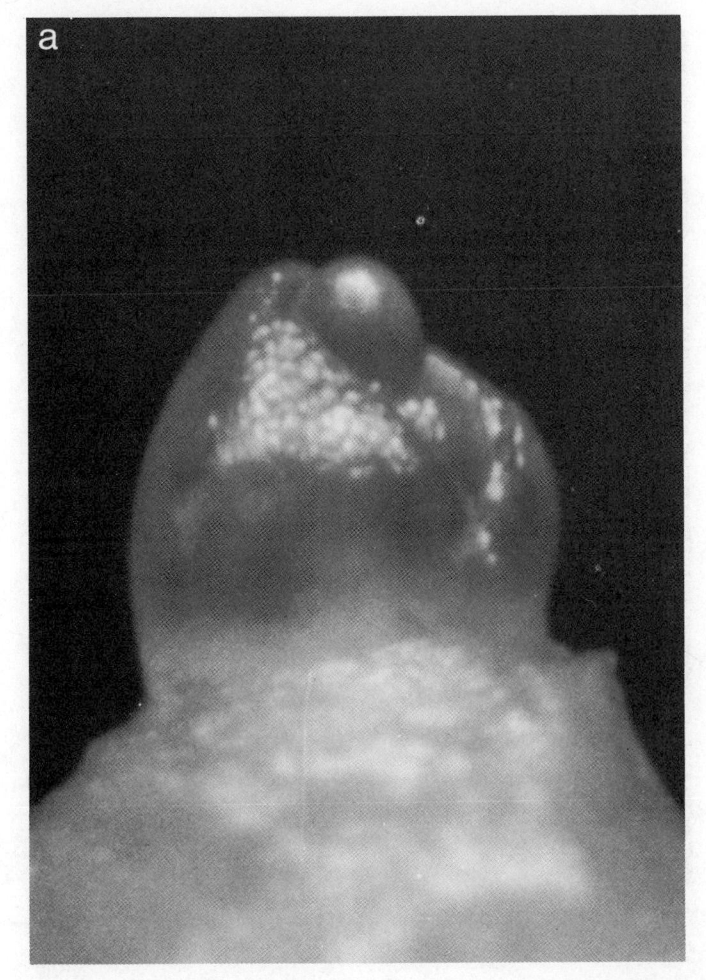

Fig. 1. Stages of development of the wheat shoot apex
(a) Vegetative stage: the apex is initiating leaf primordia

In the final phase, between anthesis and plant maturity, development and growth are restricted almost completely to the fertilized carpels in which the endosperm develops. The embryo differentiates a shoot apex which initiates leaves and a well defined radicle.

During the process of development outlined above, certain stages can be precisely defined, easily recognized by their morphological appearance, and are of fundamental physiological significance. Anthesis is one such stage. It can be recognized without difficulty because of the characteristic changes in stamen morphology, can be timed to the nearest minute in an individual floret, and marks the

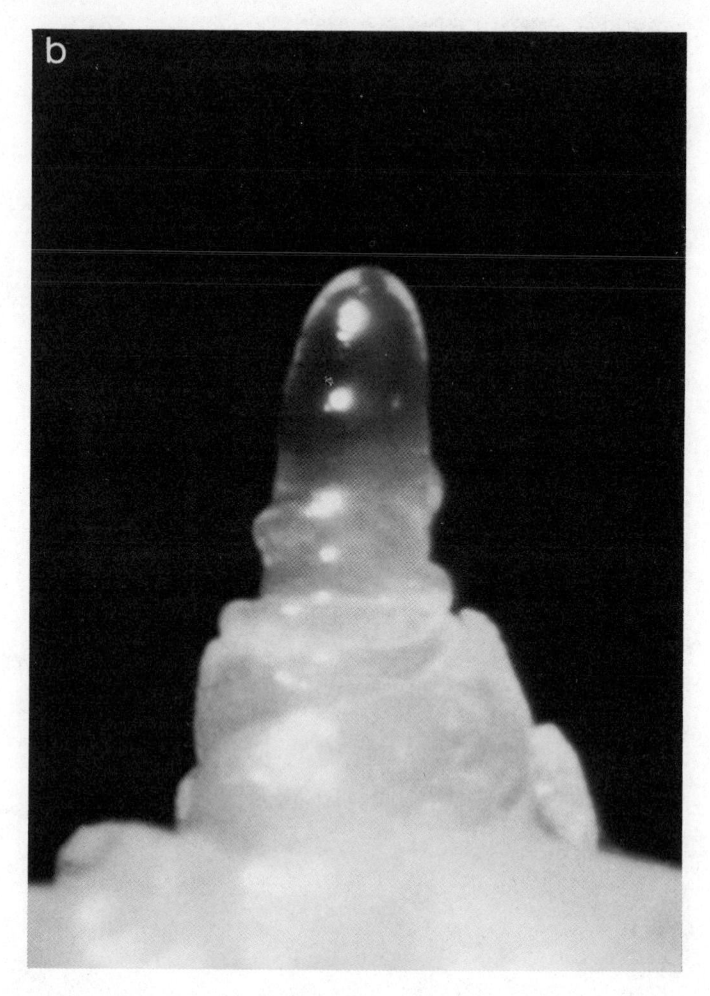

Fig. 1. (cont.) Stages of development of the wheat shoot apex
(b) Elongation of the apex: the primordia initiated at this stage will develop into spikelets

beginning of the important phase of grain growth. In other parts of the life cycle, significant changes in physiology are not associated with an easily definable, characteristic stage. For example, the transition to floral development occurs when the apex still appears vegetative in form and the changes are gradual and difficult to define. In such cases a definition based on criteria other than morphology may be necessary. To illustrate the points raised here, the process of leaf and spikelet initiation and of ear growth are analysed in more detail to identify and define possible significant stages.

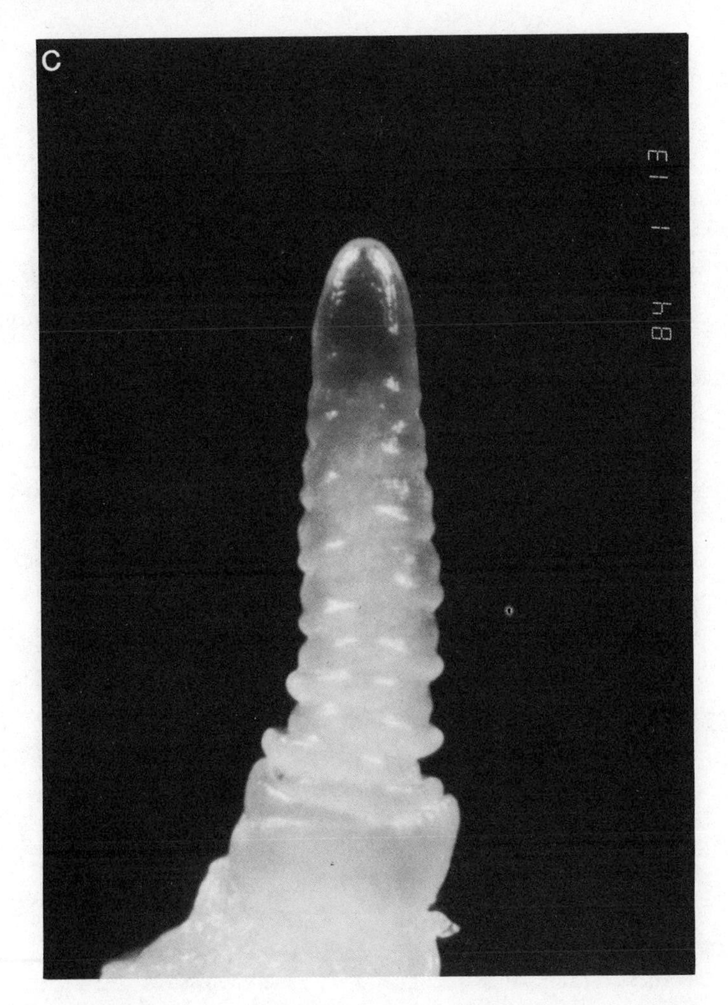

Fig. 1. (cont.) Stages of development of the wheat shoot apex
(c) Double ridge stage

APEX DEVELOPMENT IN FIELD CONDITIONS

Frequent counts of the total number of primordia initiated at the shoot apex may be plotted against time or against thermal time to show the progress of apex development. When plotted against time, the rate of primordium initiation is much affected by temperature. During the warm days of September and early October the rate of primordium initiation is relatively fast; it slows down as temperature drops during winter and increases again as spring approaches. When the variation in temperature is accounted for by plotting number of primordia against thermal time, the rate of primordium initiation is seen to fluctuate less and generally to

Fig. 1. (cont.) Stages of development of the wheat shoot apex
(d) Terminal spikelet stage

increase throughout the life cycle (Fig. 2; Baker, 1979). By linking studies of apex morphology to observation of the final destiny of the primordia which are initiated, it is possible to relate the change in primordium number to the number of leaves formed, and the time when double ridge and terminal spikelet stages are attained. The double ridge stage does not occur until about three-quarters of the total primordia have been initiated. This stage is often regarded as the first sign of floral initiation, but it is also possible to define the beginning of ear initiation as the time when the first spikelet is initiated. This can be done retrospectively, after the total number of leaves formed upon the shoot is known, although this is only possible at a fairly late stage of apex development, when the

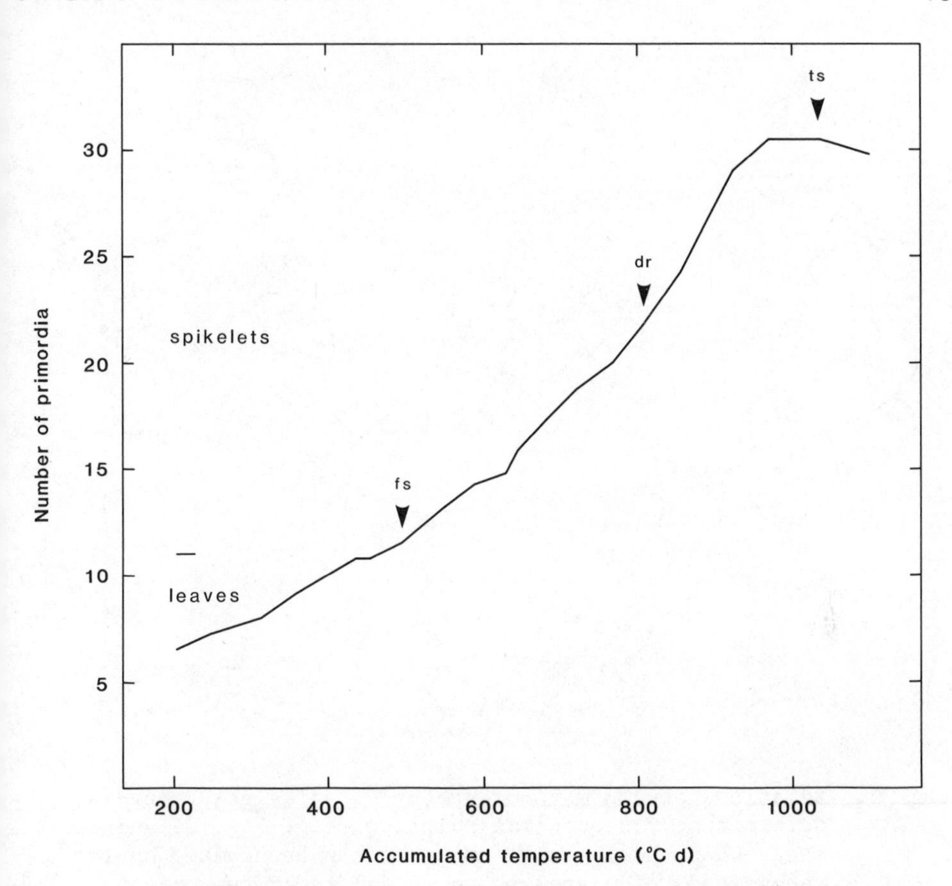

Fig. 2. Number of primordia v accumulated temperature for winter wheat cv. Norman, sown 15 October 1981. The shoot apex initiated 11 leaves. The time of initiation of the first spikelet (fs), and the times of occurrence of the double ridge (dr) and terminal spikelet (ts) stages are shown.

difference between leaf and spikelet primordia can be distinguished with confidence. The final number of leaf primordia is reached some time before the double ridge stage (Fig. 2). In winter wheat the double ridge stage usually occurs when about half the spikelets have been initiated (see also Baker and Gallagher, 1983a). Terminal spikelet stage is often noted at some time after the number of spikelets has reached its maximum (Fig. 3).

Initiation of Leaves

Leaves initiate at a more or less uniform rate (Fig. 2; Stern and Kirby, 1979). This rate depends on temperature, varying over the

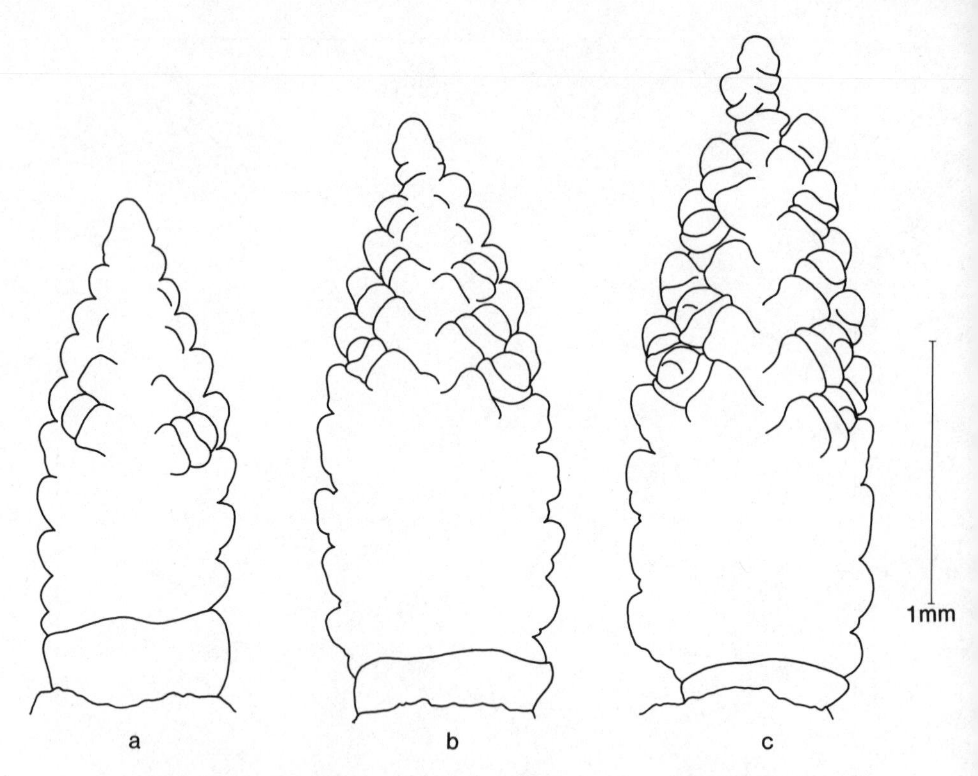

Fig. 3. Relation between maximum number of spikelet primordia, determined by sequential sampling (e.g. Fig. 2) and the stage when the terminal spikelet can be unambiguously recognised. The apex shown in (a) is at the end of spikele initiation, but the terminal spikelet is not yet distinguishable. The last primordium will form the lower glume of the terminal spikelet, but this cannot be determined morphologically. The terminal spikelet can be distinguished with reasonable confidence in the type of ape shown in (b) but the glumes are not clearly defined. The formal definition of terminal spikelet stage is shown in (c), in which the two glumes and the first floret are identifiable with certainty.

normal range experienced in the field from about zero at 0°C up to about 0.4 primordia per day at 15°C. Assuming a linear response to temperature (Baker and Gallagher, 1983b), the base temperature appears to be about zero and the rate increases by about 0.03 primordia for each degree rise in temperature. Some leaf primordia are initiated before the plant emerges so they are not influenced by daylength, and there is little or no effect of daylength on those primordia that are initiated subsequently (unpublished data and Austin, personal communication); neither has vernalization been show to affect the rate of leaf initiation. The total number of leaves is affected by vernalization and photoperiod and therefore the duration of the leaf initiation phase will depend upon the response of number of leaves to these factors.

Leaf Initiation and Leaf Emergence

The rate of leaf emergence is slower than that of leaf initiation so that the period of leaf appearance encompasses not only the leaf initiation phase but also ear initiation and stem elongation phases. The mechanism by which this difference in rates is accomplished is not solely by difference in leaf growth rates, but by each successive leaf entering a period of low growth rate. At the end of this period, the duration of which depends on leaf position, the relative growth rate increases and the leaf eventually emerges (Williams, 1975). Leaf emergence rates differ both between environments and genotypes. The rate of leaf appearance is remarkably constant through the season when expressed in thermal time and in the field may depend on rate of change of daylength at plant emergence (Baker et al., 1980; Kirby et al., 1984).

Initiation of Spikelets

Generally the transition from leaf to spikelet initiation is marked by an increase in the rate of primordium production. At a constant temperature the rate during the spikelet initiation phase is more or less uniform. The minimum temperature for initiation is about 0°C and with increasing temperature the rate increases by about 0.07 primordia per day per °C to a maximum at 20 to 25°C (Baker, 1979; Rahman and Wilson, 1978; Halse and Weir, 1974). The rate of initiation increases with increasing daylength (Fig. 4; Holmes, 1973; see also Allison and Daynard, 1976; Rahman and Wilson, 1977; Rawson, 1971). The response to daylength is dependent on genotype (Holmes, 1973). The process by which the total number of spikelets is determined has not been satisfactorily explained. In controlled environments, low initiation rates (e.g. in response to short days) are associated with high numbers of spikelets, but also with high numbers of leaves.

Floret Initiation and Death

Floret initiation begins in the spikelets in the lower mid-part of the ear before spikelet initiation is complete; by the time the last spikelet is initiated these spikelets have about two to three florets. Floret initiation continues for some time after this until about eight to ten florets are initiated and then in the period before anthesis usually more than half of these primordia die leaving

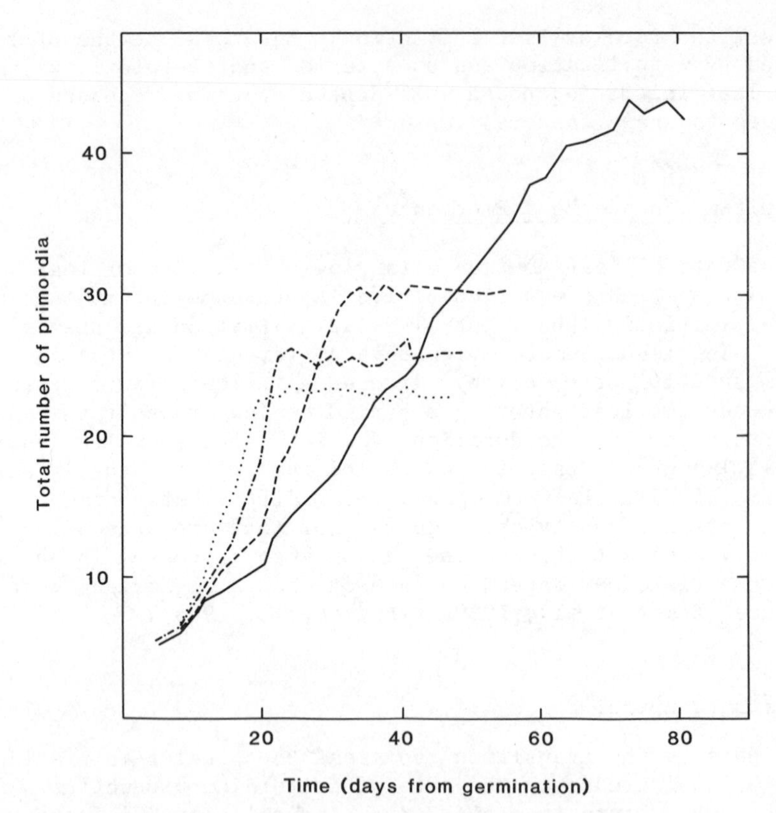

Fig. 4. Effect of daylength on primordium initiation in Marquis spring wheat. ——— 8 hr photoperiod; ----- 12 hr photoperiod; -•-•- 16 hr photoperiod; ••••• continuous light. (Modified from Holmes, 1973.)

between three and six potentially fertile florets. The effect of th environment on the initiation phase has not been investigated.

Growth Rate of the Developing Ear

There is evidence that there is an intrinsic change in the growth rate of the ear during development. This is supplemented by data from barley where it has been demonstrated that there is an increase in relative growth rate, apparently coinciding with the end of ear initiation (Kirby, 1974) and anatomical studies have shown that this may be related to changes in cell division and cell growth (Nicholls and May, 1960). Studies of ear growth in wheat by Macdowall (1973) revealed a similar change in relative growth rate but he speculated that the change in relative growth rate occurred a the double ridge stage (Fig. 5). If such an increase in relative

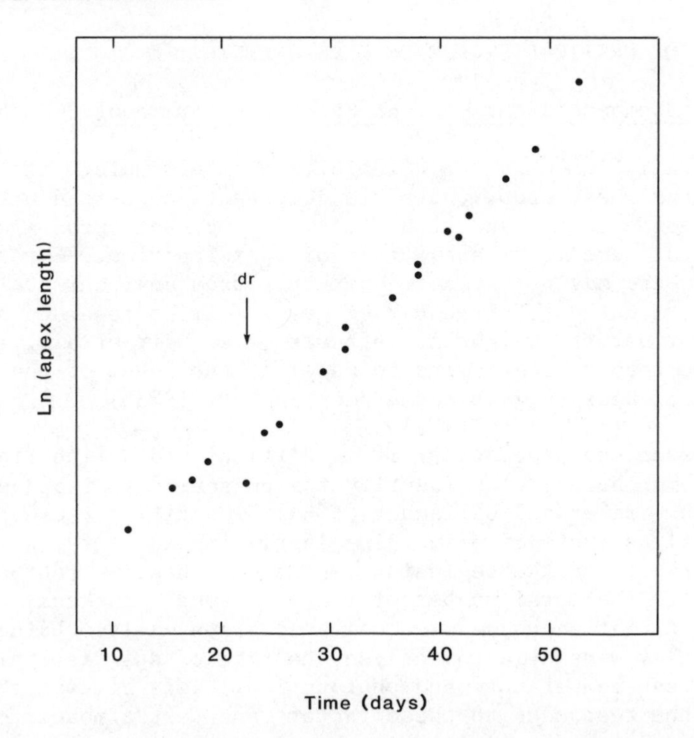

Fig. 5. Log (apex length) v time to show the increase in relative growth rate at the double ridge stage. (From Macdowall, 1973.)

growth rate takes place, it has important implications for the absolute growth rate. There will be a more marked increase after a change in relative growth rate than if no such change occurred.

Stem Elongation

The lowermost nodes of the stem bear tiller buds and the internodal regions do not elongate. Stem growth is generally confined to the last six internodes. The time at which the phase of internode elongation begins is associated with ear development, and field studies have shown that internode elongation occurs at about the lemma stage of ear development or slightly earlier (Tottman, 1977). Rapid stem growth begins at the terminal spikelet stage. Thus the ear growth phase, during which the stem is the major consumer of dry matter, occurs between the completion of spikelet initiation and anthesis.

SELECTION OF CRITICAL STAGES OF DEVELOPMENT

Predicting Numbers of Leaves and Spikelets per Shoot

Generally in any model which attempts to simulate the growth ar yield of the wheat crop, there are components concerned with the photosynthetic canopy and with yield. At present these are usually simulated at some level above that of apex function. For example, leaf emergence may be estimated directly from environmental parameters and the total number of grains may be regarded as a function of ear dry weight at anthesis (e.g. Weir et al., 1984). There have been some attempts to model at the level of the activity of the shoot apex (e.g. Baker and Gallagher, 1983b).

As described previously, it is difficult to define stages based solely on morphology that identify the phases of leaf and ear initiation. Baker and Gallagher (1983a) identified these phases frc a numerical analysis of primordium initiation identifying the end of the leaf phase and the beginning of the ear phase in retrospect, by reference to the total number of leaves formed. The end of the spikelet initiation phase was estimated graphically. Using this approach they were able to analyse the rate of spikelet initiation i terms of mean temperature and photoperiod. This allowed them to estimate the responses to these factors and test a model to estimate the rate of primordium initiation based on the multiplicative effect of daylength and temperature.

Plant Development in Relation to Numbers of Tillers and of Potential Florets

The initiation and morphology of the tiller bud has been described by Williams et al. (1975) but there are no data about the relationship between leaf growth and tiller bud growth and emergence The theoretical analysis of the relationship between leaf emergence and tiller emergence was made by Friend (1966). This assumed that a constant number of additional leaves emerged between the emergence c a leaf and the emergence of its subtending tiller (the leaf interval or phyllochron interval) and that the rate of emergence of leaves wa the same on the main shoot and on the tillers. Using this form of analysis Masle-Meynard (1978) and Kirby et al. (1984) found that, using a phyllochron interval of three, the observed number of tiller was in good agreement with prediction (Fig. 6).

Cessation of tillering is generally considered to be related to stage of development. In the data shown in Fig. 6 there was a stron correlation between the number of emerged leaves when tillering fell below potential and the number of leaves formed on the main shoot (r=0.699; 62 d.f.) and between number of leaves on the main shoot an the number of the first elongated internode (r=0.915; 62 d.f.).

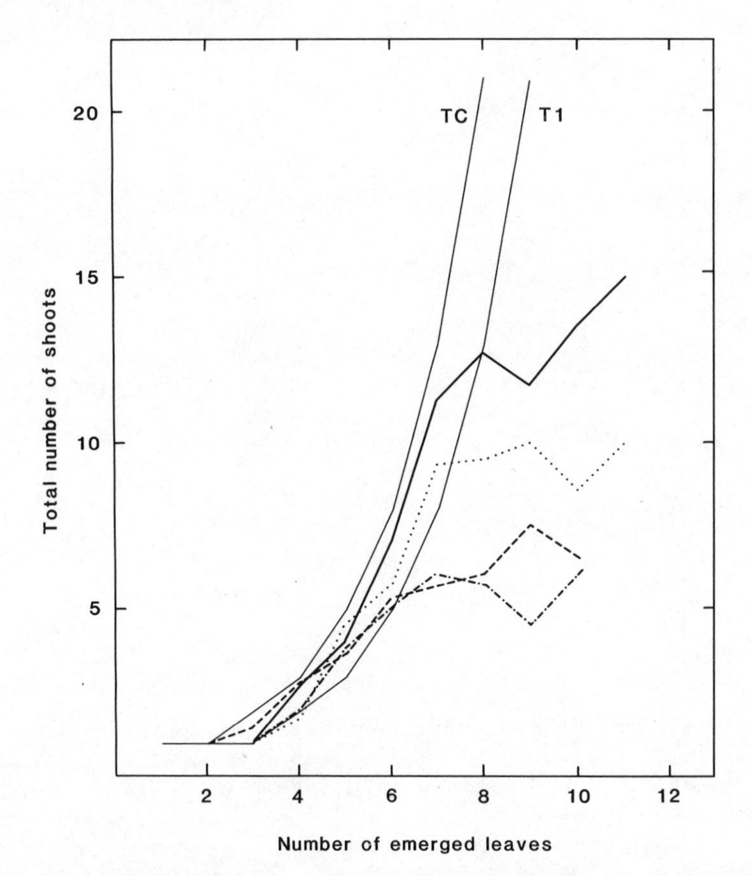

Fig. 6. Number of shoots v number of emerged leaves for cv. Maris Huntsman sown at different dates. ——— September; ----- October; -•-•- November; ••••• February. The thinner, continuous lines are the predicted number of shoots, assuming that TC or T1 were the first tiller to emerge. (See also Chapter 4.)

Baker and Gallagher (1983a) have reported that tillering ceases at the double ridge stage. Such a relationship is assumed in the AFRC model (Weir et al., 1984) but it may not apply generally. Darwinkel (1978) found that tiller production ceased at a progressively later date with decreasing rate of sowing (Fig. 7). Apex stages were not reported in his paper and although plant density is known to have some effect on cereal development (Kirby and Faris, 1970), the differences in the date at which a particular stage occurred were likely to have been small compared with the differences in the date of maximum number of tillers. In barley and wheat, tiller, floret and spikelet death may be the result of the potential growth of the plant exceeding available resources, so there is competition between

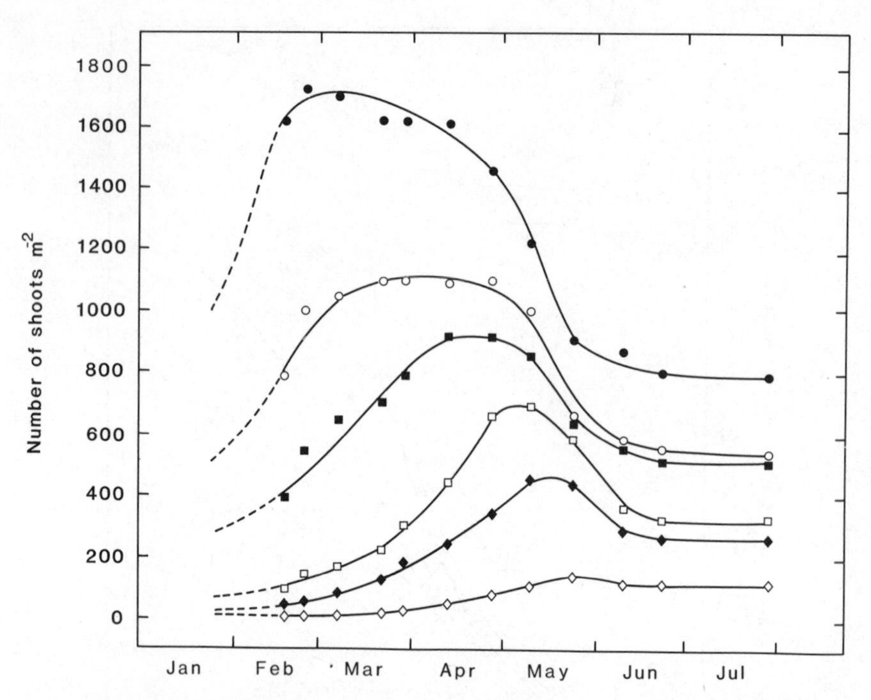

Fig. 7. Number of shoots v date for different plant populations of winter wheat. ◇ 5; ◆ 25; □ 50; ■ 200; ○ 400; and ● 800 plants m^{-2}. (Modified from Darwinkel, 1978.)

and within shoots for these resources. This hypothesis is based on the fact that the relatively high growth rate of the stem and the ea occurs at a time when the leaf area is not increasing (Fig. 8) and coincides with the death of florets and spikelets. Experiments base on reducing assimilation by shading (Fischer, 1975) or removing competing shoots (in barley) genetically or surgically (Kirby, 1974; Kirby and Jones, 1976) tend to confirm this hypothesis. Therefore the time at which tillering ceases, which determines the number of tillers produced, may not be associated with a particular developmental stage. Rather the cessation of tillering and the subsequent death of tillers and florets may depend upon environmenta and genetical factors which determine the rate of carbon assimilation, or the rate of uptake of mineral nutrients or water an the intensity of growth of the stem and the ear. A relationship of this sort has been demonstrated in an experiment in which the effect of the dwarfing gene (Rht_2/Gai_2) was investigated (Brooking and Kirby, 1981). This gene had no effect upon ear or spikelet development, which was similar in all the genotypes in the experiment, with or without the dwarfing gene. In the ear growth phase, the ear/stem growth rate ratio was on average 0.45 in those genotypes with the dwarfing gene, compared with 0.38 in the tall

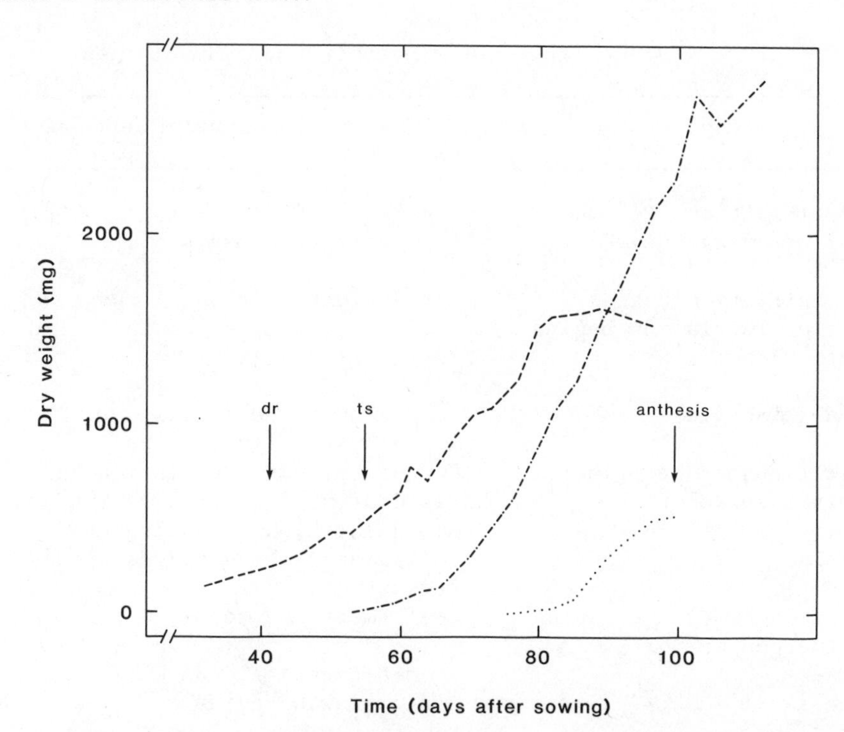

Fig. 8. Dry weight v time for leaves (- - -), stem (-•-•-) and ear (•••••) of winter wheat. The time of double ridge stage (dr), terminal spikelet stage (ts) and anthesis are shown (from Brooking and Kirby, 1981, by permission of the Cambridge University Press).

genotypes. This difference was associated with a difference in floret survival and there were more potentially fertile florets per spikelet and more grains per ear in the dwarf than in the tall genotypes.

CONCLUSIONS

The development of the wheat plant is a continuous process but a number of important stages can be identified (Table 1). These stages may refer to the morphology of the shoot apex which reflects significant changes in its growth and development or to organs, the development of which are correlated with that of the apex. Some stages involving the estimation of rate or duration may better be defined by numerical analysis rather than morphological events.

Finally, some significant stages in plant development (e.g. the onset of tiller death) may not be uniquely associated with a

Table 1. Some important stages in wheat development

Stage	Main events of development and growth
Seed imbibition	Beginning of leaf initiation phase
Seedling emergence	Leaf emergence begins
Leaf initiation ceases	End of leaf initiation phase
Spikelet initiation begins	Beginning of spikelet initiation phas
Double ridge	
Lemma primordium	Stem elongation begins
	Floret initiation begins
Maximum number of primordia	End of spikelet initiation phase
Terminal spikelet	Rapid stem elongation begins
	Rapid ear elongation begins
Meiosis	Pollen and embryo sac development begins
	Leaf emergence ceases
Ear emergence	
	Stem elongation ends
Anthesis	Grain growth begins
Physiological maturity	Grain growth ceases

particular morphological stage of the apex, but may be determined by the interaction of growth and development and assimilation processes

REFERENCES

Allison, J. C. S., and Daynard, T. B., 1976, Effect of photoperiod o development and number of spikelets of a temperature and some low-latitude wheats, Ann. appl. Biol., 83:93.

Baker, C. K., 1979, The environmental control of development in winter wheat, Ph.D. thesis, University of Nottingham.

Baker, C. K., and Gallagher, J. N., 1983a, The development of winter wheat in the field. I. Relation between apical development and plant morphology within and between seasons, J. agric. Sci., Camb., 101:327.

Baker, C. K., and Gallagher, J. N., 1983b, The development of winter wheat in the field. II. The control of primordium initiation rate by temperature and photoperiod, J. agric. Sci., Camb., 101:337.

Baker, C. K., Gallagher, J. N., and Monteith, J. L., 1980, Daylength change and leaf appearance in winter wheat, Pl. Cell Environ., 3:285.

Bonnet, O. T., 1966, Inflorescences of maize, wheat, rye, barley and oats: their initiation and development, University of Illinois College of Agriculture, Agricultural Experiment Station Bulletin, 721.

Brooking, I. R., and Kirby, E. J. M., 1981, Interrelationships between stem and ear development in winter wheat: the effects of a Norin 10 dwarfing gene, Gai/Rht_2, J. agric. Sci., Camb., 97:373.

Darwinkel, A., 1978, Patterns of tillering and grain production of winter wheat at a wide range of plant densities, Neth. J. agric. Sci., 26:383.

Fischer, R. A., 1975, Yield potential in a dwarf spring wheat and the effects of shading, Crop Sci., 15:607.

Friend, D. J. C., 1966, The effect of light and temperature on the growth of cereals, in: "The Growth of Cereals and Grasses", F. L. Milthorpe and J. D. Ivins, eds., Butterworths, London.

George, D. W., 1982, The growing point of fall-sown wheat: a useful measure of physiologic development, Crop Sci., 22:235.

Halse, N. J., and Weir, R. N., 1974, Effects of temperature on spikelet number of wheat, Aust. J. agric. Res., 25:687.

Holmes, D. P., 1973, Inflorescence development of semidwarf and standard height wheat cultivars in different photoperiod and nitrogen treatments, Can. J. Bot., 51:941.

Kirby, E. J. M., 1974, Ear development in spring wheat, J. agric. Sci., Camb., 82:437.

Kirby, E. J. M., and Appleyard, M., 1981, Cereal Development Guide, NAC Cereal Unit, Stoneleigh.

Kirby, E. J. M., Appleyard, M., and Fellowes, G., 1984, Leaf emergence and tillering in barley and wheat, Agronomie, (submitted).

Kirby, E. J. M., and Faris, D. G., 1970, Plant population induced growth correlations in the barley plant main shoot and possible hormonal mechanisms, J. exp. Bot., 21:787.

Kirby, E. J. M., and Jones, H. G., 1976, The relations between the main shoot and tillers in barley plants, J. agric. Sci., Camb., 88:381.

Macdowall, F. D. H., 1973, Growth kinetics of Marquis wheat. V. Morphogenetic dependence, Can. J. Bot., 51:1259.

Masle-Meynard, J., and Sebillotte, M., 1981, Etude de l'hétérogénéité d'un peuplement de blé d'hiver. II. Origine des différentes catégories d'individus du peuplement; éléments de description de sa structure, Agronomie, 1:217.

Nicholls, P. B., and May, L. H., 1963, Studies on the growth of the barley apex. I. Interrelationships between primordium formation, apex length, and spikelet development, Aust. J. Biol. Sci., 16:561.

Rahman, M. S., and Wilson, J. H., 1977, Determination of spikelet number in wheat. I. Effect of varying photoperiod on ear development, Aust. J. agric. Res., 28:565.

Rahman, M. S., and Wilson, J. H., 1978, Determination of spikelet number in wheat. III. Effect of varying temperature on ear development, Aust. J. agric. Res., 29:459.

Rawson, H. M., 1971, An upper limit for spikelet number per ear in wheat, as controlled by photoperiod, Aust. J. agric. Res., 22:537.

Stern, W. R., and Kirby, E. J. M., 1979, Primordium initiation at the shoot apex in four contrasting varieties of spring wheat in response to sowing date, J. agric. Sci., Camb., 93:203.

Tottman, D. R., 1977, The identification of growth stages in winter wheat with reference to the application of growth-regulator herbicides, Ann. appl. Biol., 87:213.

Weir, A. H., Bragg, P. L., Porter, J. R., and Rayner, J. H., 1984, A winter wheat crop simulation model without water or nutrient limitations, J. agric. Sci., Camb., 102:371.

Williams, R. F., 1975, The shoot apex, leaf growth and crop production, J. Aust. Inst. agric. Sci., 41:18.

Williams, R. F., Sharman, B. C., and Langer, R. H. M., 1975, Growth and development of the wheat tiller. I. Growth and form of the tiller bud, Aust. J. Bot., 23:715.

ABOUT THE LEAF-DAYLENGTH MODEL UNDER FRENCH CONDITIONS

R. Delécolle*, F. Couvreur+, P. Pluchard[x], and C. Varlet-Grancher‡

*INRA Station de Bioclimatologie
Montfavet, France

+ITCF Station Expérimentale de Boigneville
Maisse, France

[x]INRA Station d'Amélioration des Plantes Mons en Chaussée
Péronne, France

‡INRA Laboratoire de Bioclimatologie Mons en Chaussée
Péronne, France

INTRODUCTION

Leaf appearance is an important part of the pre ear-emergence phase in wheat and other cereal crops, and a knowledge of its relationship to external (chiefly climatic) factors is important if a general growth and development model is to be made.

The dependence of leaf appearance rate upon temperature is known from studies in controlled conditions (Friend et al., 1962) and from field experiments (Gallagher, 1979; Delécolle and Gurnade, 1980). It was shown that leaf appearance stage was closely and linearly related to thermal time from sowing. Thermal time is equivalent to accumulated temperature above a certain base temperature (Gallagher, 1979).

However, the slope of this linear relationship, i.e. the thermal rate of appearance, is not constant over all experiments, and Baker et al. (1980) have shown that this rate is correlated to the rate of change of daylength, measured at emergence, for cultivar Huntsman. This result has been confirmed by Kirby et al. (1982) on barley.

From field experiments in several locations at various latitudes in France, we show that a general expression for leaf appearance rate is applicable to a range of genotypes, although daylength change

Table 1. Geographical co-ordinates of stations

	Latitude (deg)	Longitude (deg)
Auzeville	43.55	-1.50
Boigneville	48.00	-2.37
Greoux	43.78	-5.87
Mons	52.90	-3.20

seems not to affect low-photosensitive genotypes in southernmost experiments.

MATERIAL AND METHODS

The leaf number on the main shoot was recorded in four locations over a range of latitudes (Table 1) and genotypes (Table 2). Three locations were supervised by ITCF during three years, and one by INRA during two years. Every year, various sowing dates were used in an attempt to represent the possible crop managements in the different areas where the experiments were sited. The dates of sowing ranged from 15 September to 22 December.

Observations were made on large plots (60 m^2) in Auzeville, Boigneville and Greoux, and nitrogen was applied twice, the amounts being computed in order to achieve a yield of 7.5 to 8 t ha^{-1}. Smaller plots (20 m^2) were used in Mons, with four replicates. Observations were made from one to three times a week, when fifteen to twenty plants were sampled on two adjacent rows, without

Table 2. Vernalization characteristics of genotypes studied; the response to vernalization varies from a strong response (++++) to no response (•).

Genotype	Response
Adam	++++
Arminda	+++
Capitole	+++
Courtot	•
Fidel	+
Maris Huntsman	+++
Talent	++
Top	++
Wattines	++
713	++++

replication at Auzeville, Boigneville and Greoux. On each plant, the main shoot was identified and the number of leaves observed: a leaf was considered as emerged when its tip was visible without unrolling the preceding leaf sheath. According to sampling studies by ITCF (Ingoat and Couvreur, 1979), the number of observed plants allows a precision of 5 to 10 per cent on leaf number determination.

RESULTS

Leaf Emergence and Thermal Time

Figure 1 shows the well-known relationship between successive average numbers of emerged leaves (leaf appearance status) and thermal time: a linear fit gives quite good correlation coefficients and the observed slope of the line is an estimate of the rate of leaf emergence (leaves (°C d)$^{-1}$). Figure 1 also illustrates the variation of this rate at Boigneville for the genotype Talent, when sown on different dates.

As mentioned above, the best way to take account of these variations is to plot the rate of leaf appearance against the rate of daylength change at emergence, which has been estimated by the difference between daylength five days before and five days after emergence, normalized to one day.

Thermal Rate of Appearance and Daylength

If the data set of available genotypes is split into two groups, low- and high-vernalization-requiring, the corresponding relationships between rate of appearance and daylength change are

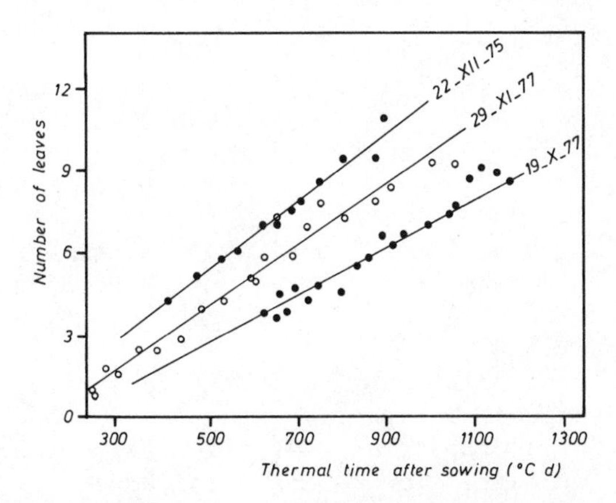

Fig. 1. Number of leaves on cv. Talent against thermal time (base 0°C) for various sowing dates at Boigneville.

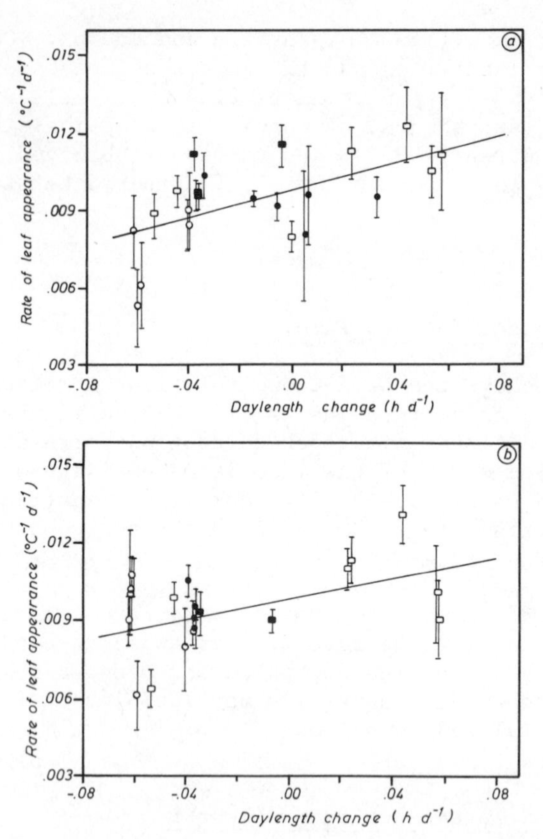

Fig. 2. Rate of leaf appearance plotted against daylength changes at emergence for (a) low vernalization genotypes (b) high vernalization genotypes: each point is plotted with it confidence interval (P=0.05).
○ = Mons □ = Boigneville ■ = Auzeville ● = Greoux
The variance accounted for, r^2, and the confidence interval of the slope (at P=0.05) are
a) $r^2 = 0.29$; b = 0.010 to 0.043
b) $r^2 = 0.40$; b = 0.003 to 0.0038

shown in Fig. 2a and b, for all available experiments. (All replicates are thus given the same weight, regardless of how each genotype is represented.)

The two sets of points may be fitted by straight lines with significant slopes (b), but poorly explained variances. From Fig. 2a, which has a high proportion of replicates of the genotype Talent, the consistency of regression relies upon results from Boigneville and Mons, as the variation in the subset from Greoux and Auzeville is not accounted for by the regression. It is clear from the confidence intervals that the slopes for the two groups of genotypes are not significantly different.

DISCUSSION

The use of field results to assess the relationship between leaf appearance and daylength change involves considerable variability. However, the individual values of leaf emergence rates are generally (but not always) precise. Observational error may have contributed to the large confidence intervals, though these may also reflect irregularities in sowing depths.

Another cause of error is the determination of the time of emergence. Depending on the time of year at which emergence occurs, the error in daylength change due to under- or over-estimating emergence time may vary dramatically. However this error is only significant around the period with minimum daylength, which corresponds to the mean point of the regression line: the importance of possible outliers is consequently reduced.

From the confidence intervals of the regression coefficients (Fig. 2a and b), it appears that the fitted lines are not statistically different and agree with the results of Baker et al. (1980) (Fig. 3). An average value over all genotypes and locations would be b = 0.03 though it should be remembered that, for French conditions, the confidence interval for this value is at least 0.01 to 0.05. In the same way, the average intercept is approximately 0.01.

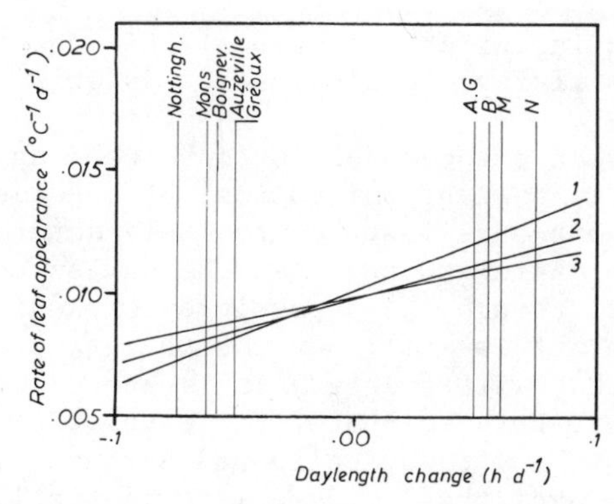

Fig. 3. Regression lines and maximum range of daylength change for the experimental stations.

1. Baker et al., 1980.
2. Low vernalization genotypes.
3. High vernalization genotypes.

Examining the results for Talent in the southernmost experiments, Fig. 2a shows that, even with generally good precision on individual regression coefficients, the results from Auzeville and Greoux would fit a line with zero slope. As these results are exclusively concerned with replicates of Talent, it could then be deduced that, for this genotype which has a low response to photoperiod (Hunt, 1979), and for latitudes where the range of daylength change is reduced (Fig. 3), no account needs to be taken of daylength in a temperature-leaf emergence model.

The lack of precision of the present results makes it inappropriate to consider in detail the best base temperature for calculations of thermal time. Increased precision on the individual leaf-thermal time slopes would not strengthen the relationship between daylength change and rate of leaf emergence over the entire set of data, so a method for determining an optimum base such as given by Kirby et al. (1982) would not be appropriate. Consequently, it is advisable to use zero as the base.

A simplified model can thus be proposed, where leaf number LN is a function of time of emergence, t_e, and thermal time $\sum_j (T_j - T_0)$; with $T_0 = 0$:

$$LN = \phi(t_e) \sum_j T_j$$

and where $\phi(t_e)$ is determined as:

$$\phi(t_e) = b\ (dD/dt)_e + a,$$

$(dD/dt)_e$ being the rate of daylength change at emergence, a = 0.01 and b = 0.03.

If applied to individual genotypes, this model should consider photoperiodic sensitivity, in a way which is still to be studied.

Finally, if such a sub-model is to be used as part of a more general model, two important points must be considered. First, the date of seedling emergence must be precisely determined, especially for periods of low daylength and when the daylength is changing rapidly. Secondly, though it is possible to predict the number of leaves at any time, it is still not possible to predict the total number of leaves that will emerge. It is known that there is no precise relationship between number of leaves and physiological stage (Joubert and Cairns, 1977; Delécolle and Gurnade, 1980) and so it is not possible to predict the limit to leaf initiation. The maximum number of leaves is genetically determined, but it is also highly dependent on environmental conditions. This point must be studied further to achieve an accurate leaf-climate sub-model.

REFERENCES

Baker, C. K., Gallagher, J. N., and Monteith, J. L., 1980, Daylength change and leaf appearance in winter wheat, Pl. Cell Environ., 3:285.

Delécolle, R., and Gurnade, J. C., 1980, Liaisons entre le développement du blé tendre d'hiver. I. Stades de développement de l'apex, apparition des feuilles et croissance de la tige, Ann. Amél. Pl., 30:479.

Friend, D. J. C., Helson, V. A., and Fisher, J. E., 1962, Leaf growth in Marquis wheat, as regulated by temperature, light intensity and daylength, Can. J. Bot., 40:1299.

Gallagher, J. N., 1979, Field studies of cereal leaf growth. I. Initiation and expansion, J. exp. Bot., 30:625.

Hunt, L. A., 1979, Photoperiodic responses of winter wheat from different climatic regions, Z. PflZucht., 82:70.

Ingoat, G., and Couvreur, A. L. P., 1979, Pour intervenir au bon stade, Perspectives Agricoles, 32:11.

Joubert, G. D., and Cairns, F., 1977, The relationship between leaf number and growth stages of different wheat cultivars, Agroplantae, 9:7.

Kirby, E. J. M., Appleyard, M., and Fellowes, G., 1982, Effect of sowing date on the temperature response of leaf emergence and leaf size in barley, Pl. Cell Environ., 5:477.

COMPETITION AMONG TILLERS IN WINTER WHEAT: CONSEQUENCES FOR GROWTH AND DEVELOPMENT OF THE CROP

J. Masle

INRA Laboratoire d'Agronomie
Paris, France

INTRODUCTION

Competition prevails during most of cereal growth. Many nutrients may be limiting in the field, depending on the main characteristics of soil, climate or cropping system. In temperate areas of Northern Europe, the most frequent limiting factors are nitrogen and water; if soil nutrients are at an optimal level and development requirements are fulfilled, competition for light is often the ultimate source of yield limitation; biomass production is then directly related to photosynthetically active radiation. Thus, in all cases, the processes which lead to the formation of ears and grains in the field are affected by competition for one or more essential growth factor.

In this paper we analyse plant and tiller growth and development in various conditions of competition in the field. From this, we infer some of the main parameters governing the competitive relations in the crop and their consequences on yield components.

VARIABILITY OF THE CROP

Competition occurs when the quantity of material tapped by a plant is, because of the presence of its neighbours, insufficient to ensure its potential growth (Clements et al., 1929; Donald, 1963). Many results show that relations between plants in a crop are largely determined by the differences in the characteristics of the individuals forming the crop (Aspinall, 1960; Donald, 1963; Harper, 1977). The first part of this paper describes the composition of a wheat crop by defining the nature of the individual plants and the elements distinguishing them.

First, we should consider not only the plants but also the different tillers as individual units. Although tillers may retain some dependence on each other, they rapidly acquire some degree of nutritional autonomy (Quinlan and Sagar, 1962). Examined at the scale of the plant or of the tiller, the wheat crop is an extremely heterogeneous community throughout its life cycle. This heterogeneity is due both to the normal process of tillering and to differences between seeds and their position in the seed bed, as described below.

The tillering process creates tiller groups which range in their stage of development and growth. At any moment in the life cycle of the crop, the classification of the different tiller groups is the same whether based on apical stage or number of leaves or dry-weight. These characters depend on the order of tiller bud initiation, and appearance (Fig. 1). This is an idealised representation of the tillering pattern followed by plants in a crop under non-limiting nutritional conditions. Sometimes plants do not have a first tiller (T1), others may have neither T1 or T2 (see below), but the order and time of appearance of subsequent tillers remains as shown in Fig. 1. The coleoptile tiller is not represented because it is rarely observed in the experimental conditions

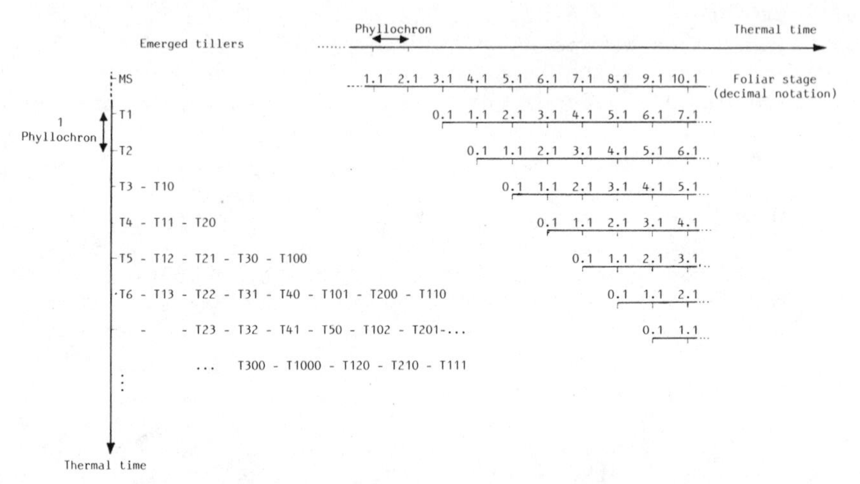

Fig. 1. Pattern of leaf and tiller appearance for a wheat plant under non-limiting nutrition (Masle and Sebillotte, 1981b). A "zero" digit refers to tillers arising from a prophyll. The foliar stage notation is as in Haun (1973); 0.1 means that the first leaf of the tiller is just arising above the ligule of the subtending leaf.

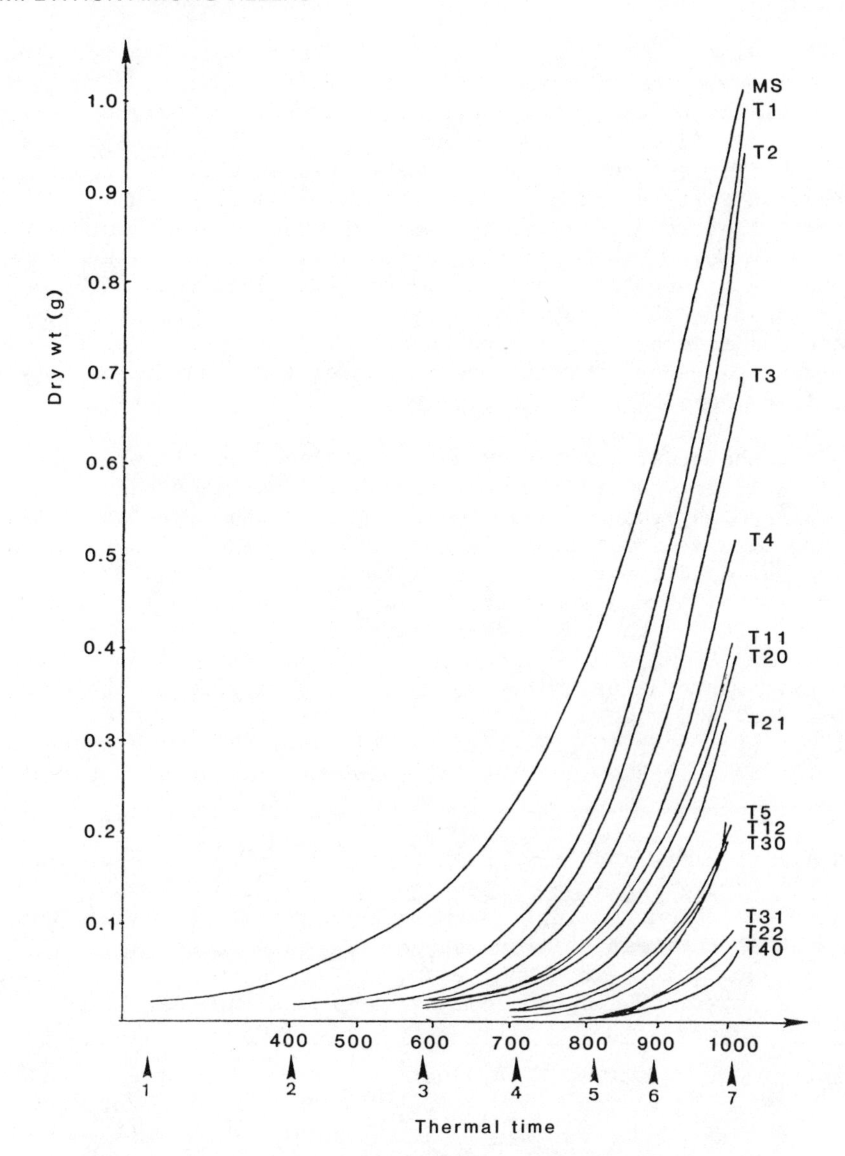

Fig. 2. Dry-weight of tillers versus accumulated temperature (thermal time) from sowing (base temperature = 0°C) for the cultivar Talent. The numbered arrows on the x-axis indicate dates of sampling. The first sample was taken at the 3 leaf stage and the last at meiosis.

described in this paper and does not have a stable development pattern (Masle, 1980; Fletcher and Dale, 1974).

The number of leaves on each tiller increases in proportion to the total number of degree days and the initial differences in leaf number related to age differences are nearly completely conserved until meiosis (Fig. 1). On the other hand, during the same period these differences are only partly reflected in the dry-weights (Fig. 2) or green surfaces of different tiller classes, which have an exponential growth curve of the same type for all tillers; $y = a \exp(b\Sigma T)$, where ΣT is thermal time (i.e. accumulated temperature). Growth is more rapid for later-produced tillers i.e. b increases (Masle, 1980). By relating growth to thermal time, differences in growth rate express an intrinsic tiller effect, and are not simply due to variations in temperature conditions to which the tillers have been subjected.

Both the order and the synchrony of leaf and tiller appearance described in Fig. 1 are remarkably stable. On the other hand, the value of the phyllochron and the relative growth rates among tillers are much more variable from one crop to another even under similar nutritional conditions; they are particularly modified by sowing date and cultivar (Table 1). Kirby et al. (1982) also noticed the effect of sowing date on the phyllochron value.

Performance of seeds may vary as a consequence of a number of factors e.g. the size of the grain which partly depends on its position in the ear (Chaussat et al., 1975), the sowing depth, and the moisture near the seed (Masle, 1980; Rickman et al., 1983).

Table 1. Comparison of the phyllochron and dry weights of tillers (expressed as a percentage of main-stem dry-weights) for different sowing dates and cultivars. The sample was taken at the "ear at 1 cm" stage. At this stage there was no competition for nutrients. The results for T2 and T4 are shown. The same tendency was found in other tillers.

Cultivar	Date of sowing	Phyllochron (°C d)	Main-stem dry-weight (g)	Relative tiller dry-weight (% main-stem) T2	T4
	1/10/80	120	0.194	55	24
Talent	20/11/80	100	0.154	40	7
	30/01/81	75	0.144	35	4
Etoile de Choisy		105	0.135	34	6
Hobbit	4/11/82	101	0.136	51	13
Cappelle Desprez		95	0.161	57	17

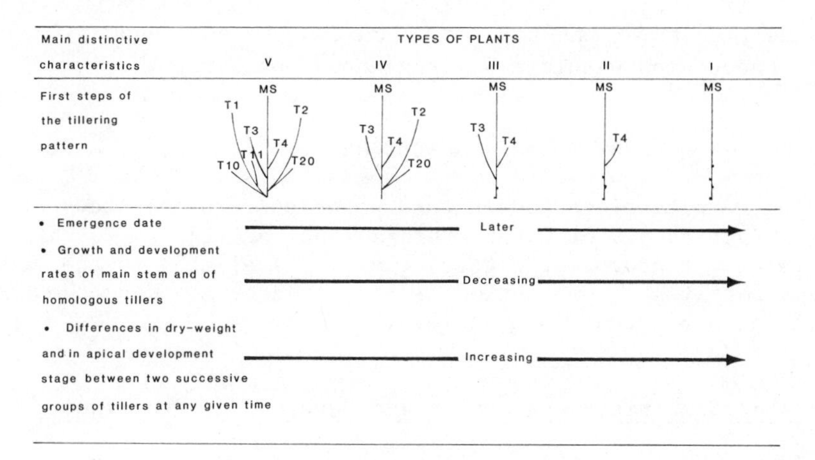

Fig. 3. Classification of the different types of plants present in a cereal crop. The differences between the types are shown by reference to the main distinguishing characteristics.

These characteristics lead to considerable variation in the time of plant emergence and thus in the rate of colonization of the space between plants (Kemp et al., 1983). They also result in plants having different tillering potentials because they influence the number of first tillers which do not grow and the behaviour of homologous tillers (Masle and Sebillotte, 1981a). Figure 3 gives the definitions and the distinctive characteristics of the 5 types of plants which can be found in a wheat or barley crop, whatever the nutritional conditions and cultivar. Generally, types I, IV and V constitute the bulk of the crop, but in different experiments the proportions vary from 5 to 90% for each type, because of the variability in sowing conditions and in the origins of commercial seeds. The genotype does not seem to exert an influence.

The variability of individual plants in the crop is thus partly due to the pattern of plant development and partly contingent on the cultivation conditions, particularly those at the time of sowing. The type of plant is therefore determined early in the life cycle but its subsequent development depends principally on the climate (especially temperature and photoperiod). In the absence of nutritional limitations, these factors determine its morphogenesis (Kirby and Riggs, 1978; Kirby and Ellis, 1980; Baker and Gallagher, 1983). At any moment in the life cycle, the plant population may be described as an aggregation of tiller groups each of which is characterized by different growth and development states. These groups are defined in a very stable way in different populations, especially those of different cultivars, by the combination of 2 principal criteria, physiological age and plant type.

Thus, whatever the time at which competition becomes important, it is competition within a heterogeneous population.

BEHAVIOUR OF TILLERS UNDER COMPETITION

The results presented in this paper concern two kinds of competition: competition for nitrogen and for light. For nitrogen the limiting factor is the total quantity available to the crop, whereas for light, competition is largely determined by the uneven distribution between canopy layers.

Effects of Competition on Growth and Development Processes

Reactions of a plant to limiting nutritional conditions. Shortage of a nutrient in relation to the potential requirement of the plant has, immediately and simultaneously, two kinds of consequences whatever the developmental stage of the plant. These are interruption of the tillering process, when axillary buds which would normally begin their growth (cf. model Fig. 1) do not elongate and reduction of the growth rate of the youngest tillers (Fig. 4).

The slackening of growth then spreads progressively from the youngest tillers to other tiller groups and eventually to the main stem. Although variable, the reduction in growth depends on the age of the tillers, being greatest in the youngest tillers. For some young tillers, growth decreases rapidly and soon no further growth i tiller size or apex differentiation can be seen and these tillers become senescent. This senescence process begins immediately for th group of tillers which emerged most recently and a little later for one or several preceding groups, in the order of their age. These effects on growth are reflected in the number of grains that a tille can set, and this yield component is strongly correlated to the vegetative dry matter measured at maturity whether for the plant or for each tiller (Masle, 1980). It is noteworthy that this relationship is constant among plants and all tillers, and also amon plants grown under different conditions of climate, and with different timing and severity of competition. This indicates that a any given stage and for any tiller the variations in growth rate can affect the processes which determine grain number.

However, except in extreme cases, nutritional conditions do not appreciably influence the development rate. For all tillers which continue to grow, the rates of leaf appearance and the differentiation of floral apices are similar to those of plants in the same environment which have adequate nutrition (Masle, 1981a and b).

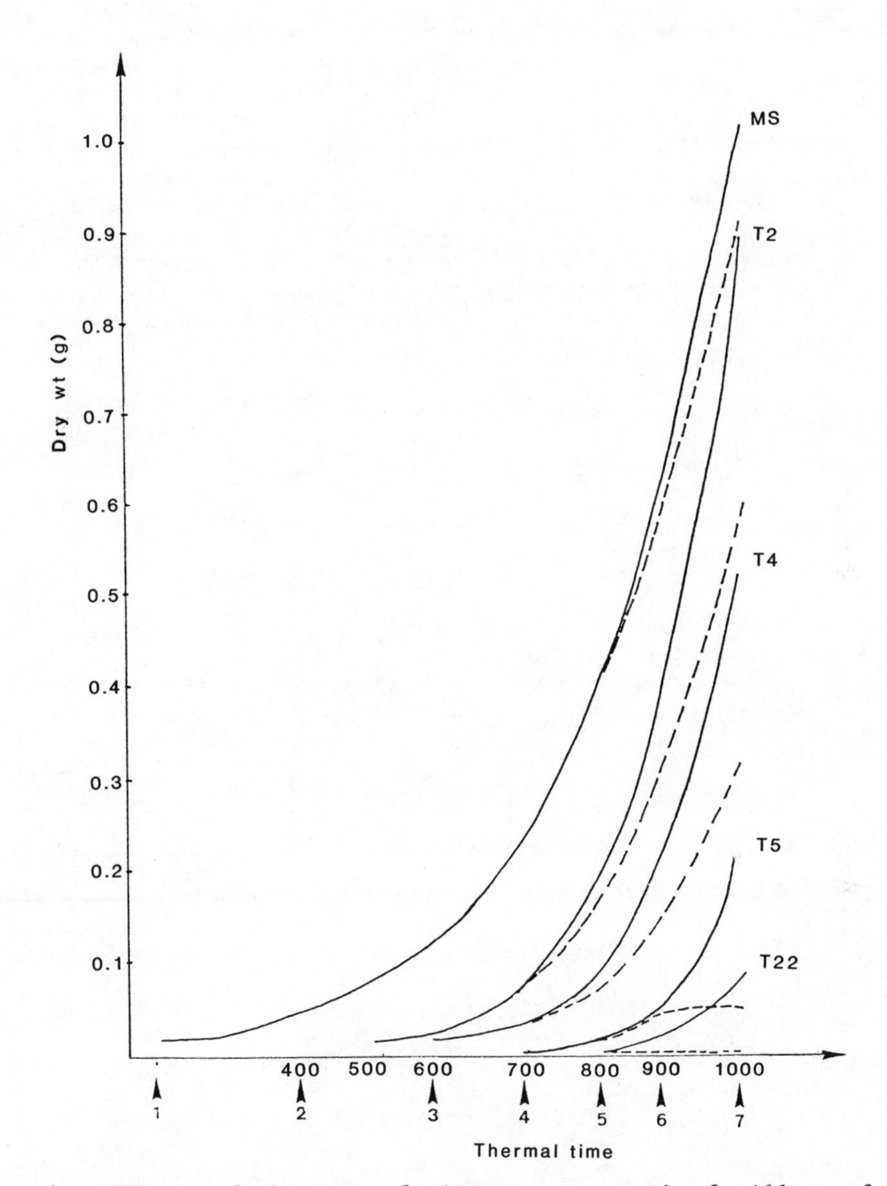

Fig. 4. Effects of shortage of nitrogen on growth of tillers of the cultivar Talent. The numbered arrows on the x-axis indicate dates of sampling (see Fig. 2). Some plants were grown with adequate nitrogen throughout their life cycle (———, the same curves shown in Fig. 2). Others (- - -) received adequate nitrogen until sample 4. After that they received a low amount of nitrogen (2.5 meq/litre) until sample 7.

Behaviour of a crop during competition. Because of their position in the field with regard to distance from neighbours, and depth of soil, the individual plants in a crop are not in equivalent environmental conditions. Moreover, they do not have the same ability to colonize the available space and, at a certain date, are not all at the same stage (see above). Thus, nutrition does not become limiting for all plants at the same moment in their life cycle; in the case of plants sown in rows, two or three phyllochrons may elapse between the first appearance of competitive effects and the involvement of the whole of the crop (Masle, 1983, 1984). Great variability results among the plants in the rank of the last tillers produced (Fig. 5), as well as in the rank of tillers which senesce;

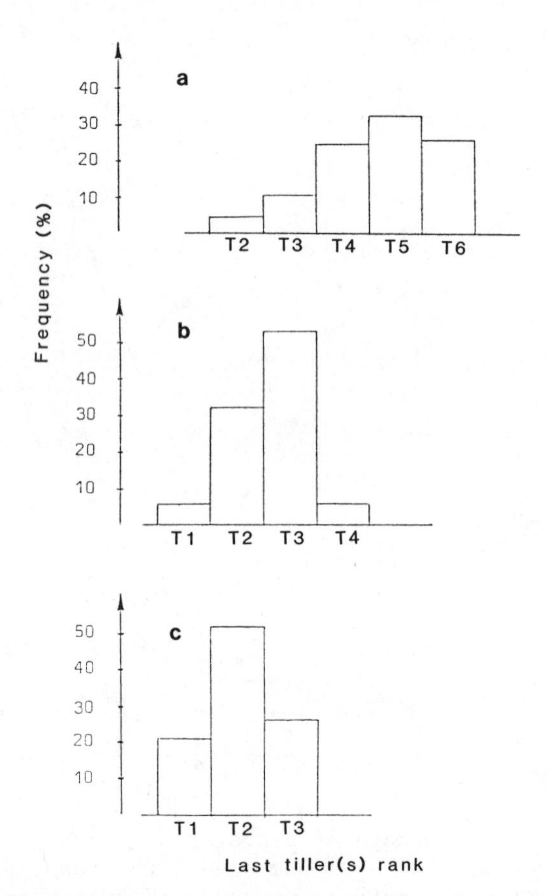

Fig. 5. Variability among plants with respect to the last tiller produced for 3 plant densities: a) 60 plants m^{-2}; b) 308 plants m^{-2}; c) 534 plants m^{-2}. The rank of the last tiller produced is indicated by the name of the primary tiller; secondary and tertiary tillers of the same age were generally also present.

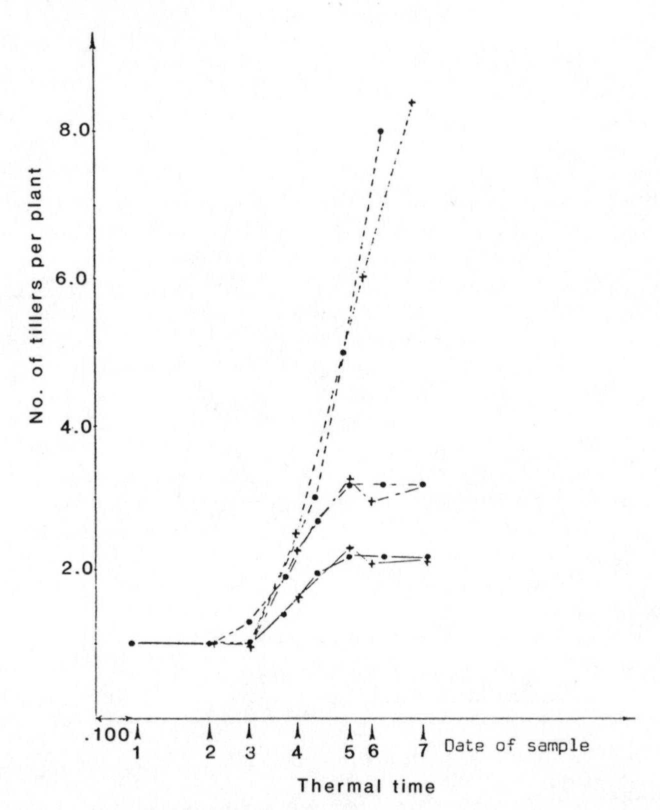

Fig. 6. Number of tillers per plant versus accumulated temperature as observed (+) or simulated (●) using the model described in Fig. 1 (phyllochron value = 100 degree-days) and the date distribution of cessation of tillering (cf. Fig. 5). The plant densities were as in Fig. 5: ----- 60 plants m^{-2}; — - — 308 plants m^{-2}; ——— 534 plants m^{-2}; competition for light on the lowest plant density had just appeared before sample date 7.

the frequency of senescence within a group of tillers of the same rank varies from 0 to 100% (Masle, 1981a).

Variability among plants apparently causes a slowing down in the average rate of tillering per plant which precedes cessation of tillering. This pattern of change, shown in Fig. 6, corresponds to results often described in the literature (Thorne, 1962b; Watson and French, 1971; Darwinkel, 1978). It has been more or less implicitly interpreted as a slower appearance of tillers under the influence of competition before the cessation of tillering. This slowing down is in fact due to the differences in the time of cessation of tillering from one plant to the next, as described above. The observations in

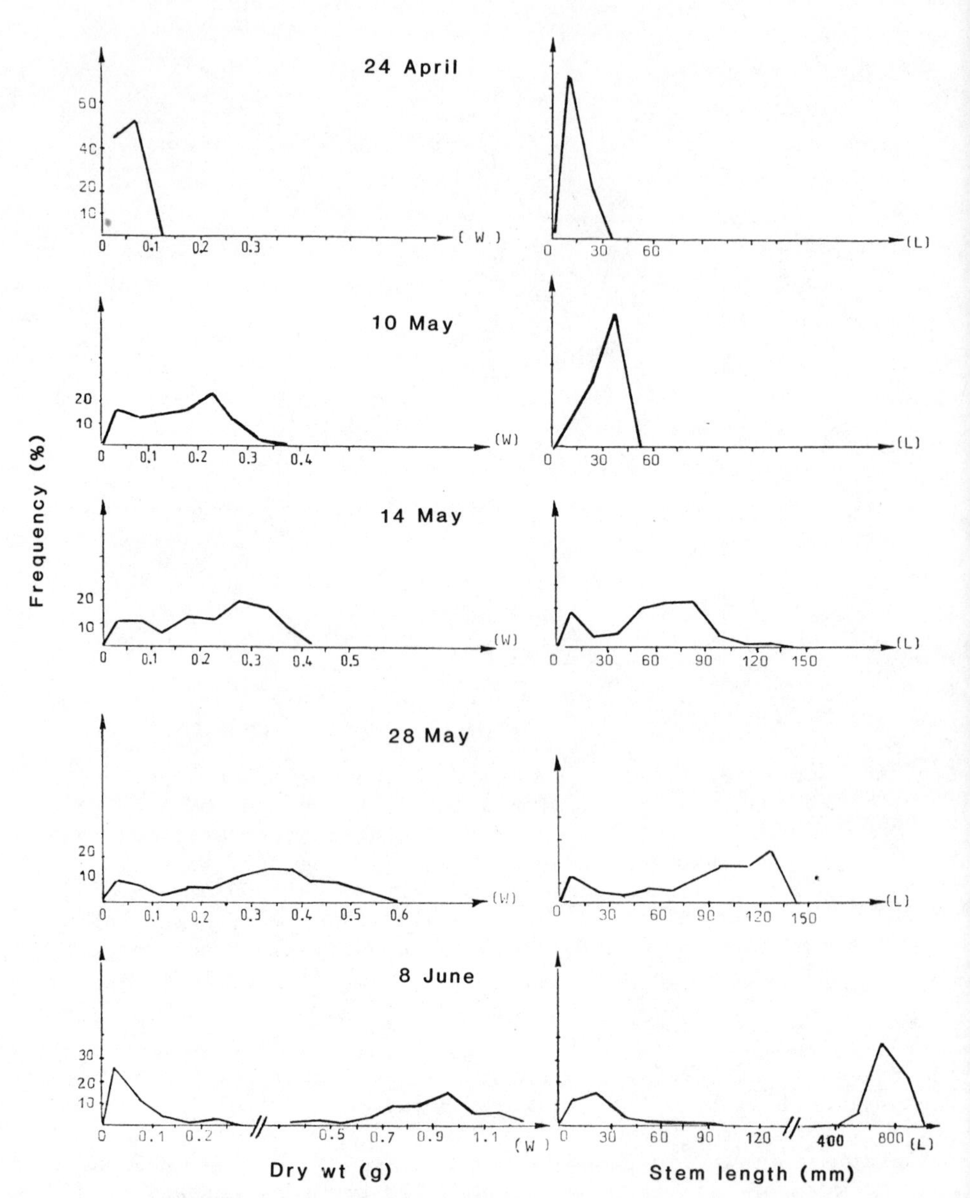

Fig. 7. Distribution of individual dry-weights and stem lengths for T2 tillers at 5 successive dates during the life cycle: on 24 April the first lemma was present on the most advanced spikelet, and the plants flowered at 10 June. The number o tillers observed at each sample varied from 50 to 100.

Fig. 6 agree closely with the curves obtained a posteriori by simulating tiller production for each plant at the potential rate predicted by the model (Fig. 1) until the rank of tillers at which cessation of tillering was observed.

There is a characteristic change in the frequency distribution of tillers, classified by growth and development stage (Fig. 7). The distribution is Gaussian until the beginning of competition, then it changes progressively until two distinct populations are apparent. In the first population of tillers, growth and apical development continue until maturity, whereas for the second population growth and development ceases. This separation into two populations occurs for each group of tillers, but the younger the group, the greater the proportion in the second population. This separation is clearly identifiable before the advent of any yellowing of the last leaf of the tiller, a criterion sometimes used for dating the beginning of senescence (Gallagher et al., 1976). The near constancy of the dry weights of the senescent tillers from the moment they stop their growth until anthesis occurs on the other tillers (Masle, 1981), indicates that they contribute very little material to the growth of the elongating tillers. Similar observations were made by Quinlan and Sagar (1962) and Bremner (1969).

Date of the beginning of competition. The occurrence of competition for nitrogen or for light thus has important repercussions on crop development and performance at anthesis. In the field, competition occurs at different times in the life cycle, depending on the experiment. Thus, in our trials, the dates of onset of competition for light are spread out, from very early stages (around the time of appearance of T2 and T3) until much later stages (near meiosis, when T5 appears). Such differences are principally caused by variation in the number of plants per square meter and, to a lesser extent, variation in sowing date. Dates of occurrence of nitrogen competition can also be very different from one plot to the next; in the case when the first application of nitrogen fertilizer is delayed until the end of February, it is not infrequent for nitrogen shortage to begin before the first tiller emerges.

Because the cessation of tillering and the senescence of tillers are immediate consequences of nutrient shortage, they are not related to a particular development stage. They occur earlier when there is less nutrient available e.g. under conditions of high plant density, poor fertilization and possibly when there is water shortage. The coincidence, observed in some experiments, of the stabilization of the number of tillers with the formation of the terminal spikelet (Riggs and Kirby, 1978), or of the start of tiller senescence with the beginning of stem elongation (Baldy, 1974), were thus fortuitous, and were due to the particular competition conditions within the experiments.

Once competition for light begins, it will continue at least until anthesis. Competition for nitrogen, on the contrary, may be remedied by fertilizer application. In this case, a resumption of tillering can sometimes be observed. Tillers which emerge as a consequence of nitrogen application correspond to buds which, according to the pattern shown in Fig. 1, are then about to elongate (Masle and Gbongue, unpublished data). This check to tillering followed by a resumption can be observed following very early nitrogen competition, or in response to early water deficits (Kleppe et al., 1982) and results in more plants without T1 and/or T2, i.e. the proportion of plants of types IV and III is greatly increased. Some resumption of tillering is also possible later (after ear emergence) when the foliage begins to senesce, if water and nitrogen are supplied at this time (Thorne, 1962a; Masle and Gbongue, unpublished).

The existence of a nutritional basis for the growth of tiller buds and the absence of any effect on the rate of leaf appearance ar consistent with the observations of other authors (Aspinall, 1961; Kirby and Faris, 1972; Fletcher and Dale, 1974; Darwinkel, 1978). I means that observation of the variability of tiller sequences among plants and crops allows reconstruction, *a posteriori*, of some events in the history of the crop and when they occured with respect to foliar stages.

Origins of differences in behaviour of tillers

The response of a tiller to competition differs depending on th factors involved. This is because of the different characteristics of the ways in which tillers capture either nutrient or light. When nitrogen is the first factor for which there is competition, and remains the dominant limiting factor throughout the life cycle, only those tillers with at least three leaves at the time nitrogen become limiting are able to maintain their growth. They are the only ones to produce ears and grains. The existence of this critical stage in relation to the nitrogen uptake of a tiller seems to correspond to a particular stage in the morphogenesis of the root system. The three leaf stage is closely correlated with the production of at least one root that is 15 to 20 mm long (Table 2; Masle, 1981a), suggesting that this is a functional root endowing the tiller with a certain autonomy *vis-à-vis* nutrients and water (Wiebe and Kramer, 1954; Rovira and Bowen, 1968, 1970). Before this stage, these elements ar provided through older tillers. Gillet (1969) and Ong (1978) indicate that this critical stage (around 3-4 leaves) also exists in rye-grass or other grasses.

When light is the cause of competition, the growth of a tiller appears to be directly related to its leaf area and the height of it leaves compared with those of the tillers in its immediate vicinity.

Table 2. Frequency distribution of root length at successive foliar stages for the cultivar Talent. All shoots are grouped together and there was no competition for nutrients. The decimal notation for foliar stage is the same as in Fig. 1.

Foliar stage of the tiller	Classes based on length (mm) of the longest root of the tiller							Number of tillers observed
	0-10	-20	-30	-40	-50	-70	≥70	
0.1 - 1.6	100							76
1.7 - 2.0	73	10	8	8				48
2.1 - 2.2	57	29	14					14
2.3 - 2.6	45	19	17	15	4			53
2.7 - 3.0	27	18	20	15	8	10	1	88
3.1 - 4.0	7	13	18	21	12	18	11	141
4.1 - 5.0		3	11	14	11	32	29	65

At any time in the life cycle, the younger a tiller, the smaller is its leaf area and the greater is the proportion of that area that is in the lower layers of the canopy (Fig. 8). As light intensity within a wheat canopy decreases exponentially from the top of the canopy (Szeicz, 1974; Gallagher and Biscoe, 1978), the quantity of energy reaching a young tiller is greatly attenuated, and all the more so when there are many older tillers or many neighbouring plants. Therefore, competition for light energy starts in the lowest layers of the canopy and affects the youngest tillers, then moves into the higher levels, reducing the growth of older tillers. The main stems have a particularly favourable position as they constitute most of the foliage in the highest layer (60 to 80% on average in the top 15 cm, whereas any other tiller corresponds to 10% at most, or exceptionally 20% (Fig. 8). The main shoots therefore play a predominant role in shading of the tillers; it is understandable why they remain unscathed by competition perhaps until 3 weeks after it has begun (Masle, 1984) and senesce only rarely, except in the case of extremely high crop densities, outside the usual agronomic range.

For all shoots, a given leaf area and leaf height in the canopy are strongly correlated with apical stage, height and dry weight of the shoot (Masle, 1984). This may be one of the reasons for increases in the potential number of ears with early sowing date, high plant density or a high proportion of plant type V. In all these cases, the population of the future fertile tillers are more homogeneous at the beginning of the elongation phase (Masle, 1984), both in stage of development and in dry weight. This is due to less marked differences between successive groups of tillers, or to a large proportion of old tillers. Therefore, within a cultivar, the greatest number of ears results when the greatest number of tillers

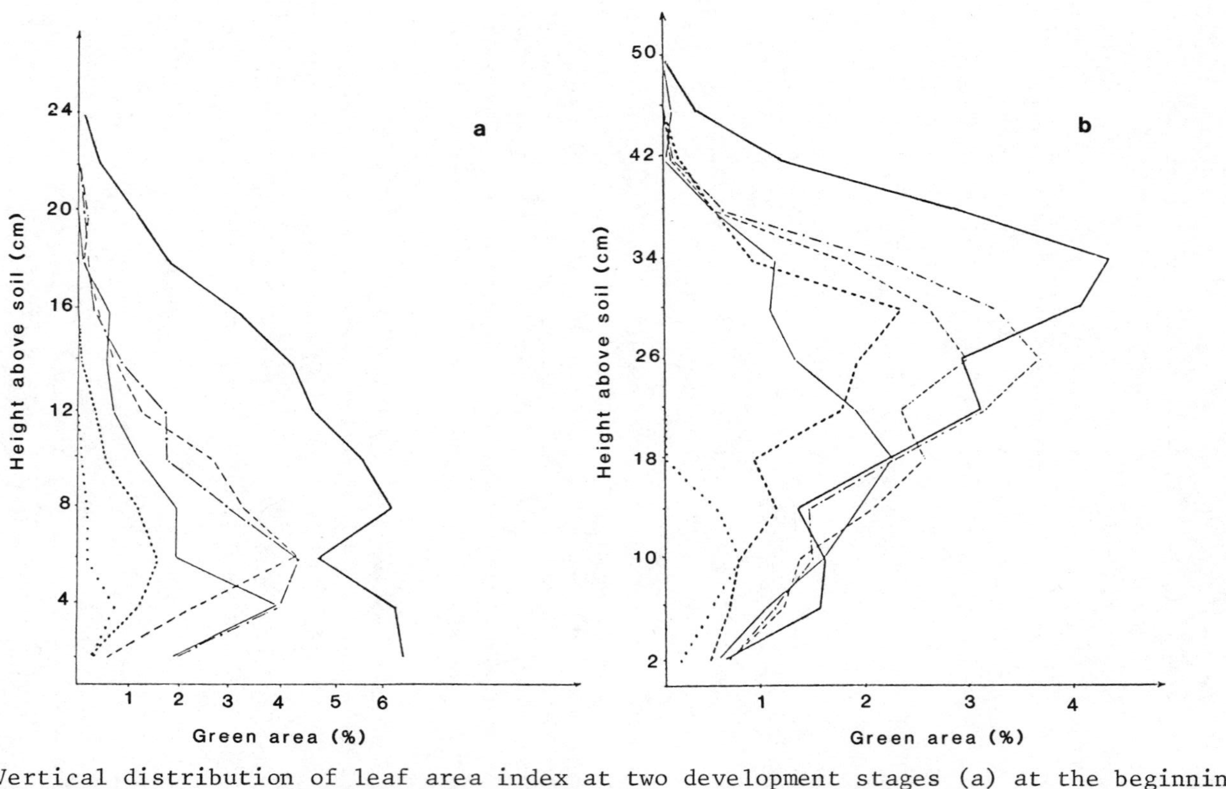

Fig. 8. Vertical distribution of leaf area index at two development stages (a) at the beginning of stem elongation, "ear at 1 cm" (———), T2 (— · —), T3 (- - -), T4 (-----) and T5 (•••••). The x-axis is the area of green leaf and sheaths per layer expressed as a percentage of the total green area of the plant.

begin their elongation phase at the same time, that is they colonize the upper layers of the canopy at the same time.

However, if the stem has grown beyond a certain length the tiller will then very rarely senesce ($\leqslant$ 5%). The critical length for the cultivar Talent is 40 to 60 mm (Masle, 1980) and similar values have been found for other cultivars. Beyond this length, although there is subsequently some overlapping of leaves which may reduce the growth of some tillers, the tillers still intercept sufficient energy to maintain a certain growth rate.

The variability in the behaviour of tillers in a crop in response to competition for nitrogen or light is due not to intrinsic functional differences between tillers, but to differences in size and position of their roots or leaves. At any time in the life cycle there will be differences, due to age, in such characters as dry weight or length of a tiller as well as number of leaves or stage of apical development which will determine their ability to compete for nitrogen or light. Because there is little transfer of material between shoots, the ability to compete will determine whether a tiller dies or produces an ear. The ability to compete for resources is determined early. For example, the ability to compete for light is defined, for the cultivar Talent, by the time the shoot is 40-60 mm long and the ear is near the terminal spikelet stage. In the case of competition for nitrogen, the critical stage is when the tiller has three emerged leaves.

ELEMENTS FOR MODELLING EAR AND GRAIN NUMBERS

The analysis of plant morphogenesis and tiller growth and development under competition provides a basis for proposing formal laws governing the number of ears and grains per square meter.

The following model is proposed to predict the number of ears m^{-2}. The model applies to situations where the quantity of nitrogen is the first factor restricting growth and then remains severely limiting. We have shown that under such conditions only tillers having 3 leaves when the limitation appears continue to grow. The model requires 3 groups of parameters; the dates at which nitrogen becomes limiting for each plant within the crop (t_c), the tiller number for each plant at that moment and the foliar stages of these tillers. The comparison of the tillering pattern of a plant with the potential pattern predicted by the model in Fig. 1 provides a means of detecting the cessation of tillering and of determining the time in the plant's life cycle when it occurred in terms of the foliar stage of the main stem. This time is practically concomitant with the intervention of the nitrogen shortage (t_c), and can be determined, for any plant, a posteriori but well before ear emergence (at least 1.5 to 2 months before anthesis, even for the latest t_c). The values for the other groups of parameters are easily obtained: the model of the potential appearance of leaves and tillers described i Fig. 1 gives the number of tillers present at t_c on any type of plant, and their leaf numbers. Thus to predict the number of ears m^{-2} we need the number of plants m^{-2}, their types and the dates at which tillering stopped. The first two are fixed by the end of winter and therefore are observable early in growth. The predictio of the number of ears is therefore possible as soon as the nitrogen competition affects all the plants within the crop.

This model and the underlying principles have been tested over several years for different soils and plant densities and over a large range of nitrogen nutrition conditions which have led to differing times of onset and severity of competition and for very different cultivars (Masle, 1982, 1983). In a majority of cases, t number of ears m^{-2} was correctly predicted even though the number of ears per plant varied from 1 to 3.6, so that the classification the ear populations were very different. When there were differenc between the observed number of ears and the number predicted by the model, this was associated with situations where the young roots of the tillers, assumed to be functional in the model, started their elongation in a dry horizon. The deeper horizons were at about fie capacity so that water did not in itself limit the growth, but the dryness of the top 10 cm of soil affected the nitrogen availability and possibly water uptake by the young tillers.

This model depends upon three main principles of construction which have a special value for predicting the development of tiller under competition. These are:

(1) The consideration of the dynamics of the onset of the competition, in relation to the crop's development cycle.

(2) The choice, as inputs to the model, of characteristics whi are related to the utilization of the factor for which there is competition. This results in the model being specific for the crop nutrition conditions, characterized by the nature of the limiting factor.

(3) The perception of the crop as a population of individuals competing to obtain these factors, and with characteristics that depend directly on the expression of morphogenesis in the heterogeneous field conditions. These characteristics can be defin for each group of tillers defined by age and plant type.

These principles are similarly the basis of a model for predicting the number of ears m^{-2} (NE) under conditions of light energy competition:

$NE = a\ W_1^{\ b}$ (Masle, 1980; Meynard, 1983)

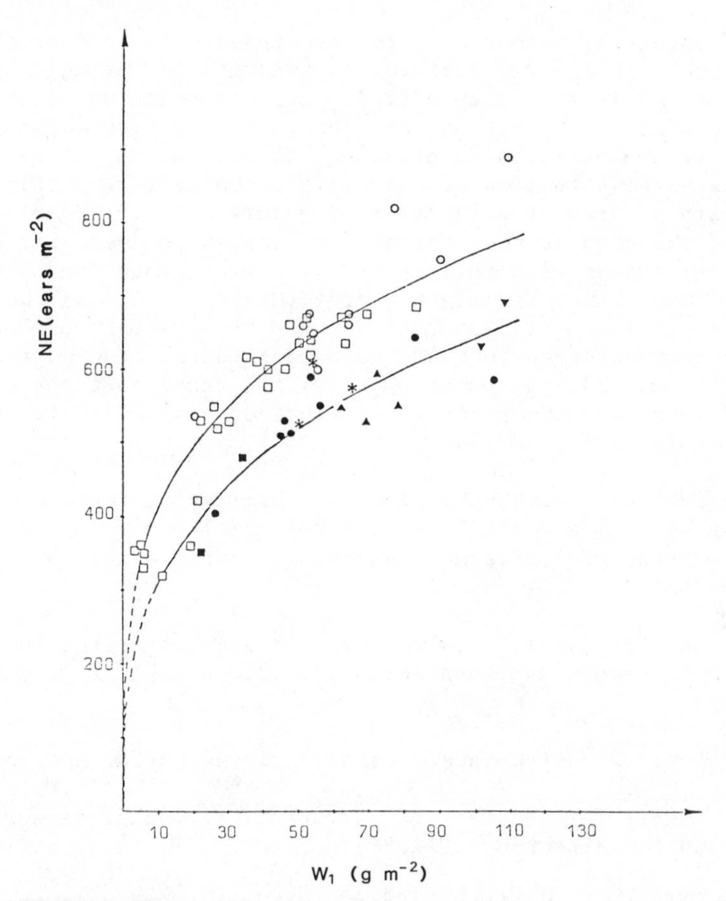

Fig. 9. Relationship between the potential number of ears (NE) and the total aerial dry weight at "ear at 1 cm" stage (W_1) for the cultivar Talent. o 1978; □ 1979 and the cultivar Lutin: ■ 1979; ▼ 1980; ● 1981; ▲ 1982; * 1983. The curve equations for the fitted curves are:

Talent: $NE = 219.8\ W_1^{0.269}$ (35 plots; Champagne Crayeuse)

Lutin: $NE = 149.6\ W_1^{0.318}$ (19 plots; Noyonnais)

where W_1 is the total dry weight per unit ground area of the aerial parts at the "ear at 1 cm" stage, (Couvreur et al., 1980), and a and b are constants. Whatever the cause of the variation in dry matter (in particular, date of sowing or number of plants), these constants are very stable from year to year and under differing cultivation conditions (Fig. 9) though they do vary with cultivar.

The potential number of ears thus appears to be directly relat to a characteristic (W_1) measured at the very beginning of stem elongation. This stage, even if it does not coincide with the beginning of the competition for light, marks a particular time in relation to the utilization of light; it corresponds to the time wh the vertical distribution of leaves is becoming well defined and wh the density of leaf area is increasing very quickly. A higher dry-matter of the crop at the "ear at 1 cm stage", caused, for example, by a higher number of plants or earlier sowing date, means an earli onset of competition for light energy (Masle, 1984) and therefore a earlier cessation of tillering. It also integrates a number of the canopy characteristics that are important in relation to the possibility of tiller elongation, i.e. the density of the different foliar layers, and the sizes and heights of the successive tiller groups (Masle, 1980, 1984).

The influence of the total aerial biomass of the crop at the beginning of stem elongation on the potential number of ears reflec the three principles presented above, in that a higher biomass corresponds to:

(1) Earlier competition for light energy, starting in the lowe layer of the canopy and then extending more quickly to the whole of the canopy.

(2) More extensive intercepting surfaces during stem elongatic

(3) A less defined vertical structure of the green surfaces ir relation to the different tiller groups.

The effect of cultivar on the constants a and b in the above equation, however, indicates that W_1 does not account for some crop characteristics which have a role in the determination of number of ears, for example leaf size and leaf posture.

By considering the laws underlying the above two models, we a able to understand why the maximum number of tillers produced is a bad predictor of the number of ears. In fact, these laws take into account many aspects of competition and the influence of characteristics which can vary greatly between crops with the same number of tillers; in particular, the physiological age distributic of tillers within the crop and the differences between tillers of different ages. The modelling of number of grains per m^2 also necessitates explanations in terms of how the different groups of plants and tillers that constitute the crop compete.

CONCLUSION

The nutritional conditions of a crop in the field vary greatly both from crop to crop and during the life cycle. These variations have an important effect on two yield components, the number of ears per m^2 and the number of grains per m^2, which both depend on growth phenomena. It has been clearly shown that the production of dry matter, and its efficiency in producing ears and grains, can be understood in terms of the relations, throughout the cycle, between the quantities of resources present in the environment and the assimilating roots and leaves. These relations depend on the morphogenetic characteristics of the crop which, for a given cultivar, are modified by the climate. These characteristics regulate the number of organs growing at a given moment, their nutritional requirements and the distribution of assimilates between organs. Also important are the changes in nutritional status in relation to the stage of development, in particular the date of the onset of competition and also the dates when there are changes in the nature of the limiting factors.

The explanation of these relations involves a wide range of knowledge concerning the laws of morphogenesis of the isolated plant, their expression within the crop as a function of variation in the quantity of nutrients, the dynamics of the growth and position of assimilating organs and the laws of their functioning, as well as the rules of distribution of the dry matter within the canopy. It is obvious that the development of this knowledge necessitates a strong coordination of disciplines and approaches concerning the phenomena of growth and development, which must be studied together.

REFERENCES

Aspinall, D., 1960, An analysis of competition between barley and white persicaria. II. Factors determining the course of competition, Ann. appl. Biol., 48:637.

Aspinall, D., 1961, The control of tillering in the barley plant. I. The pattern of tillering and its relation to nutrient supply, Aust. J. Biol. Sci., 14:493.

Baker, C. K., and Gallagher, J. N., 1983, The development of winter wheat in the field. 2. The control of primordium initiation rate by temperature and photoperiod, J. agric. Sci., Camb., 101:337.

Baldy, C. H., 1974, Sur le comportement de cultivars de blé tendre (Triticum aestivum L. em. Thell.) cultivés seuls ou en associations binaires en lignes alternées, Ann. Agron., 25:61.

Bremner, P. M., 1969, Growth and yield of three varieties of wheat, with particular reference to the influence of unproductive tillers, J. agric. Sci., Camb., 72:281.

Chaussat, R., and Bouinot, D., 1975, Hétérogénéité de la germinati
des grains de l'épi de blé (Triticum aestivum L.), C. r. hebd. Séanc. Acad. Sci., Paris, 281:527.

Clements, E., Weaver, J. E., and Hanson, H. C., 1929, Plant competition. 1. History of the competition concept, in: "Plant competition, an analysis of community functions", Carnegie Institution, Washington.

Couvreur, F., Ingoat, G., and Massé, J., 1980, Les stades du blé, ITCF, Paris.

Darwinkel, A., 1978, Patterns of tillering and grain production of winter wheat at a wide range of plant densities, Neth. J. agric. Sci., 26:383.

Donald, C. M., 1963, Competition among crop and pasture plants, Adv. Agron., 15:1.

Fletcher, G. M., and Dale, J. E., 1974, Growth of tiller buds in barley: effects of shade treatment and mineral nutrition, Ann. Bot., 38:63.

Gallagher, J. N., Biscoe, P. V., and Scott, R. K., 1976, Barley an
its environment. VI. Growth and development in relation to yield. J. appl. Ecol., 13:563.

Gallagher, J. N., and Biscoe, P. V., 1978, Radiation absorption, growth and yield of cereals, J. agric. Sci., Camb., 91:47.

Gillet, M., 1969, Sur quelques aspects de la croissance et du développement de la plants entière de graminée en conditions naturelles Festuca pratensis Huds. I. Construction de la touffe, Annls. Amél. Pl., 17:23.

Harper, J. L., 1977, The effect of neighbours, in: "Population biology of plants", J. L. Harper, ed., Academic Press, London

Haun, J. R., 1973, Visual quantification of wheat development, Agron. J., 66:116.

Kemp, D. R., Auld, B. A., and Medd, R. W., 1983, Does optimising plant arrangements reduce interference or improve the utilisation of space?, Agric. Systems, 12:31.

Kirby, E. J. M., Appleyard, M., and Fellowes, G., 1982, Effect of sowing date on the temperature response of leaf emergence and leaf size in barley, Pl. Cell Environ., 5:477.

Kirby, E. J. M., and Ellis, R. P., 1980, A comparison of spring barley grown in England and in Scotland. 1. Shoot apex development, J. agric. Sci., Camb., 95:101.

Kirby, E. J. M., and Faris, D. G., 1972, The effect of plant densi
on tiller growth and morphology in barley, J. agric. Sci., Camb., 78:281.

Kirby, E. J. M., and Riggs, T. J., 1978, Developmental consequence
of two-row and six-row ear type in spring barley. 2. Shoot ap
leaf and tiller development, J. agric. Sci., Camb., 91:207.

Klepper, B., Rickman, R. W., and Peterson, C. M., 1982, Quantitati
characterisation of vegetative development in small cereal grains, Agron. J., 74:789.

Masle, J., 1980, L'élaboration du nombre d'épis chez le blé d'hive
Influence de différentes caractéristiques de la structure de

peuplement sur l'utilisation de l'azote et de la lumiére. Thése Docteur-Ingénieur, INA-PG, Paris.

Masle, J., 1981a, Relations entre croissance et développement pendant la montaison d'un peuplement de blé d'hiver. Influence des conditions de nutrition, Agronomie, 1:365.

Masle, J., 1981b, Elaboration du nombre d'épis d'un stade critique pour la montée d'une talle, Agronomie, 1:623.

Masle, J., 1982, Elaboration du nombre d'épis d'un peuplement de blé d'hiver en situation de compétition pour l'azote. II. Modélisation du nombre d'épis, Agronomie, 2:17.

Masle, J., 1983, Comportement variétal chez le blé d'hiver en situation de compétition pour l'azote: élaboration du nombre d'épis, in: Colloques de l'INRA, in press.

Masle, J., 1984, Analyse des relations de compétition pour la lumière dans un peuplement de blé d'hiver, Agronomie, in press.

Masle, J., and Sebillotte, M., 1981a, Etude de l'hétérogénéité d'un peuplement de blé. I. Notion de structure du peuplement, Agronomie, 1:207.

Masle, J., and Sebillotte, M., 1981b, Etude de l'hétérogénéité d'un peuplement de blé. II. Origines des différentes catégories d'individus du peuplement, éléments de description de sa structure, Agronomie, 1:217.

Meynard, J. M., 1983, Perspectives pour la conduit des blés clairs, C. R. Acad. Agric. Fr., 69:830.

Ong, C. K., 1978, The physiology of tiller death in grasses. I. The influence of tiller age, size and position, J. Br. Grassld Soc., 33:205.

Quinlan, J. D., and Sagar, G. R., 1962, An autoradiographic study of the movement of ^{14}C labelled assimilates in the developing wheat plant, Weed Res., 2:264.

Rickman, R. W., Klepper, B. L., and Peterson, C. M., 1983, Time distributions for describing appearance of specific culms of winter wheat, Agron. J., 75:551.

Riggs, T.J., and Kirby, E. J. M., 1978, Developmental consequences of two-row and six-row ear type in spring barley. Genetical analysis and comparison of mature plant characteristics, J. agric. Sci., Camb., 91:199.

Rovira, A. D., and Bowen, G. D., 1968, Anion uptake by the apical region of seminal wheat roots, Nature, Lond., 218:685.

Rovira, A. D., and Bowen, G. D., 1970, Translocation and loss of phosphate along roots of wheat seedlings, Planta, 93:15.

Szeicz, G., 1974b, Solar radiation in crop canopies, J. appl. Ecol., 11:1117.

Thorne, G. N., 1962a, Effect of applying nitrogen to cereals in the spring or at ear emergence, J. agric. Sci., Camb., 58:89.

Thorne, G. N., 1962b, Survival of tillers and distribution of dry matter between ear and shoot of barley varieties, Ann. Bot., 26:37.

Watson, D. J., and French, S. A. W., 1971, Interference between rows and between plants within rows of a wheat crop, and its effects on growth and yield of differently-spaced rows, Ann. appl. Biol., 8:421.

Wiebe, H. H., and Kramer, P. J., 1954, Translocation of radio-active isotopes from various regions of roots of barley seedlings, Pl. Physiol., 29:342.

A STUDY OF THE RELATIONS BETWEEN GROWTH, DEVELOPMENT AND TILLER SURVIVAL IN WINTER WHEAT

D. W. Wood and G. N. Thorne

Rothamsted Experimental Station
Harpenden, U.K.

INTRODUCTION

Ear number is an important determinant of yield in winter wheat and it may be greatly affected by the proportion of tillers surviving to form ears. The causes of variation in tiller survival in crops where survival is not limited by nitrogen are not fully understood and were investigated in this experiment. The relation between tiller survival and development was studied in plants whose rate of development was changed by altering temperature. The relation between survival and growth was studied in plants whose growth rate was changed by altering radiation.

METHODS

Plants of winter wheat (cv. Avalon) were sown outdoors at a density of 247 m^{-2} in micro-plots made up of small square pots fed with ample water and nutrients by a capillary system. Between 17 March and 23 May, batches of plants were put in constant environment (CE) rooms for treatment for periods of 3 weeks starting at (1) the double ridge stage, (2) the terminal spikelet stage, or (3) the 2-3 node stage. At these stages the plants had on average 4.5, 4.4 or 2.7 shoots respectively. Environments used were day/night temperatures of 9/5°C and 13/9°C in factorial combination with photon flux densities of 235 and 475 $\mu E\ m^{-2}\ s^{-1}$. Daylength was the same as outdoors. After treatment, plants were returned outdoors until maturity. At intervals, batches of 36 plants per treatment were sampled. Observations were made of apical development of the main shoot and of dry weight and numbers of various plant parts.

RESULTS AND DISCUSSION

There was little interaction between the temperature and radiation treatments, so only the separate mean effects will be described (Table 1).

Plants always developed faster in the warm room than in the coo room, having 1-2 more florets per spikelet at the end of treatment. This difference in development persisted until anthesis, which was advanced by warm treatment in period 1 by 2 days, in period 2 by 1 day and in period 3 by 6 days. Warmer temperatures caused greater death of tillers during treatment. The difference in shoot number a the end of treatment in period 1 was obliterated by further tiller death in the colder treatment and final ear number was unaffected. The difference at the end of treatment in period 2 was enlarged by the greater number of shoots which subsequently died on plants that had been in the warmer room, leading to a big effect on final ear number. Tiller death had almost ceased at the end of period 3 and the small difference in ear number at the end of treatment persisted until maturity. Temperature had little effect on growth in dry weight during periods 1 or 2, but growth in period 3 was slower in the warmer temperature. Warmer conditions in period 2 or 3 decrease growth rate after treatment and resulted in smaller final dry weights. This was probably a consequence of the smaller leaf areas (leaf-area indices at anthesis, including sheath, were: period 2 - cool 10.3, warm 8.5; period 3 - cool 11.0, warm 9.8). Warmer temperatures tended to decrease the number of grains m^{-2} and grain yield.

Table 1. Yield and components in winter wheat (cv. Avalon) at maturity.

		Ears m^{-2}	Grains ear^{-1}	Grains m^{-2}	Grain DW ($g\ m^{-2}$)	Total I ($g\ m^{-}$
Mean		535	37.3	19934	753	2058
Effects of	Period 1	+14	-2.4	-883	-74	-76
warmer	Period 2	-117	+1.2	-3974	-198	-399
temperature	Period 3	-23	-4.2	-3254	-261	-328
Effects of	Period 1	-31	-2.0	-2201	-62	-43
brighter	Period 2	+56	-2.2	+1049	+35	+275
light	Period 3	+76	+4.0	+5216	+127	+432
S.E. of effect		14.4	1.06	678	33.7	71.

Giving the plants brighter light while they were in the CE rooms always increased dry weight and decreased tiller death during the treatment period. The effects induced during period 1 disappeared by maturity. The effects of treatment in periods 2 and 3 persisted until maturity, with the effects on shoot number increasing after the end of treatment. The number of grains per ear, the number of grains m^{-2} and grain yield were increased significantly only by the period 3 treatment.

Most of the effects on ear number described above were caused by differential death of the tillers in the axils of the first and especially the second leaf. This occurred mainly after the 2-3 node stage when almost all the other tillers had already died, irrespective of treatment. Surprisingly, about 25% of the main shoots died after warm treatment in period 2. The plants survived but had a smaller average ear size than plants which retained their main shoot.

Certain general relationships emerge from the data from the four separate treatment combinations in each of the three periods and from plants that remained permanently outdoors. Much of the variation in final grain yield was accounted for by variation in the number of grains m^{-2}, if plants treated in the warm room in period 3 are excluded (Fig. 1). The latter plants had small grains (mean dry weight, 28.3 mg), probably because leaf area declined very rapidly after anthesis. Grain size also varied significantly between some of the other treatments, from 34.6 to 43.1 mg. The number of grains

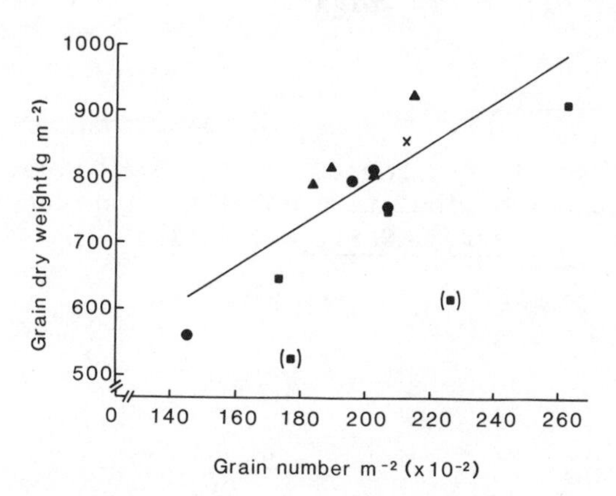

Fig. 1. Relation between grain yield (g dry weight m^{-2}) and number of grains m^{-2}. Grain yield = 166 + 0.03 (± 0.01) x grain number; 72% variance accounted for. The results from the warm room in period 3 (points in brackets) have been excluded from the regression. Data points relate to treatment period 1 (▲), period 2 (●), period 3 (■) and outdoors (X).

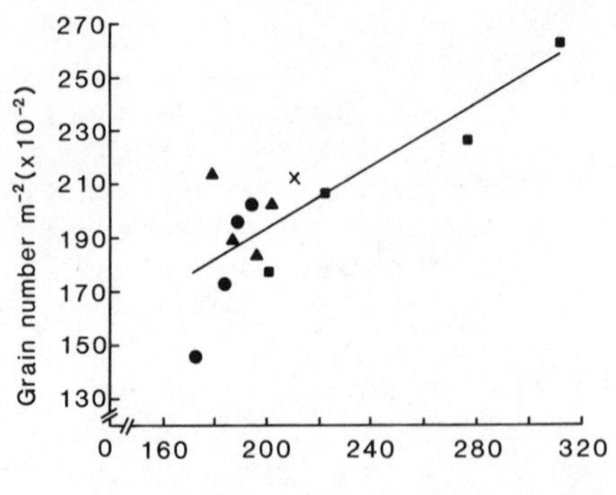

Fig. 2. Relation between number of grains m^{-2} and ear dry weight at anthesis (g m^{-2}). Grain number = 7846 + 57 (± 12) x ear dry weight; 68% variance accounted for. Data points relate to treatment period 1 (▲), period 2 (●), period 3 (■) an outdoors (X).

m^{-2} was well related to ear dry weight at anthesis (Fig. 2). Although the mean of all treatments, 20,000 grains m^{-2} per 210 g m^{-2} ear dry weight, was close to the value of 1 grain per 10 mg dry weight used in the model of Weir et al. (1984), this figure did not apply generally. There was a range from 1 grain per 8.4 mg to 1 grain per 12.2 mg ear dry weight.

REFERENCE

Weir, A. H., Bragg, P. L., Porter, J. R., and Rayner, J. H., 1984, A winter wheat crop simulation model without water or nutrient limitations, J. agric. Sci., Camb., 102:371.

PRODUCTION AND SURVIVAL OF WHEAT TILLERS IN RELATION TO PLANT GROWTH AND DEVELOPMENT

S. K. Roy and J. N. Gallagher

Lincoln College
New Zealand

INTRODUCTION

Tillering is a crucial process in determining wheat yields. It is intimately related with leaf area expansion during early growth and is the main determinant of the number of ears m^{-2} at harvest, the component most closely correlated with yield (Hampton et al., 1981; Porter, 1984). Any realistic model of wheat growth must, therefore, contain a submodel which can accurately simulate the tillering process.

The tillering of cereals and grasses has often been studied but usually by censual or demographic procedures. Such studies provided useful descriptions of tiller population dynamics but little insight into the physiological mechanisms controlling the rate and duration of tiller production and death. Detailed physiological investigations have also been made (e.g. Langer et al., 1973; Fletcher and Dale, 1977; Klepper et al., 1982) but the results from such work are not always directly applicable to crop models. In the absence of an understanding of the physiological basis of tillering some modellers have assumed that the production and retention of organs such as tillers is governed simply by plant growth or assimilate supply (Charles-Edwards, 1982; van Keulen, 1982). As a preliminary to developing a model of tillering the objectives of the present work were: (i) to investigate the factors controlling the duration of the phases of tiller production and death and (ii) to determine whether the pattern of tiller production and death is related to plant growth rate during specific developmental phases.

MATERIALS AND METHODS

Table 1 describes the essential details of the three experiment on which subsequent analysis is based. A semi-dwarf, prolifically tillering wheat (Triticum aestivum cv. Oroua; a cross of a South African cultivar, Skemer, with a CIMMYT line 66RN395) was used in al three experiments. Tiller population and plant dry matter were monitored throughout growth. It is important to compare trends in plant size and number of tillers at equivalent developmental stages. In each experiment, the thermal time (Tt) above a base of 0°C to eac stage was calculated and expressed as a percentage of Tt to anthesis This percentage is termed the developmental units after sowing, DUS.

RESULTS

Duration of Tiller Appearance and Death

The pattern of tillering in the growth cabinet was similar to that typically observed in the field (Fig. 1a). The observed tiller

Table 1. Experimental details. TS = terminal spikelet; FL = flag leaf emergence; EE = ear emergence; A+14 = fourteen days after anthesis; DCMU is N-(3,4-Dichlorophenyl)-NN-dimethyl urea.

Experiment	Sowing	Treatments	Thermal time (Tt) to anthesis (°C d
Glasshouse	200 plants m^{-2} sown in pots 5 March 1982	Drought from emergence to TS; TS to FL; FL to EE; EE to A+14; and no drought	1600
Field	Drilled in 15 cm rows 22 September 1982	In sub plots: 2 plant populations 150 and 600 plants m^{-2} In main plots: thinning no thinning and DCMU application	1100
Growth cabinet	200 plants m^{-2} sown in pots 7 July 1983	Four amounts of N: 0, 50, 100, 150 mg N pot^{-1} applied after emergence	1050(N0) t 1200(N150)

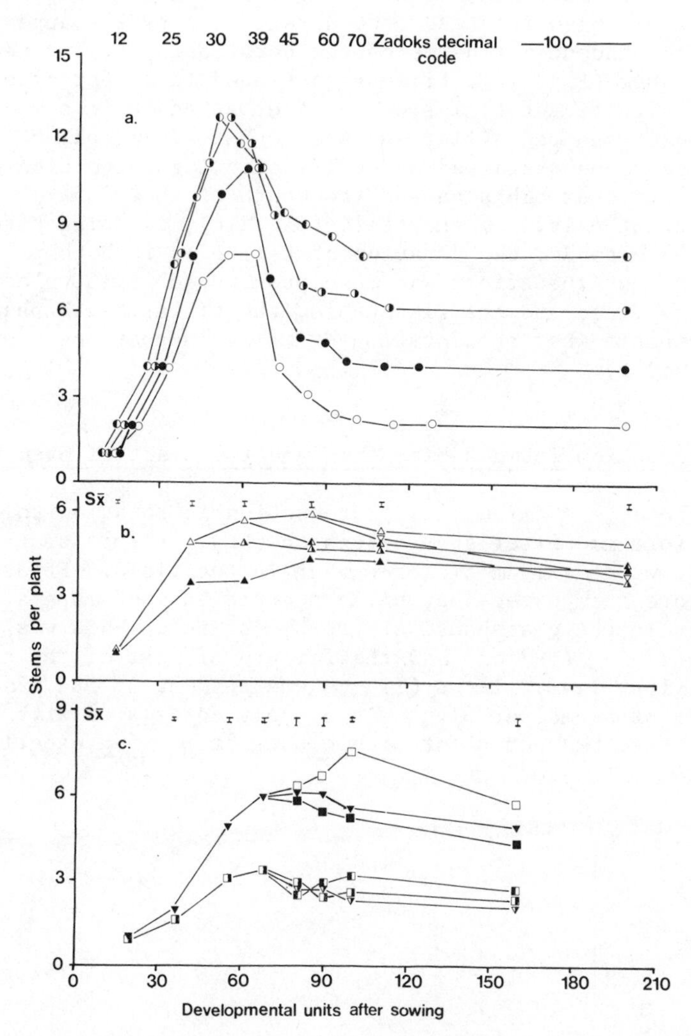

Fig. 1. The relation between stem number per plant and developmental time after sowing. (a) Growth cabinet: N0 (○), N50 (●), N100 (◑) and N150 (◐); (b) Glasshouse: Control (△), drought from; emergence to TS (▲), TS to FL (△), FL to EE (▲) and EE to A+14 (▽); (c) Field: 150 plants m^{-2} unthinned (▼); 150 plants m^{-2} thinned (□), 150 plants m^{-2} DCMU (■), 600 plants m^{-2} unthinned (◨), 600 plants m^{-2} thinned (◧) and 600 plants m^{-2} DCMU (▽).

appearance was close to the theoretical maximum (Masle-Meynard and Sebillotte, 1981b) and stopped around the time of terminal spikelet formation (TS, Zadoks 30); subsequent tiller death stopped before anthesis. Nitrogen increased the duration of tiller appearance slightly but had no influence on the duration of tiller death. In the glasshouse (Fig. 1b), tillers continued to appear after TS and there was significant tiller death after anthesis in some treatments. Early drought stopped tiller appearance between Zadoks 30 and 40 but tiller appearance restarted after TS following resumption of watering. Late droughts caused the death of about 15% of the tillers present at anthesis. In the field (Fig. 1c), thinning after booting (Zadoks 45) extended the duration of tillering in both plant populations but the effect was more pronounced at 150 plants m^{-2}. In both the field and the glasshouse, the tiller death phase extended beyond anthesis when the duration of tiller appearance was increased (Figs 1b and c).

Tiller Production Between Emergence and the Start of Stem Extension

We chose to examine tiller production in relation to plant growth before and after stem extension (SE), as this stage is often coincident with maximum tiller number in the field. Figure 1 shows that, before stem extension, tillers appeared much more slowly in the field than in the glasshouse or growth cabinet. This was associated with deep sowing (5-7 cm) and the failure of most of the coleoptile (T0) and first leaf tillers (T1) to emerge. At 45 DUS (Zadoks 32) the number of stems per plant was closely correlated with the amount of dry matter (DM) per plant <u>within</u> but not <u>between</u> experiments

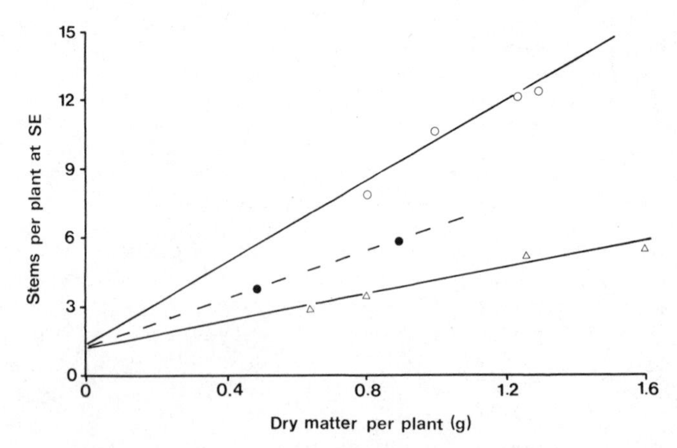

Fig. 2. The relation between stem numbers per plant and dry matter per plant at the start of stem extension (SE): growth cabinet (○); glasshouse (●); field (△). The lines were fitted by eye.

(Fig. 2). The number of stems present at a given DM in the growth cabinet was more than twice that observed in the field. This was presumably due to the absence of T0 and T1 tillers in the field which severely slowed subsequent tillering: over all experiments there was a strong overall correlation (98.6% variance accounted for) between stems per plant at 33 DUS (4-leaf stage) and 45 DUS.

Changes in Tiller Numbers After the Start of Stem Extension

The change in number of stems per plant between 45 DUS and anthesis (100 DUS) was highly correlated with a measure of the assimilate production per stem during this period, expressed as the increase in DM divided by the number of stems per plant at 45 DUS (ΔDM/N) (Fig. 3). However, the size of the response differed between environments: in the growth cabinet and in the glasshouse more stems survived (or appeared) per unit increase in ΔDM/N than in the field.

In the field and the glasshouse, the death of tillers after anthesis seemed to be associated mainly with a prolonged phase of

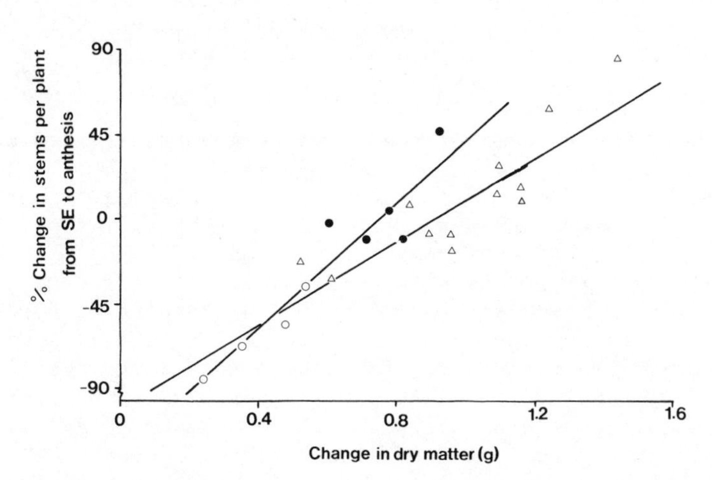

Fig. 3. The relationship between the change in stems per plant between the start of stem extension and anthesis and the change in DM per stem (ΔDM/N) during the same period: growth cabinet (○); glasshouse (●); field (△). The equations of the fitted lines are:
$y = -126$ (s.e. 12.5) $+ 167$ (s.e. 19.3)x
$y = -102$ (s.e. 23.2) $+ 111$ (s.e. 22.5)x

tiller production. Simple correlations with either plant DM or changes in DM per stem after anthesis were weak. Stepwise multiple regression also failed to reveal any combination of variables useful for predicting tiller death after anthesis.

DISCUSSION

Duration of Tiller Appearance and Death

The duration of the phases of tiller production and death was determined by factors other than the developmental stage of the plant. This agrees with results of Aspinall (1961) for barley and Langer (1963) for grasses. Developmental stage alone is not, therefore, an appropriate variable with which to predict the end of either tiller appearance or tiller death in models and other factors must be taken into account. For example, Fig. 1b shows that drought can cause a premature cessation of tillering: Aspinall (1961) found a similar cessation when nutrient supplies were withdrawn. When plants were thinned in the field, tillering was stimulated to restart in the high population and the duration of tillering was prolonged in the lower population, both responses being associated with an increase in plant growth rate (Fig. 3). Hanada (1983) found the tillering of rice to behave similarly in response to thinning.

Tiller Production Between Emergence and the Start of Stem Extension

During the early stages of development, plant growth rate and dry matter per plant are not good predictors of tillering behaviour over a range of different growing conditions (Fig. 2). Within an experiment the correlation between stems per plant and DM per plant at 45 DUS was high and this agrees with the results of Power and Alessi (1978) and Masle-Meynard and Sebillotte (1981a). However, strong correlations are to be expected because both variables are measures of plant 'size'. An important feature of the differences found between experiments was the failure of many T0 and T1 tillers to emerge in the glasshouse and in the field. Peterson et al. (1982 found that heavy seeds, shallow sowing and strong irradiance were important in controlling the emergence of T0 and T1 and that the presence or absence of these tillers was an important determinant of the subsequent tillering rate (Klepper et al., 1982). To model the growth and appearance of tillers satisfactorily a better understanding of the tillering process is needed. The size and availability of individual source leaves as suppliers of assimilate for bud growth and the availability and utilization of seed reserves will probably have to be taken into account (Dale and Felippe, 1972).

Changes in Tiller Numbers After the Start of Stem Extension

Between the start of stem extension and anthesis, tiller appearance may continue or tiller death start depending on both treatment and environment. Figure 3 supports the notion of Charles-Edwards (1982) that a certain minimum amount of growth must be made during a particular developmental phase if an organ or structure is to be retained. Thus when ΔDM/N was less than about 0.8 g, tiller death was the rule; above this amount further tillering was likely to occur. A similar relationship is implicit in van Keulen's (1982) model of tiller death in wheat. The concept is also supported by experiments with cereals showing that conditions favouring crop growth in the 30 days before anthesis also favour tiller survival (Gifford et al., 1973; Fischer, 1975; Power and Alessi, 1978). However, significantly less dry matter production was needed to ensure the survival of a tiller in an artificial environment than in the field (Fig. 3). Other workers have found markedly different relationships between the amount of structural material present in plants grown in different environments (Jones and Hesketh, 1980). The lack of a general relationship describing the response of tiller death to plant growth is disappointing and emphasises once again the problem of extrapolating from controlled environments to the field.

The severity of tiller death after anthesis was poorly predicted from the measured variates using either simple or multiple regression procedures. More careful experiments will be needed if the causes of this phenomenon are to be understood.

Predicting Ear Numbers

If ear number is to be predicted by simulating tiller production and survival, then it is clear that more attention must be paid to factors determining the appearance of early tillers. The influence of various factors on the appearance and death of tillers between the start of stem extension and anthesis seems closely related to their influence on ΔDM/N and this may form a useful basis for prediction within a given environment, i.e. artificial or field. Nonetheless, ΔDM and N have to be accurately predicted and this requires an accurate simulation of tillering and leaf area expansion up until the start of stem extension. For some types of modelling exercise it may be sufficient simply to relate ears per plant to plant growth rate ($\Delta DM/N$) from TS to anthesis or DM per plant at anthesis. In our experiments these relationships account for 78% and 75% of the variance respectively. However, such correlations are unlikely to be sufficiently general for widespread use and they will do little to enhance our understanding of how the wheat plant grows.

REFERENCES

Aspinall, D., 1961, The control of tillering in the barley plant. I. The pattern of tillering and its relation to nutrient supply, Aust. J. Biol. Sci., 14:493.

Charles-Edwards, D. A., 1982, "Physiological Determinants of Crop Growth", Academic Press, Sydney.

Dale, J. E., and Felippe, G. M., 1972, Effects of shading the first leaf on growth of barley plants. II. Effects on photosynthesis, Ann. Bot., 36:379.

Fischer, R. A., 1975, Yield potential in a dwarf spring wheat and the effect of shading, Crop Sci., 15:607.

Fletcher, G. M., and Dale, J. E., 1977, A comparison of mainstem and tiller growth in barley; apical development and leaf unfolding rates, Ann. Bot., 41:109.

Gifford, R. M., Bremner, P. M., and Jones, D. B., 1973, Assessing photosynthetic limitation to grain yield in field crops, Aust. J. agric. Res., 24:297.

Hampton, J. G., McCloy, B. L., and McMillan, D. R., 1981, Ear population and wheat production, N. Z. J. Exp. Agric., 9:195.

Hanada, K., 1983, Differentiation and development of tiller buds in rice plants, Jap. Agric. Res. Quart., 16:79.

Jones, J. W., and Hesketh, J. D., 1980, Predicting leaf expansion, in: "Predicting Photosynthesis for Ecosystem Models", Vol. II, CRC Press, Boca Raton, Florida.

Klepper, B., Rickman, R. W., and Peterson, C. M., 1982, Quantitative characterization of vegetative development in small cereal grains, Agron. J., 74:789.

Langer, R. H. M., 1963, Tillering in herbage grasses, Herb. Abstr., 33:141.

Langer, R. H. M., Prasad, P. C., and Laude, H. M., 1973, Effect of kinetin on tiller bud elongation in wheat, Ann. Bot., 37:565.

Masle-Meynard, J., and Sebillotte, M., 1981a, Etude de l'hétérogéneité d'un peuplement de blé d'hiver. I. Notion de structure du peuplement, Agronomie, 1:207.

Masle-Meynard, J., and Sebillotte, M., 1981b, Etude de l'hétérogéneité d'un peuplement de blé d'hiver. II. Origines de différentes catégories d'individus du peuplement; éléments de description de sa structure, Agronomie, 1:217.

Peterson, C. M., Klepper, B., and Rickman, R. W., 1982, Tiller development at the coleoptile node in winter wheat, Agron. J., 74:780.

Porter, J. R., 1984, A model of canopy development in winter wheat, J. agric. Sci., Camb., 102:383.

Power, J. F., and Alessi, J., 1978, Tiller development and yield of standard and semi-dwarf spring wheat cultivars as affected by nitrogen fertilizer, J. agric. Sci., Camb., 90:97.

van Keulen, H., 1982, A deterministic approach to modelling of organogenesis in wheat, in: "Simulation of Plant Growth and Crop Production", F. W. T. Penning de Vries and H. H. van Laar, eds., Pudoc, Wageningen.

APPROACHES TO MODELLING CANOPY DEVELOPMENT IN WHEAT

J. R. Porter

Long Ashton Research Station
University of Bristol
Long Ashton, Bristol, U.K.

INTRODUCTION

Most complex crop simulation models incorporate the four basic processes that are fundamental to crop growth: phenological development, canopy development, dry matter production and dry matter partitioning. Environmental factors modify each of these processes in different ways. Our understanding and ability to predict is most advanced for the dry matter production of a given crop canopy, particularly when water and nutrients are adequate. Phenological development is also well understood and, though precise definition is lacking, rules describing the dynamics of crop development in terms of the sequential production, growth and death of leaves and shoots are beginning to emerge (Baker et al., 1980; Masle-Meynard and Sebillotte, 1981a,b; Weigand et al., 1981; Willington and Biscoe, 1982; Baker and Gallagher, 1983a,b; Kirby et al., 1984a,b), but the quantitative effects of environmental factors other than temperature and radiation remain elusive.

In describing canopy development in the Agricultural and Food Research Council (AFRC) model of growth and development of wheat (Porter, 1984; Weir et al., 1984), we have viewed the crop as a population of individuals i.e. shoots, comprised of populations of organs i.e. leaves. The jump from the organ level to the field population level (Fig. 1) covers two orders of biological organisation and this explanatory bridging of two, but not more, levels is within the scope of relevant reductionist explanation as defined by de Wit (1982).

Not all crop models simulate canopy development with detailed descriptions of the production, growth and death of shoots and

ecosystem

community

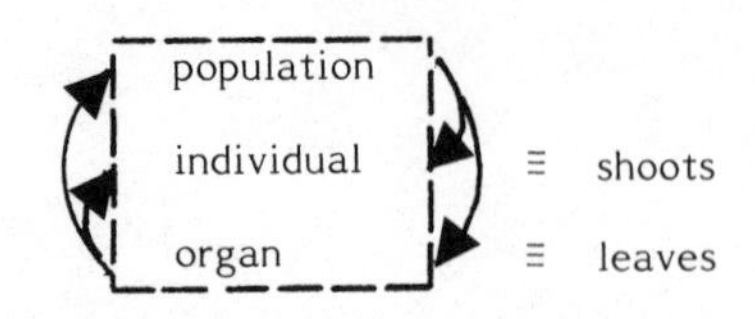

tissues

cell

organelle

molecule

atom

Fig. 1. Levels of organization in biological systems. Upward facing arrows denote that explanation of phenomena at the higher levels can be made at two lower levels. Downward facing arrows denote that concepts appropriate at a higher level are usable at lower levels.

leaves. Some earlier models, e.g. de Wit et al. (1978), and current spectral reflectance-based studies (Chapter 32) use leaf area index (LAI) as an input for simulations of dry matter production. Models of this type are useful to analyse dry matter production over large areas, where detailed simulation would be impossible and to analyse the relative efficiency of different canopies in converting CO_2 to dry matter, e.g. in the presence of disease. A second method has LAI calculated as a direct function of dry matter (van Keulen, 1975). Recently, models have been developed in which LAI is neither input nor a simple function of dry matter but a state variable that both influences dry matter production and is, in turn, affected by it (Johnson and Thornley, 1983; Porter, 1984). Models of this sort are of interest in studying the changes in leaf area with time or with specific environmental conditions or management, e.g. different sowing dates; also they provide the potential to investigate the effects of fungal disease, chemical intervention and herbicide damage on canopy performance.

There are features of wheat canopies that are easy to simulate and others that are not. Identifying them should help to focus attention on where further information is required and, as importantly, help to decide the relative priority of competing experimental approaches.

Modelling canopy development in wheat is made easier because we are dealing with a genetically homogeneous, annual and determinate crop that has no dormancy period and is grown in monoculture. Typically, in a wheat crop, the number of shoots rises to a peak at about the double ridge stage and the decline in shoot numbers commences only after this period of production has ended (Willington and Biscoe, 1982). Thus these two components of the net number of shoots can be treated separately in models of shoot population dynamics and the additional complexity of concurrent births and deaths is fortunately avoided. Furthermore, it has been shown that the maximum number of tiller production sites is related to the number of emerged leaves and can be described by a Fibonacci series (Masle-Meynard and Sebillotte, 1981; Kirby et al., 1984). Departure from this progression may indicate the onset of a shortage in mineral nutrients or carbon assimilates (Chapter 4). Finally, from a modelling viewpoint, it is useful that a single environmental variable, temperature, dominates the environmental influence on both phenological development and the production, growth and death of leaves.

On the other hand, there are a number of problems encountered when producing models of leaf area development. It is not clear whether there is a precise phenological stage associated with the cessation of shoot production. There is little information about the potential variation in the size of leaves on a shoot, although Gallagher (1979) gave values for cv. Huntsman. There is also a need for better criteria to identify those shoots that will die before crop maturity. This will require a mechanistic understanding of the processes which lead to shoot death and also the extent to which carbon and nitrogen from senescing leaves and shoots become available to those parts that continue to grow.

DESCRIPTION OF MODELS

Two models of shoot and leaf production will be described which differ in the algorithms that are used to simulate shoot production. The first model (Model 1) is essentially that described by Porter (1984) in that tiller production starts after the third leaf on the main shoot has emerged; thus the emergence of the first leaf on tiller 1 (T1) coincides with the fourth on the main shoot. The production of new shoots, in the model, continues until the double ridge stage is reached at a rate that is a linear function of temperature, and the production rate is not constrained by assimilate supply. The death of shoots starts after double ridge and continues until anthesis; the number of shoots that die in any period is dependent on shoot age and the density of the crop. The rate of leaf appearance, in thermal time, depends on the rate of change of daylength at crop emergence (Baker et al., 1980) and the end of leaf appearance is such that the last produced leaf will reach maximum

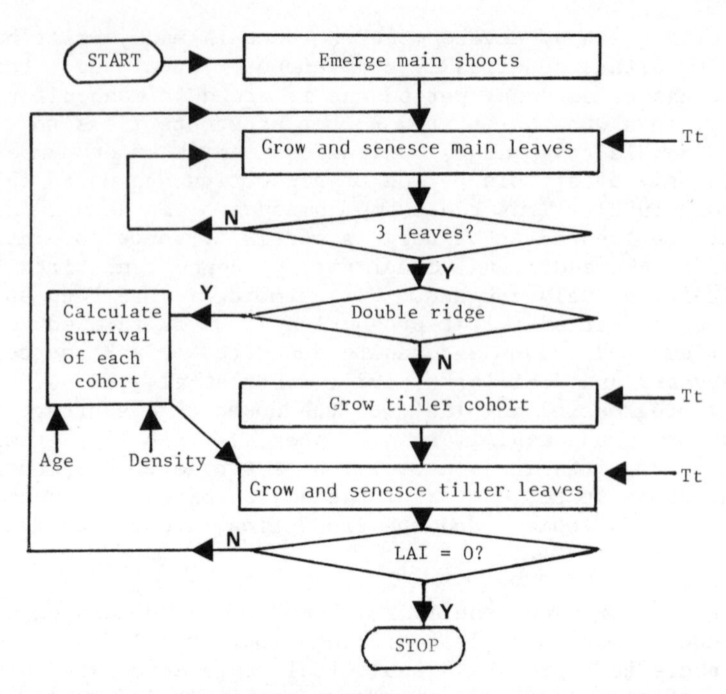

Fig. 2. A flow diagram showing the principal decisions and calculations involved in simulating a crop canopy. Tt, thermal time (degree days).

size by anthesis. This is the flag leaf. Each leaf grows to its maximum size in a thermal time period that is 1.8 times the interval between successive leaves (Gallagher, 1979). There is then a period during which it remains at its maximum size, before senescence begins. Senescence is linked to leaf production in order to maintain between three and four green leaves on a shoot (Weigand et al., 1981). Thus, for situations in which the rate of leaf production is high (for example, the spring sowing of a winter variety) the model operates so that the leaf growth and senescence rates are correspondingly greater. Figure 2 is a flow diagram showing the principal calculations and decisions taken in this model.

A second model (Model 2) uses the Fibonacci series as a basis for computing shoot production (Friend, 1965). The Fibonacci series (1, 1, 2, 3, 5, 8, 13, ..., or in general, $u_i = u_{i-1} + u_{i-2}$, $i \geqslant 3$, $u_1 = 1$) has often been used in descriptions for many species of the helical arrangement of leaves on shoots (phyllotaxy). Figure 3 shows the maximum shoot number for a given emerged plant population at any time, calculated from the leaf number on the main shoot. This potential number assumes that enough assimilate is available to produce a tiller at each potential site. Analysis of data from

Leaf number on main shoot	Maximum shoot number	Maximum number of shoots added in one leaf interval
1	n	
		0
2	n	
		0
3	n	
		n
4	2n	
		n
5	3n	
		2n
6	5n	
		3n
7	8n	

Fig. 3. The potential increase in the number of shoots according to a Fibonacci series; n = number of plants.

A.V.B. Willington and P.V. Biscoe (personal communication) suggests two rules that may constrain this potential production capability. These are that:

1) a 7.5 mg increase in dry matter in the leaf appearance interval is required for each shoot that is initiated.

2) a tillering site only remains competent during a single leaf appearance interval.

This second model can be compared with model 1 to see how changes in the rules of tiller production can affect other aspects of canopy development.

Models 1 and 2 were compared in simulations of 85, 250 or 750 plants m^{-2} for a single sowing date. The effect of sowing date on the behaviour of model 2 was assessed in a simulation of 250 plants m^{-2} for two dates, 15 September 1980 and 15 October 1980. The simulation of the earlier sowing date was compared with experimental results from Rothamsted Experimental Station (Prew et al., 1983).

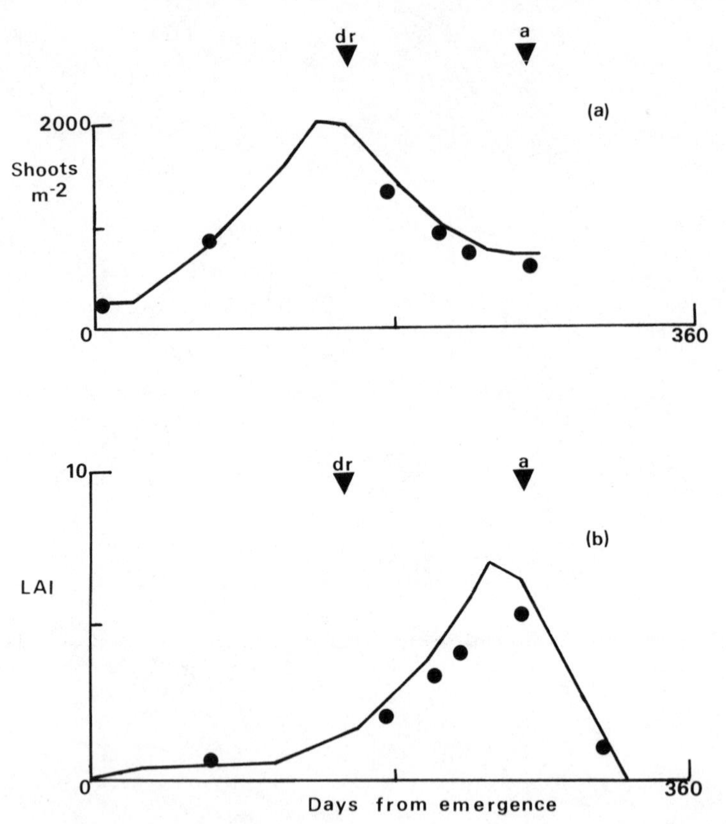

Fig. 4. Observed (●) and simulated (——) changes in (a) shoots m^{-2} and (b) LAI for a crop sown at Rothamsted in September 1980: dr, observed double ridge stage; a, observed anthesis.

SIMULATIONS

Figure 4 shows the observed and simulated (using model 2) total shoot number and LAI for the Rothamsted crop. The observations lie

→

Fig. 5. Simulated changes in shoot number m^{-2} from model 1 (linear production), a–c, and model 2 (Fibonacci production plus carbon limitation), d–f, for 3 plant densities 85 plants m^{-2} (a,d); 250 plants m^{-2} (b,e); 750 plants m^{-2} (c,f). MS, main shoot; T1, tiller 1; T2, tiller 2; RT, other tiller groups; dr, simulated double ridge stage; ts, simulated terminal spikelet stage. Closed circles are shoot number m^{-2} for production as predicted by the Fibonacci series but without any limitation by assimilate supply.

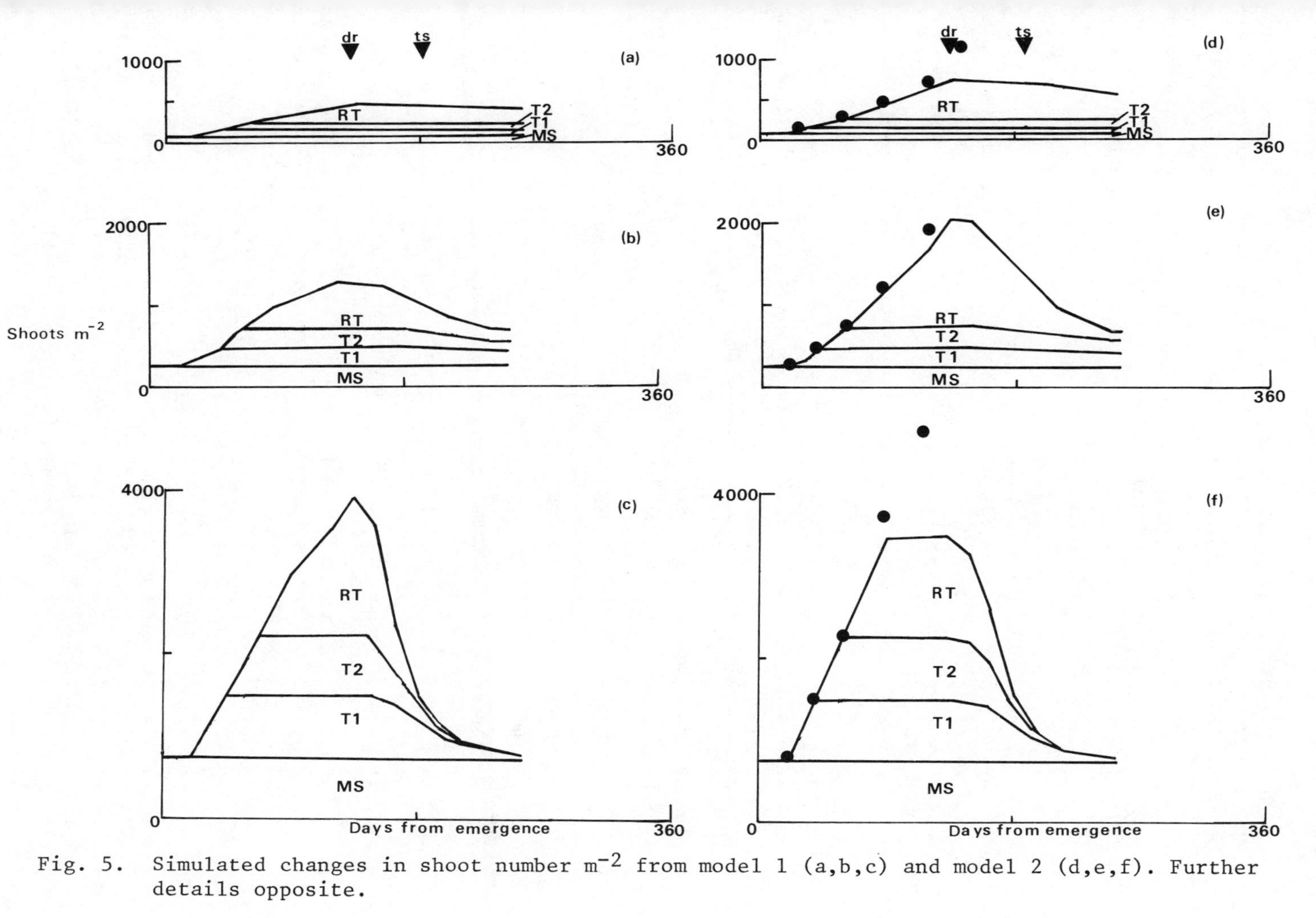

Fig. 5. Simulated changes in shoot number m^{-2} from model 1 (a,b,c) and model 2 (d,e,f). Further details opposite.

close to the simulated curve in both cases, though there are no data at the time of simulated maximum shoot number. The predicted number of shoots in the decline phase is rather larger than the observed number. The model overestimates the rate of increase and the maximu value of the crop LAI but the rate of decline is well simulated. Given the complexity in the model, which represents the overall behaviour of about 15 x 10^3 to 20 x 10^3 organs m^{-2}, the results represent the build-up of the canopy in a way that is close to observation.

Comparison of the performance of the two models (Fig. 5) shows that at 85 plants m^{-2} the Fibonacci-based shoot production (model 2) predicts many more higher order tillers (RT) than model 1. Shoot number in model 2 increases at the Fibonacci rate (indicated by closed circles) until the double ridge stage, when shoot production was terminated in the model. A similar distinction between the models is seen in the simulated case of 250 plants m^{-2}.

In contrast, at 750 plants m^{-2} the maximum shoot number is higher in model 1 than model 2. This is the result of the constrain to production in model 2 associated with the assimilate supply. Thi constraint comes into operation before the double ridge stage and causes the cessation of shoot production. Therefore in model 2, although there is a rule which links the cessation of shoot production to a particular phenological stage, this rule does not always take effect as shoot numbers may be limited by a constraint which operates via the supply of dry matter. In addition, as plant density increases, so the model predicts that maximum shoot number will be reached earlier in crop growth, as observed by Darwinkel (1978).

Differences in the time course of LAI between the models are no as evident as those for shoots (Fig. 6). The most striking effect i that as plant density increases there is a proportional increase in the contribution of LAI from main shoots to the total LAI and hence proportional decrease in the contribution from tillers.

The simulations using model 2 for two sowing dates 30 days apar show total LAI for both sowing dates rising to similar maxima and declining to zero at the same rate (Fig. 7a). However the date on which complete crop cover is achieved (approximately at LAI = 3) is delayed by a month, and this would substantially delay the onset of the rapid increase in dry matter in the spring. The differences between the different sowing dates in the timing of the phenological stages double ridge and terminal spikelet is much less than a month (Weir et al., 1984) but the difference in vegetative development is maintained. In large part, the difference in early leaf area betwee the two simulations is associated with the presence of leaves on the T2 and higher order shoot classes (RT) (Fig. 7b-e).

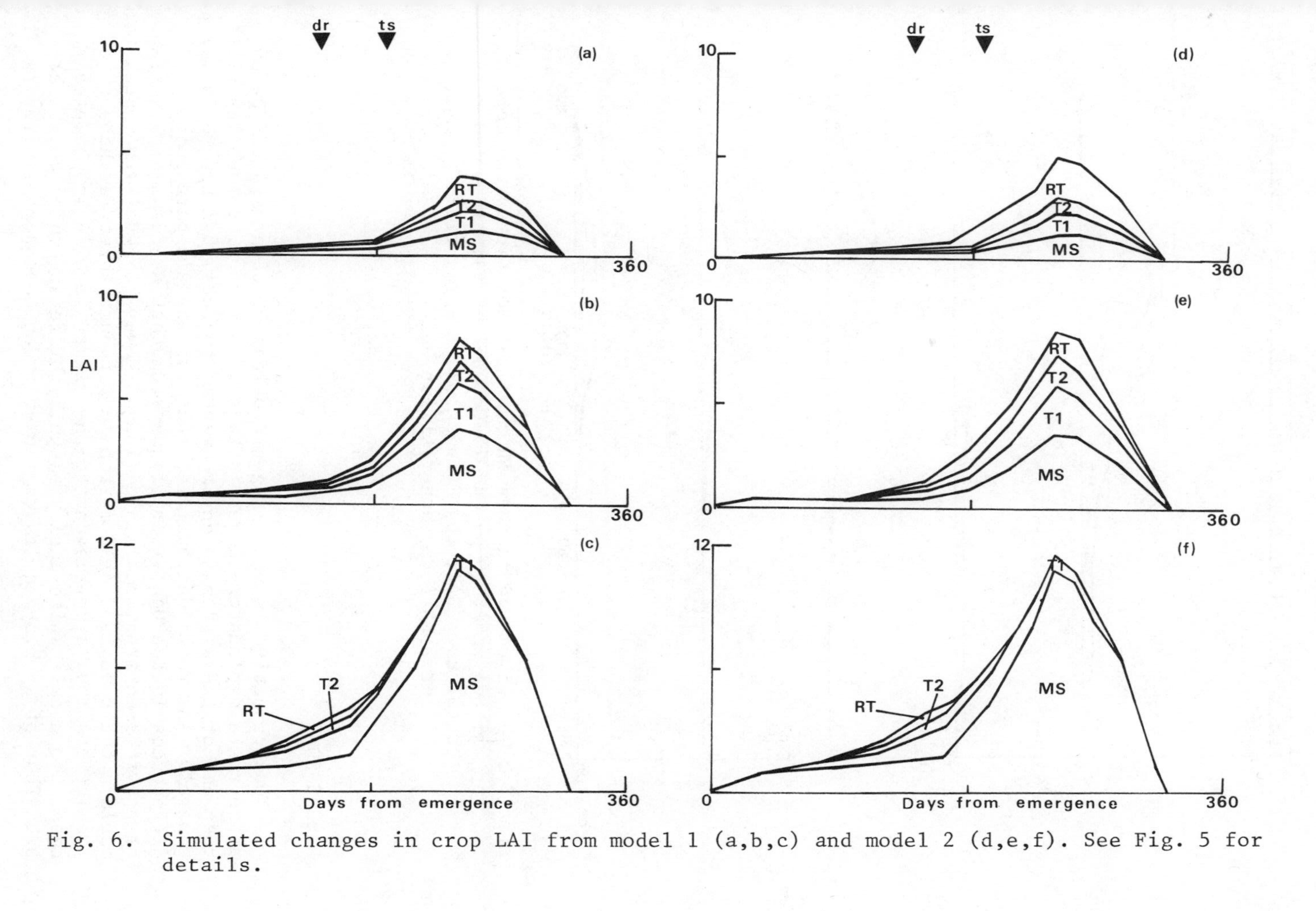

Fig. 6. Simulated changes in crop LAI from model 1 (a,b,c) and model 2 (d,e,f). See Fig. 5 for details.

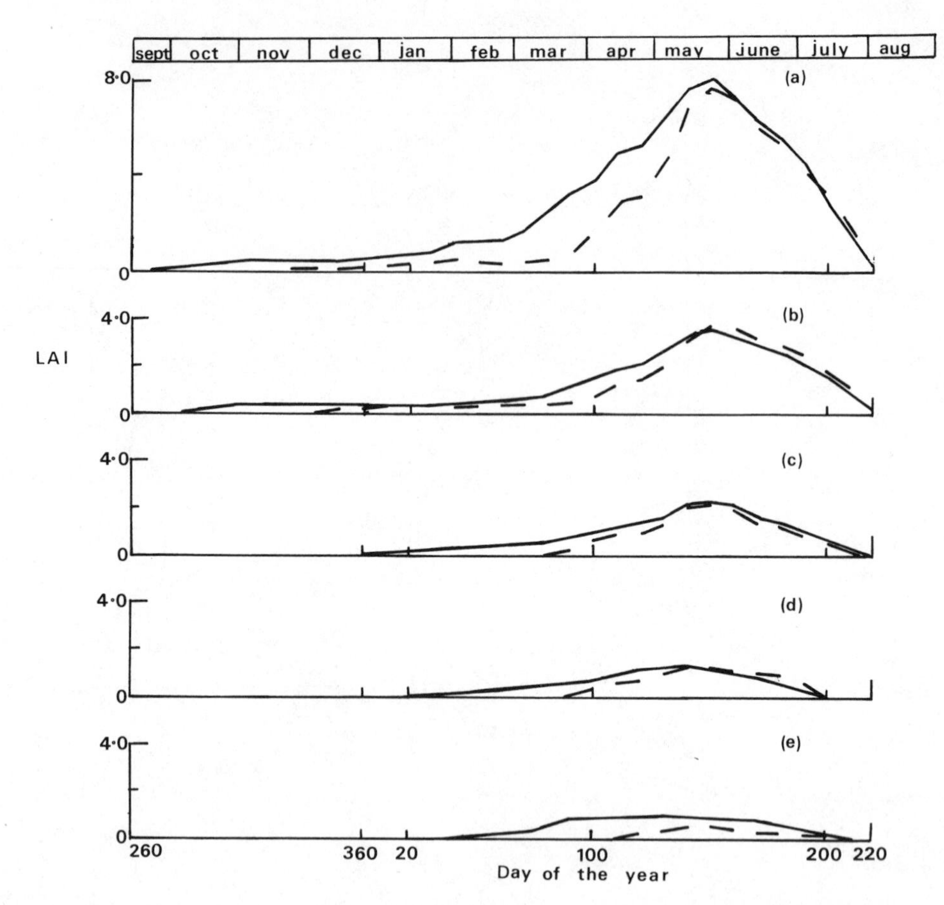

Fig. 7. The simulated changes in crop LAI for two sowing dates, 15 September (——) and 15 October (- -), 1980. (a) total total crop LAI; (b) main shoot LAI; (c) tiller 1 LAI; (d) tiller 2 LAI; (e) other tiller LAI.

DISCUSSION

One objective in modelling biological or other systems at an appropriate explanatory level is to set up general principles of the response of the system to environmental factors. We try to formulate general rules for model building that can explain particular differences in the behaviour of a system that are caused by current or previous environmental conditions. However, models that seek to explain the behaviour of a system in a reductionist way have to offer predictions that are relevant to the higher level. In the present case, two models which employ slightly different rules about the behaviour of shoots and leaves (lower order components) have predicted that as plant density increases the maximum shoot number peaks earlier and fewer higher order tillers are produced; that the main shoot contribution to crop LAI increases with density; and that for a crop sown early the potential increase in radiation capture, compared to that for a later-sown one, is by the extra higher order shoots that the earlier sown crops has developed.

Whilst rules for shoot and leaf production can be defined, we are still unclear as to the reasons for shoot death and how it should be recognised. We are not even sure which factor the plant responds to when regulating shoot survival: is it carbon or nitrogen or both, interactively or additively, or even other nutrients? However, it may be possible to identify general observations that might enable us to model the survival/death of shoots based on more appropriate mechanistic principles. It has been observed in crops where tillers are dying that:-

- the youngest/smallest shoots die first (Thorne, 1962 (barley); Masle-Meynard, 1981a; Fraser et al., 1982)
- high temperatures increase the rate of loss of shoots and leaves (Thorne et al., 1967 (barley); Thorne et al., 1968)
- an increase in the rate of death of shoots occurs at terminal spikelet (Willington and Biscoe, 1982)
- provision of extra nutrients prevents or delays shoot death (Bremner, 1969; Masle-Meynard, 1981a; Fraser and Dougherty, 1982)
- development of adequate nodal roots and/or a threshold number of leaves makes shoots less likely to die (Masle-Meynard, 1981b).

Using such observations, it is possible to postulate a model of shoot death which defines a minimum mean dry weight to be achieved by terminal spikelet for the survival of a group of shoots, after which main shoots obtain all the necessary assimilate to maintain their potential growth. Failure to reach the threshold levels means that members of a shoot group are vulnerable should the assimilate supply become limiting. Actual shortage of assimilate would reduce the number of shoots in those groups whose mean dry weight is furthest below the threshold. Similar ideas have been used by Dayan et al. (1981) in modelling shoot turnover in perennial Rhodes grass (Chloris gayana Kunth.).

The detailed models described in this paper are not appropriate for all situations in which crop physiologists, modellers or agronomists wish to understand and predict the routes to given dry matter production and yield. However, where interest is centred on the understanding of the build-up of green area by a wheat crop in the field, the synthesis of information that a model achieves by its ability to describe unequivocally the processes of phenological development, dry matter production and canopy development should represent a critical and expanding complement to well-defined experimental work.

ACKNOWLEDGEMENT

I thank Dr. M. Shaw for helpful suggestions and for reading the manuscript critically.

REFERENCES

Baker, C. K., and Gallagher, J. N., 1983a, The development of winter wheat in the field. 1. Relation between apical development and plant morphology within and between seasons, J. agric. Sci., Camb., 101:327.

Baker, C. K., and Gallagher, J. N., 1983b, The development of winter wheat in the field. 2. The control of primordium initiation rat by temperature and photoperiod, J. agric. Sci., Camb., 101:337.

Baker, C. K., Gallagher, J. N., and Monteith, J. L., 1980, Daylength change and leaf appearance in winter wheat, Plant, Cell and Environ., 3:285.

Bremner, P. M., 1969, Effects of time and rate of nitrogen application on tillering, sharp 'eyespot' (Rhizoctonia solani) and yield in winter wheat, J. agric. Sci., Camb., 72:273.

Darwinkel, A., 1978, Patterns of tillering and grain production of winter wheat at a wide range of plant densities, Neth. J. agric. Sci., 26:383.

Dayan, E., van Keulen, H., and Dovrat, A., 1981, Tiller dynamics and growth of Rhodes grass after defoliation: a model named TILDYN, Agroecosystems, 7:101.

de Wit, C. T., 1982, Simulation of living systems, in: "Simulation of Plant Growth and Crop Production", F. W. T. Penning de Vries and H. H. van Laar, eds., Pudoc, Wageningen.

de Wit, C. T., et al., 1978, "Simulation of Assimilation, Respiratio and Transpiration of Crops", Pudoc, Wageningen.

Fraser, J. and Dougherty, C. T., 1982, Effects of sowing rate and nitrogen fertiliser on tillering of 'Karamu' and 'Kopara' wheats, Proc. Agron. Soc. New Zealand, 7:81.

Fraser, J., Dougherty, C. T., and Langer, R. H. M., 1982, Dynamics o tiller populations of standard height and semi-dwarf wheats, New Zealand J. Agric. Res., 25:321.

Friend, D. J. C., 1965, Tillering and leaf production in wheat as affected by temperature and leaf intensity, Can. J. Bot., 43:1063.

Gallagher, J. N., 1979, Field studies of cereal leaf growth. 1. Initiation and expansion in relation to temperature and ontogeny, J. exp. Bot., 30:625.

Johnson, I.R., and Thornley, J. H. M., 1983, Vegetative crop growth model incorporating leaf area expansion and senescence, and applied to grass, Pl. Cell Environ., 6:721.

Kirby, E. J. M., Appleyard, M., and Fellowes, G., 1984a, Effect of sowing date and variety on leaf emergence and number of leaves of barley and wheat, Agronomie, (in press).

Kirby, E. J. M., Appleyard, M., and Fellowes, G., 1984b, Leaf appearance and tillering in barley and wheat in response to variety and date of sowing, Agronomie, (in press).

Masle-Meynard, J., 1981a, Relations entre croissance et développement pendant la montaison d'un peuplement de blé d'hiver. Influence des conditions de nutrition, Agronomie, 1:363.

Masle-Meynard, J., 1981b, Elaboration du nombre d'épis d'un peuplement de blé d'hiver en situation de competition pour l'azote. Mise en evidence d'une stade critique pour la montée d'une taille, Agronomie, 1:623.

Masle-Meynard, J., and Sebillotte, M., 1981a, Etude de l'hétérogénéité d'un peuplement de blé d'hiver. I. Notion de structure du peuplement, Agronomie, 1:207.

Masle-Meynard, J., and Sebillotte, M., 1981b, Etude de l'hétérogénéité d'un peuplement de blé d'hiver. II. Origine des differentes catégories d'individus du peuplement: elements de description de sa structure, Agronomie, 1:216.

Porter, J. R., 1984, A model of canopy development in winter wheat, J. agric. Sci., Camb., 102:383.

Prew, R. D., Church, B. M., Dewar, A. M., Lacey, J., Penny, A., Plumb, R. T., Thorne, G. N., Todd, A. D., and Williams, T. D., 1983, Effects of eight factors on the growth and nutrient uptake of winter wheat and on the incidence of pests and diseases, J. agric. Sci., Camb., 100:303.

Thorne, G. N., 1962, Survival of tillers and distribution of dry matter between ear and shoots of barley varieties, Ann. Bot., 26:37.

Thorne, G. N., Ford, M. A., and Watson, D. J., 1967, Effects of temperature variation at different times on growth and yield of sugar beet and barley, Ann. Bot., 31:71.

Thorne, G. N., Ford, M. A., and Watson, D. J., 1968, Growth, development and yield of spring wheat in artificial climates, Ann. Bot., 32:425.

van Keulen, H., 1975, "Simulation of Water Use and Herbage Growth in Arid Regions", Pudoc, Wageningen.

Weigand, C. L., Gerbermann, A. H., and Cuellar, J. A., 1981, Development and yield of hard red winter wheats under semitropical conditions, Agron. J., 73:29.

Weir, A. H., Bragg, P. L., Porter, J. R., and Rayner, J. H., 1984, A winter wheat crop simulation model without water or nutrient limitations, J. agric. Sci., Camb., 102:371.

Willington, V. B. A., and Biscoe, P. V., 1982, Growth and development of winter wheat, I.C.I. Agricultural Division Financed Research Programme Annual Report 2, Broom's Barn Experimental Station.

DEVELOPMENTAL RELATIONSHIPS AMONG ROOTS, LEAVES AND TILLERS IN WINTER WHEAT

R. W. Rickman*, B. Klepper*, and R. K. Belford ‡

*USDA-ARS, Columbia Plateau Conservation Research Center Pendleton, Oregon, USA

‡AFRC Letcombe Laboratory Wantage, UK

INTRODUCTION

Most of the effort on modelling the development of cereals has gone into research on shoots. Little has been done to elucidate the developmental history of root systems. As a result, below-ground aspects of cereal yield modelling are too crude to be useful, especially if models purport to respond to environmental variables which influence growth and yield through their action on the root system. Variables such as soil water and nutrient supply, aeration, and levels of microbial inoculants can be affected by management. Manipulations of these variables to benefit the plant could be done more reliably if more information about the root system were available.

Seminal roots arise from primordia which are already present in the seed (Esau, 1965). This root system usually comprises three to six root axes and their branches (Percival, 1921). MacKey (1973) found that cultivar and seed size were the main factors which affected number of seminal roots in the germ. The position of the grain in the ear determined both the grain weight and the number of root primordia (MacKey, 1978). Apart from laboratory-based research on seedlings however, few investigators have followed the later development and depth of penetration of seminal roots as compared to crown roots. In most field studies, a mixture of seminal, coleoptilar and crown roots have been sampled as one population. Additionally, little effort has been made to differentiate crown roots coming from the main axis from those attached to tillers, although one can surmise that, on average at least, tiller roots should penetrate to shallower depths than main-stem roots.

The work reported here was undertaken to discover, for cereal roots, developmental information similar to that already described for shoots (Klepper et al., 1982).

EXPERIMENTAL METHODS

Experiments were done in field plots at the Columbia Plateau Conservation Research Center near Pendleton, Oregon, on a 1.5 to 2.0 m deep Walla Walla silt loam (mixed mesic family of typic Haploxerolls). The plots had been planted to peas for the previous three years. The Research Center is located on the border between areas of wheat-pea and wheat-fallow rotations in north-eastern Orego and receives about 400 mm annual precipitation.

Prior to planting, pea vines from the crop of the previous spring were removed. The plots were disc ploughed twice on 9 September, cultivated with a spring-tooth harrow on 5 October, and sown with winter wheat (*Triticum aestivum* L. em. Thell. cv. Stephens - a soft white wheat) on 14 October with a double-disc Melroe end-wheel drill; the seed depth was 4 to 5 cm, with 17.8 cm row spacing. The experiment consisted of 12 plots: four treatments each replicated three times. Three of the treatments were imposed b broadcasting 90 kg ha^{-1} of nitrogen (as $(NH_4)_2SO_4$) either on 9 October 1981 (preplant), 24 February 1982 (double ridge) or 2 Apri 1982 (terminal spikelet); the fourth was an unfertilized control. Plots were 3.5 m wide and 12 m long. The dates for spring fertilization were determined after microscopic examination of dissected main-stem apices.

After seedling emergence, metal tubes 17.8 cm square and 1.1 m long were pushed into the soil so that the root systems of about 6-8 plants were totally confined to the soil within the tube (Belford et al., 1982). This permitted extraction of the intact root-soil mass during the following spring on 23 February, 17 March, 19 April and 1 June 1982. A removable side of the tubes permitted soil to be washed from the intact roots after extraction of the tube. Each roo axis was then identified (Klepper et al., 1984) and the following measurements were made: degree of branching, maximum depth of root axis, number of branches per cm on the axis within each 10 cm depth increment and total root length present at each depth. The total living root length present at each depth was determined by counting the grid intersections of roots stained red by Congo Red (Ward et al., 1978) either by eye or photographically with transects scanned by an automated system (Voorhees, 1980).

Plants from two 0.5 m lengths of row were removed on 23 February, 22 March, 19-20 April, 1-4 June and on 10 May 1982, and analysed for tiller identification, leaf development (Haun, 1973) on each shoot and evidence of tiller abortion.

ROOT NAMING SYSTEM

Cereal plant development can be described using a naming system that gives a unique number to each node on the plant (Klepper et al., 1982). The first foliar node, N1, bears the first true leaf (L1) and in the axil of L1 is a tiller bud (T1). The second foliar leaf (L2) is found on the opposite side of the stem from L1 and has a tiller (T2) associated with it. Below the crown are three more nodes, the coleoptilar node, and the epiblast and scutellar nodes in the seed. These are numbered 0, -1 and -2, respectively.

The root-naming system (Fig. 1) extends this description by assigning numbers to root axes with respect to the node from which they arise, and describing the direction of growth of the root with respect to the midrib of the leaf at the node, with X (towards), Y (away from), A (to the left), and B (to the right) (Klepper et al.,

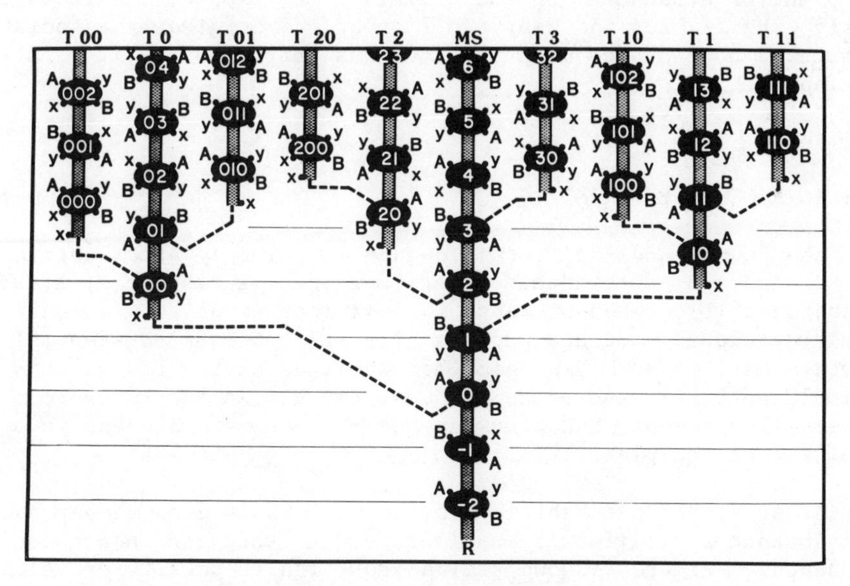

Fig. 1. Spatial and temporal relationships of seminal roots (negative node numbers), nodal roots produced at the coleoptilar node (designated by "0"), and crown roots on main stems (single digits), primary tillers (double digits), and secondary tillers (triple digits). Roots are represented by dots. (Taken from Klepper et al., 1984.)

1984). For example, the earliest crown roots to be seen are usually 1A (left of midrib) and 1B (right of midrib) which appear simultaneously on either side of the stem at about the same time tha T1 elongates. Note that the reference point turns 180° with each successive node so that the 2A root is directly above 1B. Thus the roots associated with the main shoot can be referred to as seminal (from nodes -1, -2), coleoptilar (from node 0), or crown (from nodes 1, 2, 3, etc.).

The first seminal root to appear is the radicle called R. The rest of the seminal roots arise from the scutellar (-2) and epiblast (-1) nodes (Percival, 1921; McCall, 1934; Boyd and Avery, 1936). They are -2A, -2B, -2Y, -1A, -1B and -1X, i.e. seven identifiable roots. The -2Y root rarely develops.

In like manner, nodes on primary tillers have two digit numbers and secondary tillers have a three digit number (Klepper et al., 1983). The prophyll has a designation 0. Thus, the prophyll node c T1 would be Node 10 and would bear T10 in its axil. The first folia leaf on T1 is L11 and it bears T11 in its axil. Similarly, the prophyll tiller on T10 is called T100. The roots associated with tillers include those arising from the nodes on the tiller as well a the X-root at the parent node. This root develops immediately below the point of attachment of the tiller. For example, the roots on a T2 might be 2X, 20A and 21B; the 2X root was originally associated with the 2 node on the main stem but was "captured" by T2 as it developed.

SEMINAL ROOT DEVELOPMENT

The R, -2A and -2B roots elongate immediately after germination The -1A and -1B roots appear after emergence, in the early stages of expansion of L1. When present, the -1X root is often a strong and highly developed root and appears after L1 has made substantial growth. Earlier work has described wheat as having from 3 to 6 seminal roots. In the present study, the R, -2A and -2B were universally present, and roots at the -1 node were present in only about 75% of the plants.

After seedling establishment, seminal roots grow deeper and their branches proliferate and increase the root length present at all depths. Figure 2 shows seminal root length density profiles for four sampling times for plants from the four treatments. Note that the seminal root system continues to proliferate well into late spring. Fall-applied nitrogen increased the overall seminal root length density during early spring by about 15%. There were no significant effects of spring-applied fertilizer on the seminal root systems (Fig. 2).

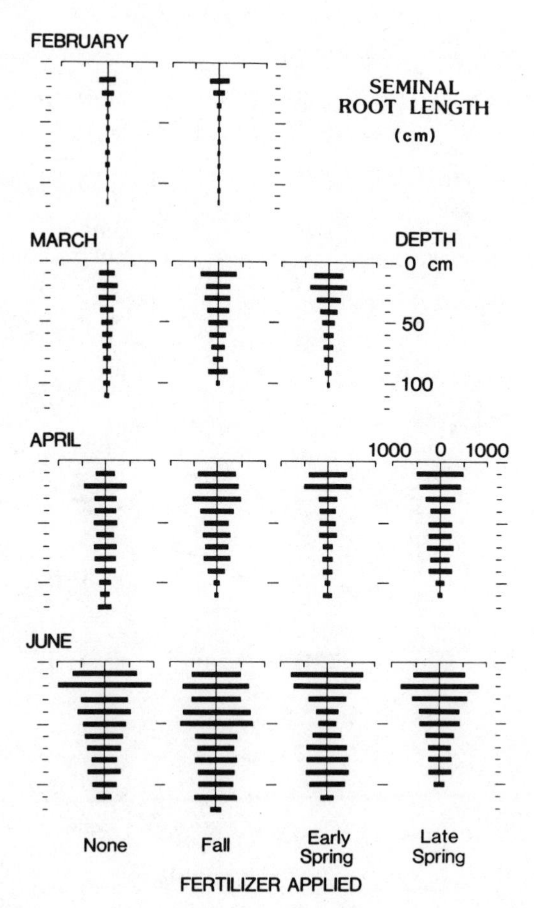

Fig. 2. Seminal root axis length at four sampling times as related to timing of nitrogen application. The horizontal bars show the length of roots in each 10 cm layer of soil, cross sectional area 317 cm^2.

DEVELOPMENT OF ROOTS AT THE COLEOPTILAR NODE

Roots at the coleoptilar node are sometimes confused with seminal roots because the internode between the epiblast node and the coleoptilar node does not elongate in wheat. The OA and OB root were commonly found on plants with three or more main stem leaves. The OX root appeared if a healthy tiller was produced at the coleoptilar node (TO) and after TO had at least two leaves. The OY root was rare.

Figure 3a shows the developmental history of the root systems growing from the coleoptilar node for the four treatments. The solid bar shows the contribution of the OA and OB roots and the dashed bar

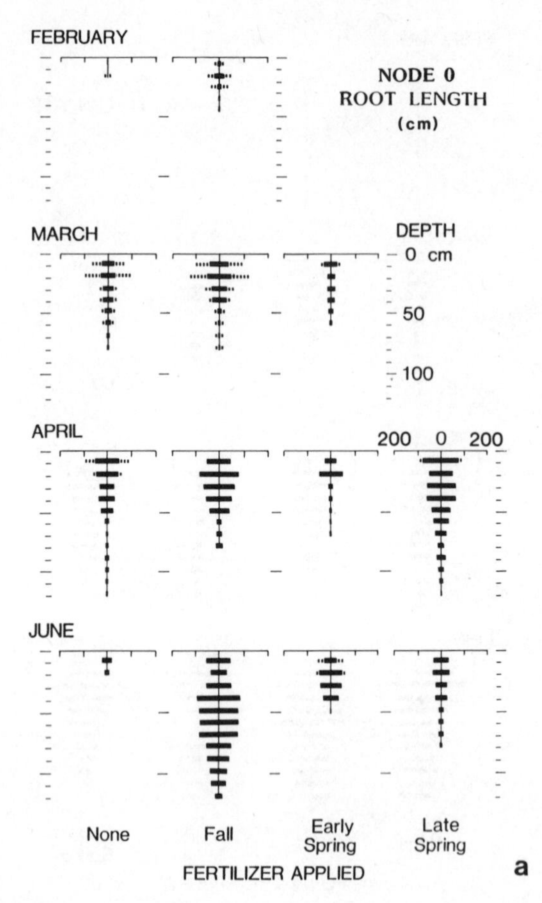

Fig. 3. Development of roots at (a) the coleoptilar node and at fou sampling times as related to timing of nitrogen application. The solid horizontal bars show the length of and B roots and the dashed bar of X and Y roots in each 10 cm layer, cross sectional area 317 cm^2.

gives OX and OY contributions. Absence of fertilizer caused a marked depression in rooting from this node throughout the growing season, and application of spring fertilizer did not increase the rooting from this node significantly. Root development at the coleoptilar node is more erratic than that at any other node.

DEVELOPMENT OF MAIN STEM CROWN ROOTS

Each main-stem node develops an A and a B root about three phyllochrons after the leaf at that node first appears. If a tiller is produced at the node, an X root and sometimes a Y root will

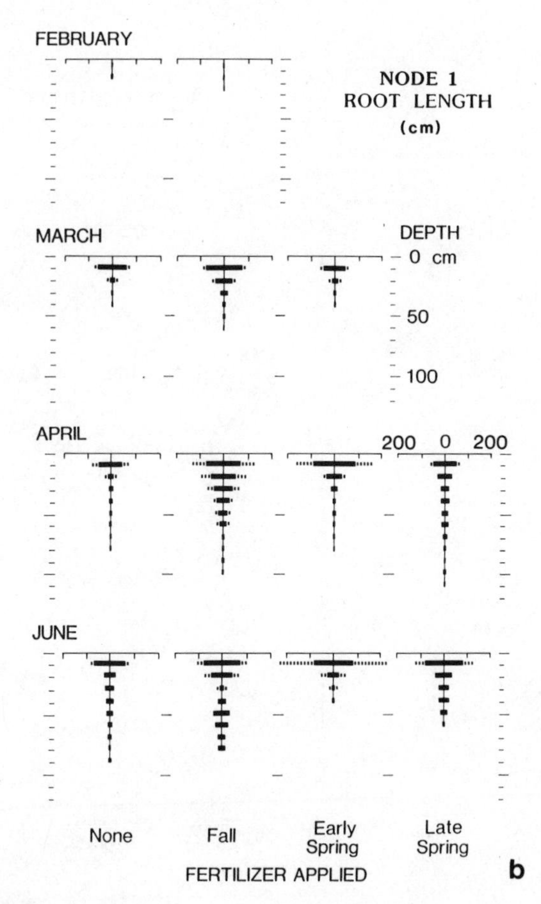

Fig. 3. Development of roots at (b) the first crown node (N1) and at four sampling times as related to timing of nitrogen application. The solid horizontal bars show the length of A and B roots and the dashed bar of X and Y roots in each 10 cm layer, cross sectional area 317 cm^2.

elongate about 4.5 phyllochrons after the first appearance of a leaf at the node. For example, the 2A and 2B roots generally begin to elongate about four phyllochrons after emergence and the 2X and 2Y roots about 5.5 phyllochrons after emergence.

Root length profiles from axes associated with nodes 1, 2 and 3 are shown in Figs. 3b, c and d, respectively. The early-spring nitrogen was applied when main stems had about 4.8 leaves; late-spring fertilization coincided with about 6.9 leaves on main stems. Therefore, the early fertilizer was applied after the A and B roots at the 1 and 2 nodes had elongated but at almost the same time as the X and Y roots appeared at node 1. The A and B roots at node 3

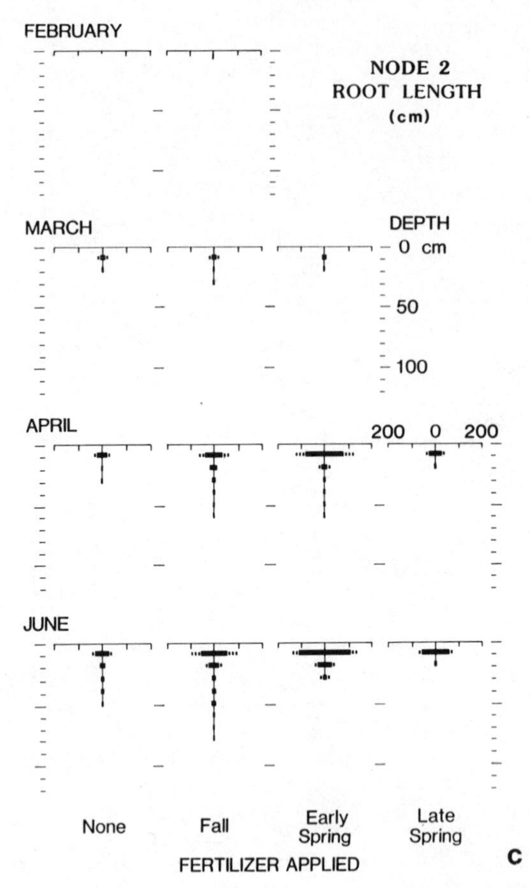

Fig. 3. Development of roots at (c) the second crown node (N2) and at four sampling times as related to timing of nitrogen application. The solid horizontal bars show the length of and B roots and the dashed bar of X and Y roots in each 10 cm layer, cross sectional area 317 cm^2.

and the X and Y roots at nodes 2 and 3 elongated between the two fertilizer application times.

Effects of the early spring fertilization can be seen in the increase in root length on X and Y roots at nodes 1, 2 and 3 and on the A and B roots at nodes 2 and 3. The length of main-stem crown roots on the plants which received fertilizer in late spring are not significantly different from those in the unfertilized control treatment.

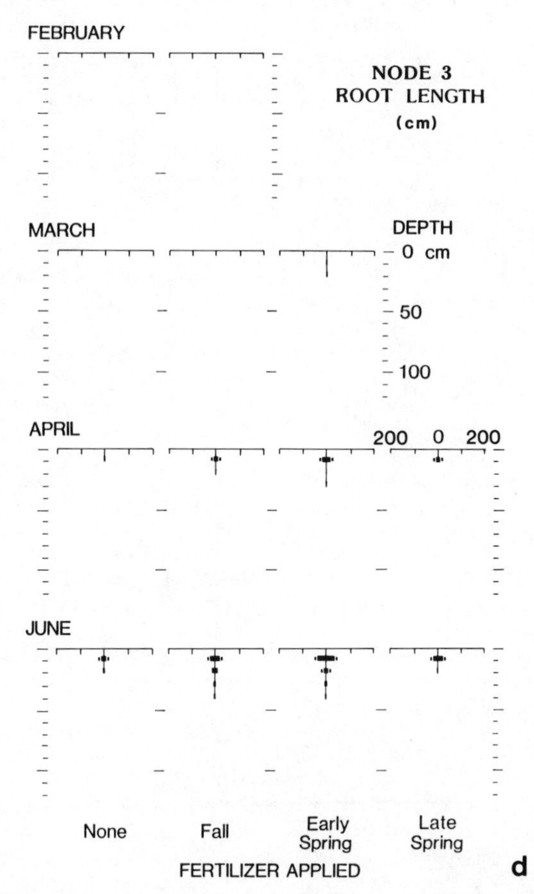

Fig. 3. Development of roots at (d) the third crown node (N3) and at four sampling times as related to timing of nitrogen application. The solid horizontal bars show the length of A and B roots and the dashed bar of X and Y roots in each 10 cm layer, cross sectional area 317 cm^2.

ROOT DEVELOPMENT ON TILLERS

Generally, tiller roots begin to grow after the tiller has at least two leaves. Figure 4 shows the generalised pattern of tiller root development for a tiller, related to the number of phyllochrons since it appeared. Just as on the main stem, the root axes on tillers elongate in an orderly and predictable way with respect to a phyllochron time scale. When total length of roots on a tiller was plotted as a function of the number of emerged leaves, it was found that the length increased approximately exponentially once the tiller had three emerged leaves (Fig. 5). The depth of rooting appears to

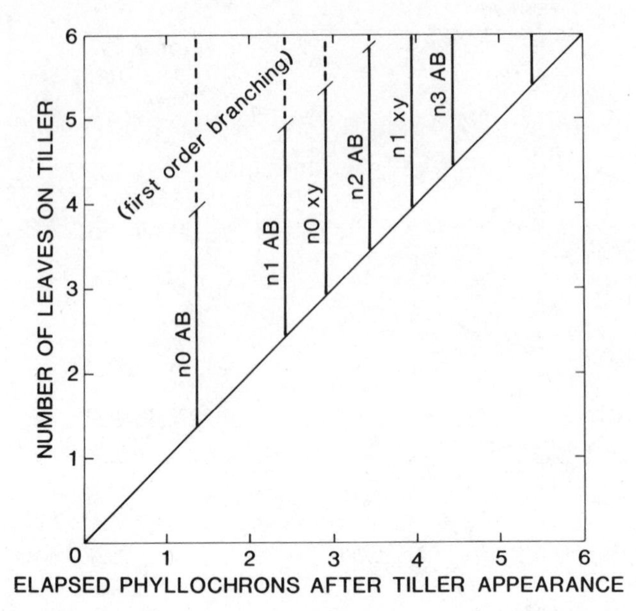

Fig. 4. Relationships among leaf, tiller and root development at main stem nodes. (Taken from Klepper et al., 1984.)

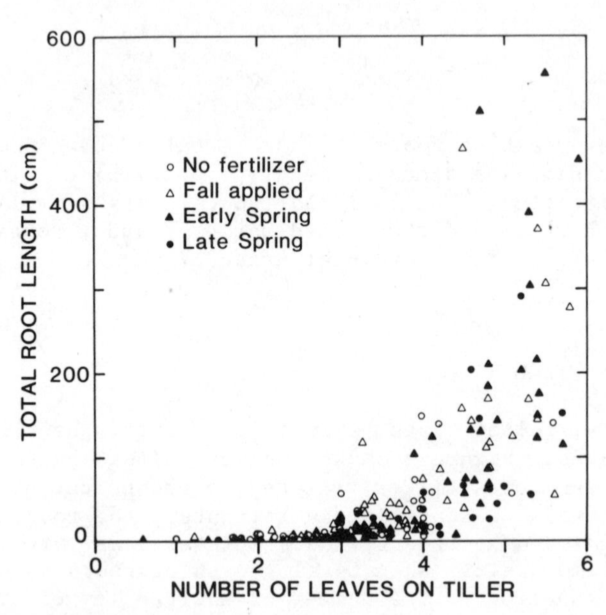

Fig. 5. Relationship between total root length and leaf number for tillers on plants from all four treatments in March and April.

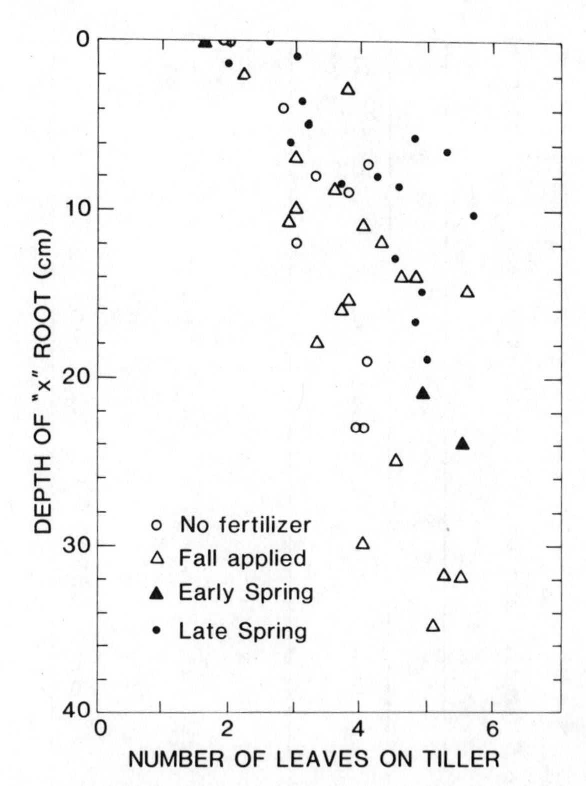

Fig. 6. Depth of deepest root on a tiller as related to the number of leaves on the tiller. Only unbroken roots were included here.

increase linearly with leaf number (Fig. 6). (Depth of rooting will of course be strongly influenced by the specific soil profile under study.)

Table 1 shows the April abortion patterns of primary tillers in the four treatments. These data show that applying nitrogen in the late spring (at terminal spikelet on the main stem) did not change the tiller loss pattern, but adding it at double ridge resulted in fewer of the primary tillers aborting. The average root length for a tiller class was not correlated with abortion of that tiller class (Table 1). The only treatment which had significant secondary tiller abortion in April was the fall-fertilized treatment (data not shown). Table 2 summarizes the condition and number of leaves on tillers in April samples. The unfertilized plants had ceased tillering as shown by the small numbers of living tillers with low leaf numbers. In contrast, the plants fertilized in fall and early spring had large numbers of living tillers with small leaf numbers. Notice that the

Table 1. The number, average Haun stage values (leaf number) and average total root length of aborted tillers on sets of 14 plants in April.

Fertilizer treatment	T0			T1			T2			T3			T4		
	n	Haun	RL cm	n	Haun	RL cm	n	Haun	RL cm	n	Haun	RL cm	n	Haun	RL cm
None	2	2.0	0	8	4.4	59.0	6	3.3	13.8	3	1.3	0	0	-	-
Fall	0	-	-	2	3.1	45.2	3	3.4	10.7	4	2.7	15.2	4	1.7	5.5
Early	0	-	-	0	-	-	1	2.0	2	3	2.7	0.9	0	-	-
Late	1	1.0	0	6	3.8	53.3	5	3.0	12.0	3	2.2	0.7	0	-	-

Table 2. Tiller conditon as related to leaf number on plants in April for the four fertilizer treatments, none (N), fall (F), early spring (E) and late spring (L) applied nitrogen. The tillers include secondary tillers not mentioned in Table 1. On all treatments, 14 plants were sampled.

Treatment	Tiller number	per cent living or dead	Number of leaves on tiller				
			0-1.9	2-2.9	3-3.9	4-4.9	5-5.9
N	12 Living	39	1	1	8	0	2
	19 Dead	61	3	1	5	10	0
F	37 Living	65	4	6	9	11	7
	20 Dead	35	7	7	4	2	0
E	39 Living	91	7	3	5	14	10
	4 Dead	9	0	2	2	0	0
L	20 Living	57	0	3	5	8	4
	15 Dead	43	2	1	8	4	0

plants from plots fertilized in early spring have much less tiller abortion than those fertilized in fall or in late spring.

LEAF, TILLER AND ROOT DEVELOPMENT

The idealized developmental pattern of the main stem leaves, tillers and roots of an unstressed wheat plant, related to phyllochron units, is shown in Fig. 7. This graph allows one to estimate the number of leaves to be expected on the tillers as well as the state of branching on those root axes which are borne on the main stem. For example, an unstressed plant with 5.5 leaves on the main stem would have about 3.5 leaves on T0, 2.6 leaves on T1, 2.1 leaves on T2, and 1.1 leaves on T3. This shoot information is obtained by laying a straight-edge vertically at 5.5 phyllochrons and reading the leaf number for each culm from the ordinate. Similarly a straight-edge placed horizontally at 5.5 leaves will show the order of branching of particular root axes. There would be second-order branching on the seminal roots, but the crown roots would be mostly unbranched with only the 1A and 1B roots showing first-order branching. (Each axis branches about 2.5 phyllochrons after it has started to elongate and this same interval is required before the branches produce laterals.)

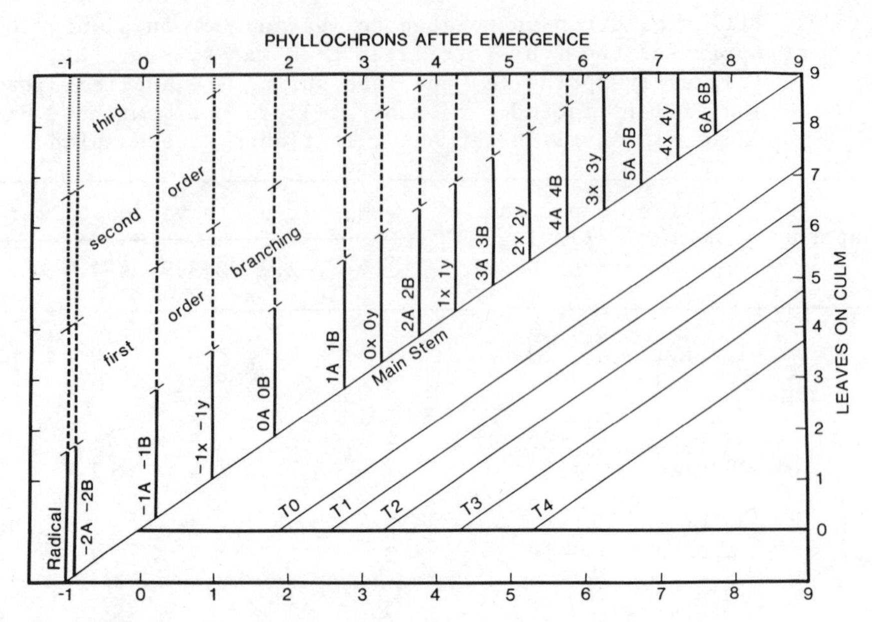

Fig. 7. Phyllochron relationships of developmental patterns of leaves and roots on a tiller. To identify roots, substitute the tiller identification number for n.

Figure 8 shows the relative depth of rooting and numbers of root axes to be found on the main stem and tillers of a typical wheat plant in mid-spring developed in a non-restrictive soil. The oldest and largest shoots have the deepest and largest root systems.

CONCLUSION

The cereal root naming system used in this work has permitted the orderly development sequence of root axes to be analysed and related to the developmental history of shoots. The same time scale based on the phyllochron can be applied to shoot development and to the production of root axes and the branching history of at least the shallower parts of these axes. This information should be of use in modelling root axis appearance and branching pattern of cereals in a uniform, deep friable soil. However, it is important to understand that the information summarised in Figs 4 and 7 merely specifies a sequencing of events, i.e. the appearance of plant morphological units relative to one another. There is no prediction of the relative sizes of these morphological units.

For cereal root growth models to apply to local conditions, information must be obtained on relationships between physical and chemical conditions of specific soil profiles and root growth rates

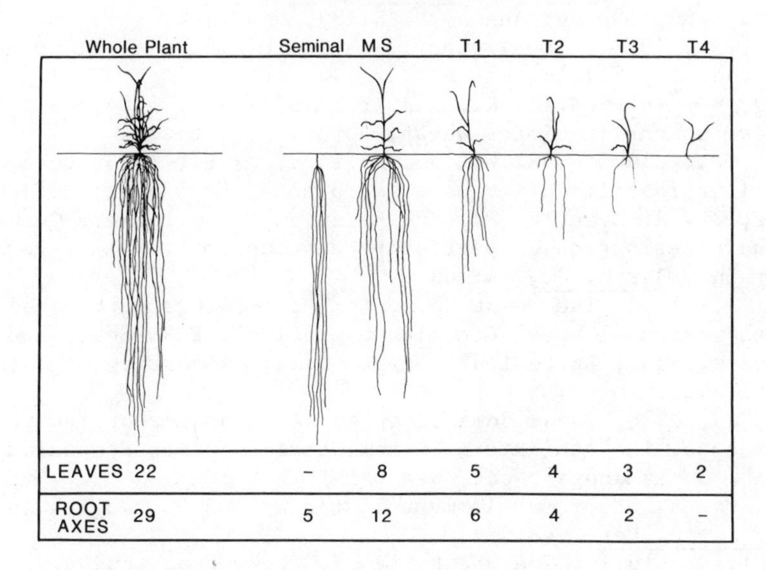

Fig. 8. Relative shoot development and root axis development for the component shoots of a wheat plant.

and branching habits. Development of this information from several sites will provide the basis for a generic model which will eventually permit generalised predictions of root length density profiles. This paper gives the timing of axis appearance which will make such model development feasible.

ACKNOWLEDGEMENTS

The authors wish to thank W. R. Warn for assistance with the design and construction of the metal extraction tubes, P. E. Rasmussen and D. E. Wilkins for use of plants from their fertility trials for part of this study and T. R. Toll, G. E. Fischbacher, K. Skirvin and L. Baarstad for skilled technical assistance. Award of a Nuffield-Leverhulme Civil Service Fellowship to R. K. Belford permitted this study to be done while he was on study leave in Oregon.

REFERENCES

Belford, R. K., Rickman, R. W., Klepper, B., and Allmaras, R. R., 1982, A new technique for sampling intact shoot-root systems of field grown cereal plants, Agron. Abstr., 11.

Boyd, L., and Avery, G. S., 1936, Grass seedling anatomy; the first

internode of *Avena* and *Triticum*, *Bot. Gaz.*, 97:756.
Esau, K., 1965, "Plant Anatomy", Wiley, New York.
Haun, J. R., 1973, Visual quantification of wheat development, *Agron. J.*, 65:116.
Klepper, B., Belford, R. K., and Rickman, R. W., 1984, Root and shoot development in winter wheat, *Agron. J.*, 76:117.
Klepper, B., Rickman, R. W., and Belford, R. K., 1983, Leaf and tiller identification on wheat plants, *Crop Sci.*, 23:1002.
Klepper, B., Rickman, R. W., and Peterson, C. M., 1982, Quantitative characterisation of vegetative development in small cereal grains, *Agron. J.*, 74:789.
MacKey, J., 1973, The Wheat Root, *in* "Proceedings of the Fourth International Wheat Genetics Symposium", E.R. Sears and L.M.S. Sears, eds., Agricultural Expt. Sta., University of Missouri, Colombia.
MacKey, J., 1978, Wheat domestication as a shoot-root interrelation process, *in* "5th International Wheat Genetics Symposium, Vol. II", S. Ramanujam, ed., New Delhi.
McCall, M. A., 1934, Developmental anatomy and homologies in wheat, *J. Agric. Res.*, 48:283.
Percival, J., 1921, "The Wheat Plant", Duckworth, London.
Voorhees, W. B., 1980, Root length measurement with a computer-controlled digital scanning microdensitometer, *Agron. J.*, 72:847.
Ward, K. J., Klepper, B., Rickman, R. W., and Allmaras, R.R., 1978, Quantitative estimation of living wheat-root lengths in soil cores, *Agron. J.*, 70:675.

MEASUREMENT OF THE GROWTH OF WHEAT ROOTS USING A TV CAMERA SYSTEM IN THE FIELD

R. K. Belford and F. K. G. Henderson

AFRC Letcombe Laboratory
Wantage, U.K.

INTRODUCTION

Modelling the growth of wheat requires knowledge of the growth and distribution of the whole root system. The phenology of root development, and the relations between leaves, shoots and roots of winter wheat are now understood (Klepper et al., 1984 and Chapter 8); it is therefore desirable that field measurements can distinguish between seminal, nodal and lateral roots, and to know how individual roots respond to stresses in the soil environment. The aim of field measurements of roots must be towards prediction of the water and nutrient uptake capabilities of the whole root system based on the length, distribution and function of component axes within the soil profile.

Many techniques have been described for the study of root systems (Bohm, 1979), and several groups now use coring or profile wall procedures for routine field examination of roots. However, such techniques have yielded little of the detailed information specified above because they are destructive and disrupt the continuity of measurement at any sampling site, making identification of axes difficult. Such techniques are also very labour-intensive, which precludes the frequent sampling within a season which is desirable to complement sampling of shoots.

In situ observation of root growth is non-destructive, minimises the point-to-point variation that is inherent with other techniques and is potentially very rapid. Observation tubes ("mini-rhizotrons") which allow study of root growth using simple mirrors and a light source (Bohm, 1974) are inexpensive. They have given reasonable correlations between washed root lengths and *in* situ measurements of root length, in determinations of root growth and distribution (Gregory, 1979; Bragg et al., 1983). However, simple equipment is awkward to use to obtain quantitative data in all field conditions. Recently, several authors have described technical improvements to minimise these practical difficulties; Sanders and Brown (1978) used a 35 mm camera to record data, while Upchurch and Ritchie (1983) described a borescope/video system to obtain a permanent record of root images.

This paper describes a lightweight TV camera/recorder system developed at Letcombe Laboratory to study root growth of winter wheat on a range of soil types.

TECHNIQUE

Transparent tubes, 5 cm outside diameter and up to 2 m long, are installed at an angle of 45° between and parallel to rows of wheat 17 cm apart; the tubes are marked with grids of either 4 x 2.5 or 4 x 5 cm. Holes are made by an auger mounted in a jig, just at the time of crop emergence; 45° installation is used to ensure that roots intercepting the tube have grown through undisturbed soil above the tube, and to minimise the subsequent tracking of roots along the interface between tube and soil. This can be a problem in heavy soils, where vertical tubes create a plane of weakness for rapid root exploration, giving rise to atypical root distributions adjacent to the tube (Bragg et al., 1983). Both glass and acrylic (plexiglass) tubes have given similar root information, but plastic is preferred for routine work because of its greater flexibility in swelling/shrinking clay soils.

A black and white television camera, 40 mm in diameter[1], is lowered inside the tube by hand to scan the upper surface of the tube; the camera is stopped briefly at each grid position. The camera is connected by a cable to a control unit; this contains a small monitor screen to check image quality before recording on a portable video recorder. The complete camera/control unit/recorder assembly is battery powered, and weighs 18 kg. Setting up, scanning and recording root images from one tube takes less than 5 minutes, with two operators.

Recorded images are transcribed in the laboratory. Counts of the number of roots, or measurements of the total length of root which have intercepted the tube within each grid, are made from a monitor screen. The field of view of the camera includes one 4 x 2.5 cm grid, and on the monitor screen this is effectively

---

[1] Rees R93/03 camera, Rees Instruments, Old Woking, Surrey, England

magnified ten times. Thus, lateral roots of 0.1 mm diameter can be easily seen, and root hairs also identified.

Knowing the times of initiation of roots in relation to shoot development, it is also possible to distinguish between seminal and nodal axes, and lateral roots arising from these axes. Further work on the analysis of TV images will help discrimination and classification of roots.

We have found that the data from 12 to 24 tubes extending to 1.40 m depth can be collected and transcribed by two people in one day, though this time does depend on the distance of the site from the laboratory, and on root density. By comparison, collection of a similar number of soil cores, and washing out and measuring roots takes 18-30 man days. With the 5 cm or 2.5 cm grids, the camera system can define root density in 3.5 cm or 1.8 cm vertical soil layers, which is better than can be obtained routinely with cores.

CALIBRATION

Relative lengths and distribution of roots to compare treatments, or occasions within a treatment, are available directly as counts or estimated lengths from the monitor screen. To calculate water and nutrient fluxes, however, absolute root lengths must be derived; this requires calibration of camera data against root lengths washed from adjacent cores.

As with all root measurement techniques, point-to-point variability is high even when sampling is consistent with respect to row position and crop density. Coefficients of variation for six replicate profiles for either camera root counts or washed lengths in silt loam over clay are typically 30% for total root length, and up to 100% for roots in the most variable subsoil horizons. If replicate values are combined, however, there are good correlations between root lengths, washed from cores and measured by an intercept method (Tennant, 1975), and either estimated camera root length ($r^2 = 0.87$) or camera root counts ($r^2 = 0.91$). On the screen, counts can be made much more rapidly than estimates of length, and a typical relationship to root length density is shown in Fig. 1.

This relationship holds to within 10 cm of the soil surface. For reasons which are not fully understood, observation techniques consistently underestimate root length close to the soil surface (Gregory, 1979; Bragg et al., 1983; Upchurch and Ritchie, 1983). Contributory factors in our studies with winter wheat included the relatively low surface root densities between crop rows where the tubes are placed, and the difficulty of identifying fine lateral roots against the background of dead root and other organic debris

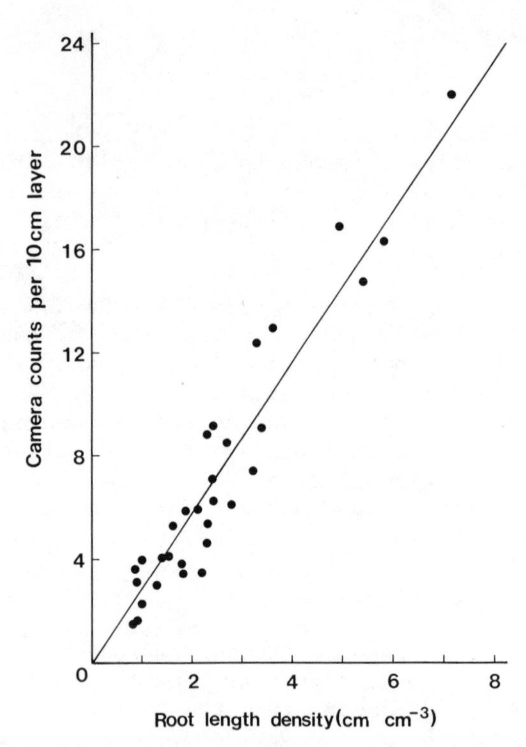

Fig. 1. Calibration of camera root counts per 10 cm depth against washed root length density of winter wheat grown on Harwell series silt loam soil (each point is the mean of 6 values). The equation for the regression line is $y = 2.94x$ ($r^2 = 0.91$).

close to the surface. Alternative techniques must be used for the upper 10 cm of soil.

ROOT DISTRIBUTION

The potential of the camera system to generate consistent data is shown in Fig. 2; this shows root distributions of winter wheat between January and July in a silt loam soil over greensand. The average rate of root elongation prior to January was 9 mm day^{-1}; from January to March this rate slowed to approximately 2 mm day^{-1} below 80 cm depth, probably because of the large resistance to penetration caused by a very high abundance of sandstone in a clay matrix beyond this depth. The sequence of profiles shows the rapid increases in root length at depths above 40 cm between April and May and between May and June (anthesis). Root length density declined from anthesis to July; similar changes have been noticed by other

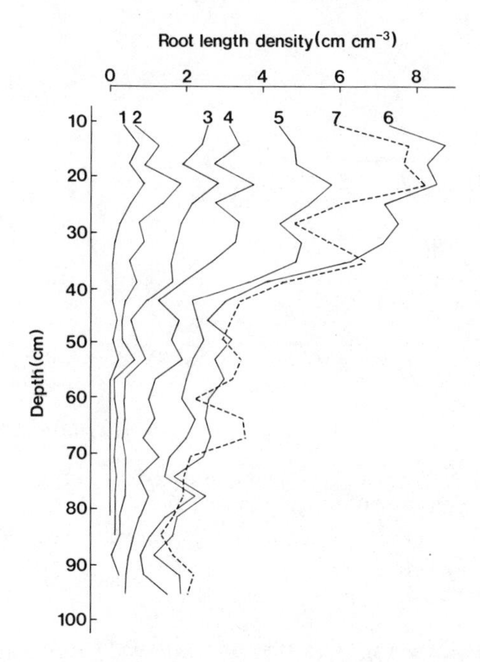

Fig. 2. Distribution of roots of winter wheat cv. Avalon on Harwell series silt loam soil, from January to July. Root length density was derived using the relationship given in Fig. 1, from camera counts made in six observation tubes on 11 January, 16 February, 8 March, 13 April, 12 May, 13 June and 19 July (1 to 7 in diagram).

workers (e.g. Bragg et al., 1984), but with the camera system it was clear that lateral roots were lost whilst the main axes of seminal and nodal roots remained intact and visible throughout the season. Below 40 cm depth, root growth continued until July, although more slowly and irregularly after anthesis than in the preceding two months; this probably reflects the variable distribution of available water in the stone/clay subsoil.

ROOT GROWTH

The ability to make repeated measurements with little labour simplifies measurement of the penetration growth of roots. For example, the effect of tillage on the rate of penetration of seminal root axes was determined on a Denchworth series clay soil in 1983 (Fig. 3). In a dry autumn, good continuity of pores in the direct-drilled soil encouraged more rapid elongation of main root axes (13 mm day^{-1}) than in the ploughed soil (9 mm day^{-1}). However, total length of roots (determined by summation of root counts within each profile) did not differ significantly between treatments,

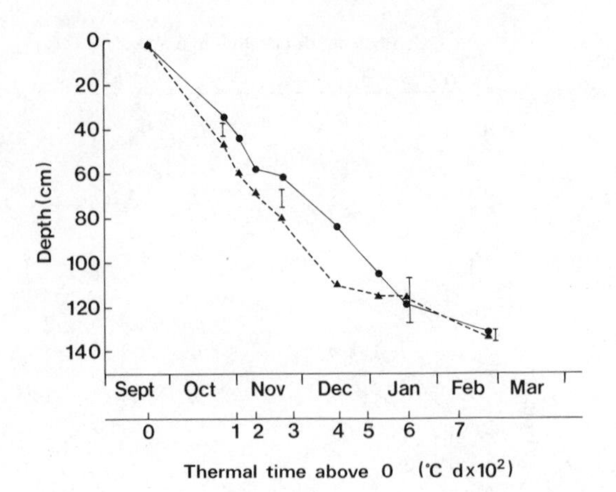

Fig. 3. Depth of penetration of roots of winter wheat (cv. Avalon) established on Denchworth clay by ploughing ●——● or direct drilling ▲- -▲. Vertical bars denote L.S.D. (P = 0.05).

because there was greater growth of lateral roots in the loosened topsoil of the ploughed plots than in the more compact direct-drille soil.

CONCLUSIONS

This paper outlines a procedure to look at root growth of wheat *in situ*, which also allows some aspects of the development of individual roots to be followed during the season. The system has the potential for rapid and frequent measurement of root length and distribution to match the observations that can be made, using existing techniques, of soil moisture extraction and nutrient uptake and shoot development and growth. More work is needed on calibratio of the system to determine absolute root lengths for a wider range o soils and root densities than so far considered; to resolve the problem of near-surface measurement of root length; and to gain a better understanding of the information contained in the recorded images in terms of root branching and senescence.

ACKNOWLEDGEMENTS

We thank Dr. R. Q. Cannell for encouragement, Dr. P. L. Bragg for early development work, and Miss J. Godding for root measurement during calibration of the system.

REFERENCES

Bohm, W., 1974, Mini-rhizotrons for root observation under field conditions, Z. Acker Pflanzenbau, 140:282.
Bohm, W., 1979, "Methods of studying root systems", Springer-Verlag, Berlin.
Bragg, P. L., Govi, G., and Cannell, R. Q., 1983, A comparison of methods, including angled and vertical minirhizotrons, for studying root growth and distribution in a spring oat crop, Pl. Soil, 73:435.
Bragg, P. L., Rubino, P., Henderson, F. K. G., Fielding, W. J., and Cannell, R. Q., 1984, A comparison of the root and shoot growth of winter barley and winter wheat, and the effect of an early application of chlormequat, J. agric. Sci., Camb., (in press).
Gregory, P. J., 1979, A periscope method for observing root growth and distribution in field soil, J. exp. Bot., 30:205.
Klepper, B., Belford, R. K., and Rickman, R. W., 1984, Root and shoot development in winter wheat, Agron. J., 76:117.
Sanders, J. L., and Brown, D. A., 1978, A new fiber-optic technique for measuring root growth of soybeans under field conditions, Agron. J., 70:1073.
Tennant, D., 1975, A test of a modified line intersect method of estimating root length, J. Ecol., 63:955.
Upchurch, D. R., and Ritchie, J. T., 1983, Root observations using a video recording system in mini-rhizotrons, Agron. J., 75:1009.

PHOTOSYNTHESIS, CO_2 AND PLANT PRODUCTION

J. Goudriaan,* H. H. van Laar,* H. van Keulen,‡ and W. Louwerse‡

*Agricultural University
Wageningen
The Netherlands

‡Centre for Agrobiological Research
Wageningen
The Netherlands

INTRODUCTION

Agricultural production can be increased through better plant characteristics obtained either through breeding or through better growing conditions, both in the soil and above ground. In the chain of events necessary for plant growth, photosynthesis stands at the beginning as the primary conversion of light energy to chemical energy stored in organic substances. This paper deals with the influence of photosynthetic performance on the eventual dry matter production of plants. The approach used is mechanistic and quantitative, and, because the number of interacting factors is large, a simulation method is used. Our simulation model is essentially BACROS (de Wit et al., 1978), modified to the present (1983) version in a number of ways indicated below. We consider characteristics of C_3 plants only.

The main emphasis of this paper is on the effect of atmospheric CO_2 concentration on dry matter yield and water consumption. The results of our simulations indicate that this relation can be described by a simple response function, and we compare simulations with empirical results. The part of the effect of CO_2 concentration that is mediated by photosynthesis can be compared with the effects of internal changes in photosynthetic properties. Some reasons for differences will be discussed.

A potentially high rate of photosynthesis is not sufficient for a high plant growth rate. It is equally important that this rate is realised and that the photosynthetic products are used in the desired fashion. In the case of a sub-optimal nutrient supply, growth is impaired and then the rate of crop photosynthesis will inevitably fall. Although we consider only situations of optimum supply of nutrients and water, our results are relevant to poorer conditions because they define the potential demand for other growth factors.

PHOTOSYNTHESIS AT THE LEAF LEVEL

Photosynthesis

The properties of leaves that affect photosynthetic performance can be separated into those that describe the light reactions, and those that describe the dark reactions. We use a single parameter for the light reactions: the quantum use efficiency ε_o (mol CO_2 E^{-1} photosynthetically active radiation (PAR)). According to Farquhar and Von Caemmerer (1981), a minimum of 8.4 photons are required for the reduction of 1 molecule of CO_2. At the average spectral composition of solar radiation this can be converted into 25×10^{-9} kg CO_2 J^{-1} (PAR). However, measured maximum values of ε have rarely exceeded 17×10^{-9} kg CO_2 J^{-1} (12 Einstein per mole), probably because of absorption in non-chlorophyllic tissue and of some other energy-requiring processes. Therefore we set ε_o at 17×10^{-9} kg CO_2 J^{-1} bearing in mind that this efficiency is further reduced by photorespiration.

We use three parameters to describe the dark reactions.

The carboxylation conductance g_x (m s^{-1}). This parameter can be used directly to describe the maximum rate of assimilation, F_m, under low CO_2 levels and sufficient irradiance. Then the assimilation rate is proportional to the CO_2 level, and a resistance scheme can be used (Fig. 1). Between the outside air and the chloroplast, two resistances of a physical nature can be distinguished, the leaf boundary layer resistance, r_b, and the stomatal resistance, r_s, and one resistance of a chemical nature, the carboxylation resistance, r_x. Under optimum water supply, the stomatal resistance drops to about 130 s m^{-1} for H_2O, equivalent to about 200 s m^{-1} for CO_2. The carboxylation resistance is about 250 s m^{-1}. With a typical leaf boundary layer resistance of about 20 s m^{-1}, the CO_2 use efficiency is the conductance of the total resistance chain and equals about 1/470 m s^{-1}. With a CO_2 concentration difference of 500 mg CO_2 m^{-3} (about 270 ppmv) between ambient air and the chloroplast, the calculated maximum assimilation rate is 1.06 mg CO_2 m^{-2} s^{-1} (38 kg CO_2 ha^{-1} h^{-1}). In the BACROS model, r_s is computed from the water status of the the crop and r_b is computed on the basis of micrometeorological considerations (Goudriaan, 1977). The carboxylation conductance, g_x is made dependent on temperature by a multiplication factor that increases linearly from zero to unity between temperatures 5 and

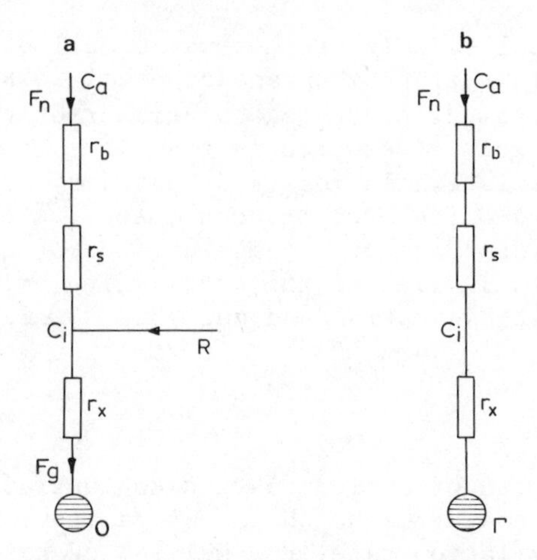

Fig. 1. Two equivalent resistance schemes for the CO_2 assimilation of a leaf under high radiation and low CO_2 concentration. The respiration rate, R, (left) can be taken into account by an apparent CO_2 concentration in the chloroplast at the compensation concentration, Γ (right). C_a is the ambient CO_2 concentration; r_b the boundary layer resistance; r_s the stomatal resistance, r_x the carboxylation resistance, C_i the sub-stomatal CO_2 concentration, F_g the rate of gross photosynthesis, and F_n the rate of net photosynthesis.

13°C. Moreover, if the plant does not have sufficient sinks, a reserve level rising above 20% on a dry weight basis will gradually diminish the carboxylation conductance to zero when reserves reach 25%. The justification for this procedure is that reserve levels higher than 25% on a dry matter basis are hardly ever found. An alternative possibility would be for additional respiration to remove surplus assimilates (Lambers, 1982); we do not include such respiration.

The CO_2 compensation point Γ (mg m^{-3}). This is the CO_2 concentration at the bottom of the resistance chain (Fig. 1). The maximum assimilation rate, F_m, is linearly related to the ambient CO_2 concentration, C_a, with an intercept at C_a = Γ. It appears that the value of Γ is almost entirely determined by photo-respiration, and is equal to about 100 mg CO_2 m^{-3} at 20°C. Measurements show a strong rise with temperature (Bykov et al., 1981) and with water stress (Lawlor and Pearlman, 1981).

Dark respiration rate R_d(mg m^{-2} s^{-1}). The dissimilation rate in the dark, R_d, per unit leaf area is estimated as the maintenance respiration of the whole crop multiplied by the leaf weight ratio and divided by the leaf area, to give a value of about 1 kg CO_2 ha^{-1} h^{-1}

at 20°C. Theoretically, this process may contribute to the value of the CO_2 compensation point by an amount equal to R_d times r_x (about 5 mg CO_2 m^{-3} or 3 ppmv). According to Peisker et al. (1983) this respiration is largely superseded in the light by direct ATP consumption, but this conclusion is disputed by Azcon-Bieto and Osmond (1983). The difference between these two possibilities is equivalent to a difference of 3 ppmv in value of the CO_2 compensation point. In view of the conflicting evidence we have chosen the simplest assumption and neglected the influence of R_d on Γ

Photorespiration

In an atmosphere of ambient oxygen concentration, and at a constant leaf temperature, the photorespiration rate, R_f, is equal to the gross assimilation rate, F_g, multiplied by Γ/C_i (Laing et al., 1974):

$$R_f = \frac{\Gamma F_g}{C_i}, \tag{1}$$

where C_i is the CO_2 concentration in the substomatal cavities. This simple equation is valid over the full range of combinations of light and CO_2 concentration.

In conditions of high light and low CO_2, the dark respiration, R_d, has vanished and the net assimilation rate, F_n, is equal to $F_g - R_f$. Because the gross assimilation rate, F_g, is given by C_i/r_x and the photorespiration rate is then equal to Γ/r_s, the net assimilation rate, F_n, is:

$$F_{n,c} = (C_i - \Gamma)/r_x. \tag{2}$$

The subscript c indicates that this is CO_2 limited F_n. In BACROS, Equation (2) is used to compute the carboxylation resistance, r_x, from a measured light-saturated assimilation rate. A fundamental assumption is that r_x itself is independent of the CO_2 concentration.

In conditions of high CO_2 and low light it is important to consider the energy requirement of photorespiration as well. According to data of Peisker and Apel (1981) twice as many photons are required to regenerate PGA in the photorespiratory cycle as in the reducing cycle of photosynthesis, where both are expressed per mole of CO_2:

$$2 R_f + F_g = \varepsilon_o H, \tag{3}$$

where H is absorbed radiation (PAR) per unit leaf area. In low light, dark respiration is not suppressed and the net assimilation,

F_n, is equal to $F_g - R_f - R_d$. When this is combined with Equation (3) we find that:

$$F_{n,l} = \varepsilon H - R_d, \tag{4}$$

where subscript l indicates light-limited F_n, and ε contains the photorespiratory losses,

$$\varepsilon = \frac{(C_a - \Gamma)}{(C_a + 2\Gamma)} \varepsilon_o. \tag{5}$$

The CO_2 concentration in this equation should be the internal CO_2 concentration, C_i, but since the equation applies close to the light compensation point, C_i has about the same value as the ambient concentration, C_a.

The fraction of assimilate lost via photorespiration is usually more under low light than under high light conditions, because of the energy requirement of photorespiration. Equation (5) calculates the loss at about 35% under low light, whereas according to Equation (2) the loss is only about 25% under high light conditions.

Interaction of light and CO_2

The transition between the light- and the CO_2-limited region of the photosynthesis-light response curve can be either sharp, or more smooth (Fig. 2). A broad transition zone ("shoulder") can be caused by spatial heterogeneity in the leaf. In addition the extinction of radiation inside the leaf leads to a broadening of the response. From biochemical considerations, the degree of coupling between NADPH supply and PGA regeneration is expressed in the sharpness of the transition. Another explanation for a sharp shoulder would be a high transport resistance in the mesophyll in combination with a low carboxylation resistance (Prioul and Chartier, 1977). However, this possibility must be discarded in view of biochemical evidence, and of data on discrimination between C^{13} and C^{14} (Björkman, 1981).

For our purpose all the factors influencing the shape of the shoulder are contained in an empirical relationship that gives a reasonable description of photosynthesis, such as the asymptotic exponential:

$$F_n = (F_{n,c} + R_d)\ (1 - \exp\ (-\varepsilon H/(F_{n,c} + R_d)) - R_d) \tag{6}$$

with ε given by Equation (5).

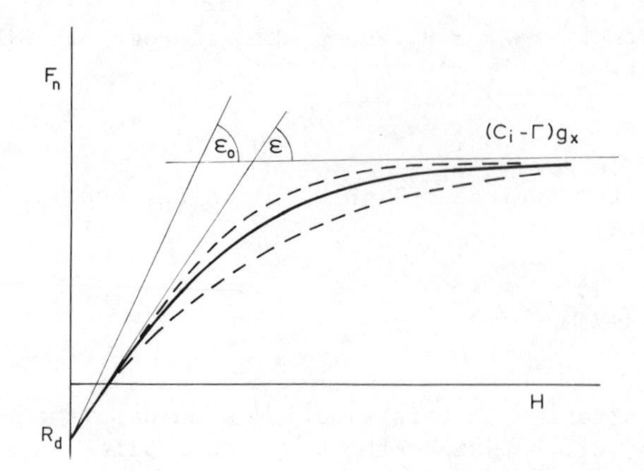

Fig. 2. Typical relationship between net CO_2 assimilation rate, F_n, and absorbed photosynthetically active radiation, H, both expressed per unit leaf area. The dashed lines indicate alternative relations which have the same asymptotes but differ in the 'shoulder' region.

STOMATAL REGULATION

Variability in stomatal conductance between leaves under the same irradiance is much larger than variability in photosynthesis. Even so, there is ample experimental evidence to indicate a linear relationship between stomatal conductance and assimilation rate at least when light is varied (Louwerse, 1980; Bell, 1982; Wong, 1979; Goudriaan and van Laar, 1978). When ambient CO_2 concentration is raised the ratio of conductance to assimilation is decreased. The difference in CO_2 concentration across the leaf boundary layer and stomatal resistances is proportional to the ambient CO_2 level, and can be described by:

$$C_i = f\ (C_a - \Gamma) + \Gamma. \tag{7}$$

The value of f is between 0.7 and 0.8 for C_3 plants. Equation (7) is a key equation for computing $F_{n,c}$ (Equation (2)), net assimilation F_n (Equation (6)) and the associated stomatal conductance. It not only generates the internal CO_2 concentration as a supply level for assimilation, but also the difference $C_a - C_i$. The magnitude of this difference, in combination with the assimilation rate, determines the value of the stomatal conductance. This description assumes that stomatal conductance follows assimilation rate. This may not be a direct cause-and-effect relationship, but an intricate simultaneous reaction to irradiance and CO_2. Such stomatal behaviour results in increasing CO_2 assimilation and decreasing transpiration per unit leaf area with rising ambient CO_2 levels.

It is well-known that this relationship between stomatal conductance and assimilation can be reversed by water shortage. The water status of the crop may put an upper limit to stomatal conductance that is lower than the value computed above. In this case we assume that the effect of water shortage becomes dominant and assimilation is calculated for the new value of r_s. Effects of water shortage on the carboxylation resistance or CO_2 compensation point have been ignored in the present model.

There are situations in which the observed stomatal conductance considerably exceeds the value needed for the realised assimilation rate. Usually this happens under ample water supply, and in that situation there is virtually no response of stomata to CO_2 concentration. It is probable that this non-regulated behaviour is more characteristic of phytotron-grown plants than of field-grown plants, and is therefore not simulated here.

PUTTING THE MODEL TOGETHER IN SPACE AND TIME

For the transition from leaf level to crop level and from seconds to a growing season, a quantitative synthesis of the constituent processes is necessary. This integration in space and in time requires careful consideration of numerous interactions.

Integration in Space

The contribution of various leaves to crop assimilation and respiration must be integrated. Leaves partly shade each other, and are at different angles with respect to incoming light, so that their radiation environment must be modelled (Goudriaan, 1977). In addition, the leaves modify their aerial environment by the release of heat and moisture. A detailed evaluation of these micro-meteorological effects requires the solution of a complicated scheme of resistances and fluxes, but a great simplification is possible by neglecting the profiles of temperature and moisture inside the leaf canopy. For the computation of the fluxes of transpiration and assimilation, it is sufficient to lump the ventilation by turbulence into just one resistance connecting canopy space and crop atmosphere (Goudriaan et al., 1983).

Integration in Time

Because the time span covered by the present model is a growing season, the amounts of dry matter in the organs are important state variables. Although the most natural time step of integration is one day, we have chosen much smaller time steps, one hour, to allow for diurnal courses. The non-linearity of the photosynthesis-light response curve, and a possible afternoon depression in assimilation resulting from water stress can then be taken into account. Moreove the reserve level may fluctuate considerably during a 24 hour perio as the balance between carbohydrate production and consumption alters.

The concept of feedback is central in temporal integration. A mentioned before, we need a method to ensure an upper limit to reserve levels, especially in periods of high radiation and low temperature. For such a stabilisation, one or more negative feedbacks are required. Such stabilisation can occur by an increas in the rate of conversion of assimilates into structural material with increasing reserve level or by a decrease in assimilation rate. We have used an instantaneous response of the carboxylation resistance to fluctuating reserve levels but a better description is badly needed. In the early spring period of growth we often simulat a high assimilation rate in the morning, consequently the reserve level builds up and severely reduces photosynthesis for the rest of the day. These unlikely results suggest the existence of a more subtle feedback with a much longer time-constant. Experimentally these effects may be confounded with a simultaneous afternoon depression due to water stress. Both water stress and high reserve levels are induced by high radiation levels. Manipulation of air humidity may be the key to distinguishing these two possible causes of an afternoon depression in photosynthesis.

ACTIVE AND PASSIVE GROWTH RESPONSE

When assimilation is stimulated by a higher level of CO_2, the reserve level is increased and, as a consequence, so is the growth rate of the plant. Therefore, increased CO_2 will always be beneficial for plant production until the upper level of reserves ha been reached. However, its impact on future growth depends on whether the additional carbon gained is stored as passive material such as starch, or is used productively to increase leaf area. In the passive response mode, the leaf area growth is independent of th CO_2 concentration, even though the weight of the plant and of its organs may have increased. In the opposite type of response, which can be called the active response mode, the additional assimilates are not distinguished from the "reference assimilates" and are converted into active plant material as well. In that situation the time course of leaf area/plant weight ratio is not affected and is identical to the control. In this way the positive feedback loop from assimilation to leaf area to assimilation causes a steeper response to increased CO_2 than in the passive response mode.

In the real world, these response modes are often mixed. The active response mode can be expressed by increased tillering (wheat: Lemon, 1983), branching (faba bean) or increased individual leaf siz (poplar: Goudriaan and de Ruiter, 1983). In the vegetative growth phase, plants have more opportunities to respond actively than after initiation of the storage organs. We have investigated both types of response by simulation and have used either a fixed time course of leaf area index, LAI, to represent the passive response mode, or a fixed time course of leaf area ratio, LAR, to represent the active response mode.

We started with a simulation of a field experiment on a C_3 species (faba bean) grown at Wageningen in 1979, to verify that the model behaved realistically under normal conditions. With the measured time course of LAI as input, we obtained reasonable results (Fig. 3). In the field, the maximum crop growth rate was about 140 kg ha^{-1} d^{-1}. We did not change the value of any parameter to improve the fit, because we felt that these results could serve as a point of reference for further investigation of the effect of CO_2 concentration. Preferably, the response curves for the active and the passive modes should intersect at normal ambient CO_2. This can only be achieved by using the same time course of LAR as simulated in the passive mode, and not the measured one.

The effect of CO_2 is now easily investigated by a sensitivity analysis. We changed the CO_2 level in the model and plotted the simulated above-ground dry matter and total transpiration at the end of the season as a function of CO_2 level (Fig. 4). Not surprisingly, the results show a diminishing response of dry matter

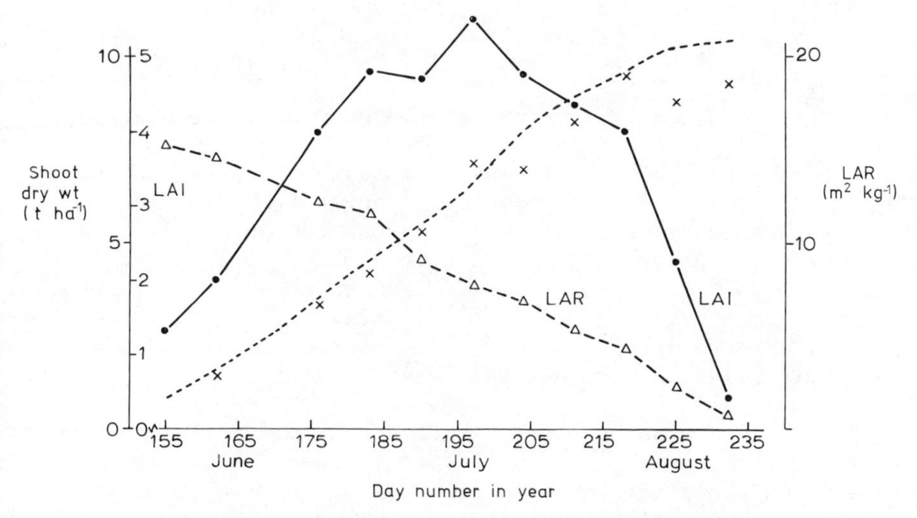

Fig. 3. Measured time courses of leaf area index (LAI, ●——●), leaf area ratio (LAR, Δ- -Δ), and of shoot dry weight (X) of faba bean. Shoot dry weight was also simulated (-----), using the measured LAI as an input.

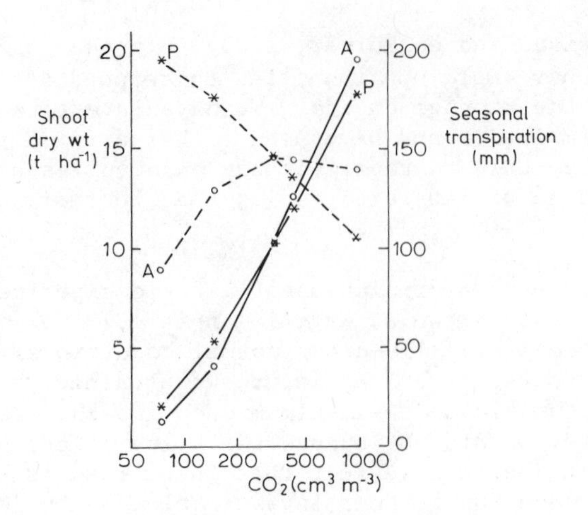

Fig. 4. Simulated shoot dry weight (———) and total transpiration (-----) at the end of the season as a function of ambient CO_2 level. In the passive response mode (P) LAI is a fixed function of time, in the active response mode (A) LAR is a fixed function of time.

with increasing CO_2. Plotting log (dry matter) against CO_2 concentration gives a remarkable linearity in response for both type of behaviour over the range from 200 to 1000 ppmv. Hence, the model results can be well represented by the equation:

$$Y = Y_o \ (1 + \beta \ln \ (\frac{C}{C_o})), \tag{8}$$

where Y_o is the total dry matter at ambient CO_2 concentration, C_o.

This equation not only shows diminishing returns, but also allows for a non-zero CO_2 compensation point (Gifford, 1980). Graphically, the value of the CO_2 compensation point can be read from Fig. 4 by extrapolation to a zero yield. It is clear that the intercept is higher than the single leaf value of 50 ppmv, presumabl because of respiration in other plant parts. However, there is also some upward curvature at the lower end of the graph, and the logarithmic relation does not seem valid in this region.

It is, of course, possible to find an alternative equation for the description of the CO_2 response of the model crop. The Michaelis-Menten equation (hyperbola) is attractive, because it allows for a saturation level. Within the range of 200 to 1000 ppmv however, its behaviour is almost identical to that of the logarithmi equation. Because the logarithmic relation requires one parameter less, it is preferable for descriptive purposes. Moreover, the meaning of β is straightforward and can be interpreted as the relative sensitivity to CO_2 around the ambient concentration.

The concept of a logarithmic response is widely used in models for the global carbon cycle, more by virtue of its simplicity than because of its physiological meaning. In these studies the parameter β is termed "biotic growth factor", but the more neutral "relative sensitivity to CO_2" seems more appropriate. Our study gives a theoretical basis for the validity of the logarithmic equation. This relation was not introduced a priori, but was established a posteriori as a result of complex model relationships. It is valid, both in the active and in the passive response modes, although the values of β differ.

In the range of 200 to 1000 ppmv, β has a value of about 0.65 in the passive response mode and about 0.77 in the active one. With these relative sensitivities, a doubling of atmospheric CO_2 level will result in an increase of dry matter yields by a factor of 1.45 and 1.53 respectively. Such increases are not unrealistic and have been found under experimental conditions (Goudriaan and de Ruiter, 1983; Gifford and Evans, 1981; Lemon, 1983; Combe, 1981).

TRANSPIRATION AND ROOT/SHOOT RATIO

Both short- and long-term feedback mechanisms influence the relation between transpiration and root/shoot ratio. In the short term, water stress may limit transpiration. Then the root/shoot ratio constrains the transpiration rate. This effect only shows up during a part of the day, but its duration and intensity decreases with higher CO_2 levels. Water stress changes the partitioning of dry matter in favour of the root system, so that a long-term negative feedback mechanism accounts for some adaptation of the root/shoot ratio to the transpirational demand. In the simulated example, the root/shoot ratio dropped from 0.19 at 200 ppmv to 0.12 at 1000 ppmv in the active growth response. In reality increased CO_2 does not always lead to lower root/shoot ratio. Sometimes the opposite happens, possibly because of earlier saturation of shoot compared to root growth by the accumulation of reserves. Such a preferential demand of the shoot was not modelled because of lack of data. An interesting result of the simulation is the calculated maximum value of total transpiration at around 300 ppmv of CO_2, in the active response mode. Leaf transpiration per unit area decreases with increasing CO_2 level because of stomatal closure, but this effect is more than compensated for by an increase in leaf area index. Only above 300 ppmv does total crop transpiration decrease again. Of course, the ratio of transpiration to dry matter production declines sharply over the entire range of CO_2 levels.

INTERNAL AND EXTERNAL STIMULATION OF PHOTOSYNTHESIS

Stimulation of photosynthesis by external factors, such as radiation or CO_2, almost invariably results in increased plant production and dry matter accumulation. Although some problem areas remain, such as the effect of CO_2 on partitioning, the main features of model calculations and CO_2 experiments agree, showing that assimilate supply does limit growth. Once the assimilates are in the reserve pool, it does not seem to matter whether they have been brought there by improved external conditions or by an improved photosynthetic response. Therefore our model seems ideally suited to evaluate the growth response to genetic (internal) changes of the leaf photosynthetic properties. We have investigated the effect of changes in the potential light use efficiency, ε_o, and the carboxylation conductance, g_x. In the control situation the values of these parameters are 17×10^{-9} kg CO_2 J^{-1} and 4×10^{-3} m s^{-1} respectively. The results of the model computations are presented in Fig. 5. It is immediately clear that the relative effect of changes in light use efficiency is much larger than that of carboxylation conductance. The relative sensitivity to ε_o is about 0.7 and to g_x only about 0.25. These results suggest that a much larger potential exists in breeding for improved quantum use

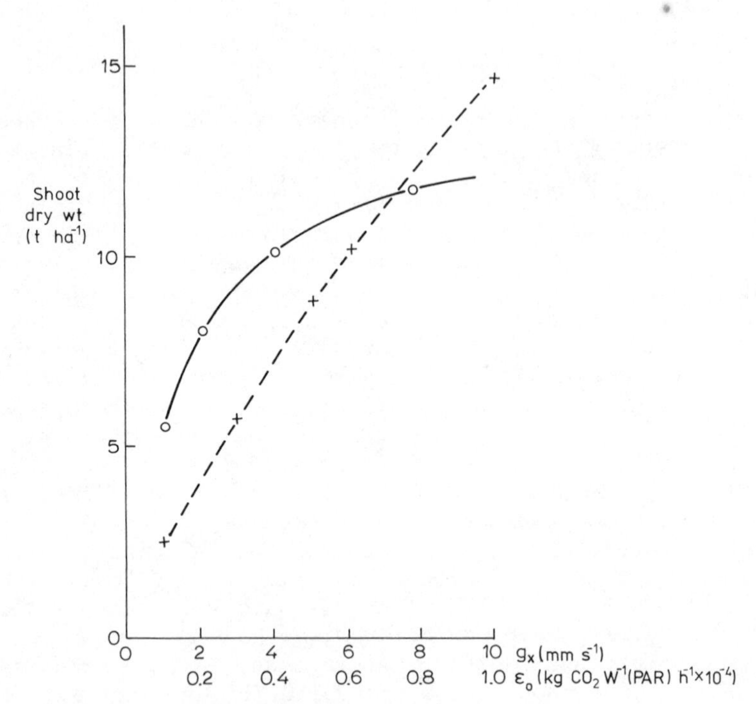

Fig. 5. Simulated shoot dry matter at the end of the season as a function of carboxylation conductance, g_x (———), and of light use efficiency, ε_o (-----).

efficiency rather than for improved carboxylation properties. Austin (1982) found a higher sensitivity to $F_{n,max}$ but this value includes the resistance of the stomata. The idea of improving photosynthetic properties has been evaluated by many plant breeders working with a large number of species. It is surprising and theoretically challenging that so far the practical results of such efforts have been disappointing. Gifford and Evans (1981) mentioned in a review on this subject that

a) The relationship between assimilation and yield is tenuous,

b) No indirect increase of maximum assimilation rate has occurred during domestication and improvement of wheat in the past (Khan and Tsunoda, 1970) nor in other species (maize, sorghum, millet, sugarcane, cotton, cowpea),

c) Breeding for lower photorespiration (CO_2 compensation point) has been unsuccessful.

They conclude that selection for photosynthetic properties may be failing because of counterproductive associations. If this statement is true then our next task is to find these associations and to devise ways to counter them.

DISCUSSION

Both internal and external stimulation of photosynthesis lead to the same end products, so that feedback mechanisms through end-product inhibition or through increased respiratory losses cannot explain the difference between the apparent lack of success in breeding for improved photosynthesis and the effect of CO_2 enrichment. Presumably the crucial difference is the high cost of building the photosynthetic structure itself.

The light-saturated rate of assimilation ($F_{n,c}$) is strongly related to leaf thickness (Louwerse and van der Zweerde, 1977), to leaf nitrogen content (van Keulen, 1984) and to RuBP carboxylase/oxygenase concentrations. RuBP carboxylase is known to account for up to half of the leaf protein, and therefore it is understandable that photosynthetic capacity is related to both nitrogen and RuBP concentrations. This brings us to the agriculturally trivial conclusion that nitrogen uptake must be improved, or that the efficiency of nitrogen use must be improved by a better distribution. But this latter conclusion poses difficulties because a relatively small increase in nitrogen involved in photosynthesis implies a relatively large decrease in other nitrogenous compounds. If there is little scope for changing the allocation of nitrogen from non-photosynthetic functions to photosynthetic ones, another possibility is a more efficient, but perhaps uneven, redistribution between the leaves in the canopy. The photosynthetic capacity of single leaves may then be quite variable, because a locally high rate of photosynthesis can only be achieved at the expense of a lower value somewhere else. Since leaf position is correlated with leaf age, this would also imply that high rates cannot be maintained for a lon time in the same leaf. Measurement of single leaf photosynthesis then requires intensive sampling, and it is probably necessary to measure whole canopy photosynthesis as well.

The remarkable constancy of the quantum use efficiency also indicates physical and biochemical limits. Osborne and Garrett (1983) found no worthwhile variation with ploidy in the quantum use efficiency of absorbed radiation. Optical properties of the leaves did vary slightly, but because of counteracting effects of light loss to soil and sky, and of improved light distribution, the overal effect was small (Goudriaan, 1977). Even small improvements in this area are attractive, however, because of the high sensitivity of plant production to this factor. The steepness of the transition between the light-limited and the CO_2-limited part of the photosynthetic response curve can influence assimilation. With equal ε and $F_{n,c}$, crop photosynthesis can still be improved by a steeper transition. The extreme effect is obtained by replacing Equation (6 by a Blackman-type curve. The effect of such a change is a calculate increase of final dry matter of almost 10%. Experimentally, the detection of differences between leaves with respect to the shape of the photosynthesis curve requires measurements at a number of light levels and is therefore labour intensive. The scope for genetic improvement in this characteristic is only modest.

Finally, it is interesting to return to the response to changes in CO_2 concentration, because it partly consists of an effect through ε (Equation (5)) and partly through $F_{n,c}$ (Equation (2)). Using these equations, together with the chain rule for derivatives and the sensitivities just mentioned, we find that the CO_2 effect of 0.77 is composed of an ε effect of 0.28, a carboxylation effect o 0.3, and a remainder of 0.19. The latter may be attributed to improved water use efficiency, leading to more prolonged periods fre of water stress, and to a shift in partitioning in favour of the shoot. Nutrient-imposed limitations reduce the CO_2 response of the crop but there is no reason to abandon the concept of a logarithmic response. The value of the relative sensitivity will fall, dependen on the degree and character of the nutrient shortages. An important prediction is that water shortage should not reduce the CO_2 response, because the water-use efficiency increases and there is some evidence that the efficiency of nitrogen use but not that of phosphorus slightly improves at higher CO_2 concentrations (Wong, 1979; Goudriaan and de Ruiter, 1983).

ACKNOWLEDGEMENT

Thanks are due to Mrs. Uithol-van Gulijk for the typing of the manuscript.

REFERENCES

Austin, R. B., 1982, Crop characteristics and potential yield of wheat, J. agric. Sci., Camb., 98:447.

Azcon-Bieto, J., and Osmond, B. C., 1983, Relationship between photosynthesis and respiration, Pl. Physiol., 71:574.

Bell, C. J., 1982, A model of stomatal control, Photosynthetica, 16:486.

Björkman, O., 1981, Responses to different quantum flux densities, in: "Encyclopaedia of Plant Physiology, New Series Vol. 12B, Physiological Plant Ecology II", O. L. Lange, P. S. Nobel, C. B. Osmond and H. Ziegler, eds., Springer-Verlag, Berlin.

Bykov, O. D., Koshkin, V. A., and Catský, J., 1981, Carbon dioxide compensation concentration of C_3 and C_4 plants: dependence on temperature, Photosynthetica, 15:114.

Combe, L., 1981, Effet du gaz carbonique et de la culture en climat artificiel sur la croissance et le rendement d'un blé d'hiver, Agronomie, 1:177.

de Wit, C. T., et al., 1978, "Simulation of assimilation, respiration and transpiration of crops", Simulation Monosgraphs, Pudoc, Wageningen.

Farquhar, G. D., and von Caemmerer, S., 1981, Modelling of photosynthetic response to environmental conditions, in: "Encyclopaedia of Plant Physiology, New Series Vol. 12B, Physiological Plant Ecology II", O. L. Lange, P. S. Nobel, C. B. Osmond and H. Ziegler, eds., Springer-Verlag, Berlin.

Gifford, R. M., 1980, Carbon storage by the biosphere, in: "Carbon dioxide and climate: Australian Research", G.I. Pearman, ed., Australian Academy of Sciences, Canberra.

Gifford, R. M., and Evans, L. T., 1981, Photosynthesis, carbon partitioning and yield, A. Rev. Pl. Physiol., 32:485.

Goudriaan, J., 1977, "Crop micrometeorology: a simulation study", Simulation Monographs, Pudoc, Wageningen.

Goudriaan, J., and van Laar, H. H., 1978, Relations between leaf resistance, CO_2 concentration and CO_2 assimilation in maize, beans, lalang grass and sunflower, Photosynthetica, 12:241.

Goudriaan, J., and de Ruiter, H. E., 1983, Plant growth in response to CO_2 enrichment, at two levels of nitrogen and phosphorus supply. I. Dry matter, leaf area and development, Neth. J. agric. Sci., 31:157.

Goudriaan, J., van Laar, H. H., van Keulen, H., and Louwerse, W., 1983, "Simulation of the effect of atmospheric CO_2 on assimilation and transpiration of a closed crop canopy", in: Proceedings of Symposium "Biophysik Pflanzlicher Systeme",

Akademie der Wissenschaften DDR, Berlin.
Kahn, M. A., and Tsunoda, S., 1970, Evolutionary trends in leaf photosynthesis and related leaf characteristics among cultivat wheat species and its wild relatives, Jap. J. Breed, 20:133.
Laing, W. A., Ogren, W. L., and Hageman, R. H., 1974, Regulation of soybean net photosynthetic CO_2 fixation by the interaction of CO_2, O_2 and ribulose-1,5-diphosphate carboxylase, Pl. Physiol., 54:678.
Lambers, H., 1982, Cyanide-resistant respiration: a non-phosphorylating electron transport pathway acting as an energy overflow, Physiol. Plant., 55:478.
Lawlor, D. W., and Pearlman, J. G., 1981, Compartmental modelling o photorespiration and carbon metabolism of water stressed leave Pl. Cell Environ., 4:37.
Lemon, E. R. (ed.), 1983, "CO_2 and Plants", Westview Press, Colorado.
Louwerse, W., and van der Zweerde, W., 1977, Photosynthesis, transpiration and leaf morphology of Phaseolus vulgaris and Zea mays grown at different irradiances in artificial and sunlight, Photosynthetica, 11:11.
Louwerse, W., 1980, Effects of CO_2 concentration and irradiance on the stomatal behaviour of maize, barley and sunflower plants i the field, Pl. Cell Environ., 3:391.
Osborne, A., and Gerrett, M. K., 1983, Quantum yields for CO_2 uptake in some diploid and tetraploid plant species, Pl. Cell Environ., 6:135.
Peisker, M., and Apel, P., 1981, Influence of oxygen on photosynthesis and photorespiration in leaves of Triticum aestivum L. 4. Oxygen dependence of apparent quantum yield of CO_2 uptake, Photosynthetica, 15:435.
Peisker, M., Tichá, I., Catský, J., Kase, M., and Jank, H. W., 1983, Dependence of carbon dioxide compensation concentration on photon fluence rate in French Bean leaves and its relation to quantum yield and dark respiration, Photosynthetica, 17:344.
Prioul, J. L., and Chartier, P., 1977, Partitioning of transfer and carboxylation components of intra-cellular resistance to photosynthetic CO_2 fixation: a critical analysis of the methods used, Ann. Bot., 41:789.
van Keulen, H., 1984, "Simulation of water use, nitrogen nutrition and growth of a spring wheat crop", Simulation Monographs, Pudoc, Wageningen (in press).
Wong, S. C., 1979, Elevated atmospheric partial pressure of CO_2 and plant growth. I. Interactions of nitrogen nutrition and photosynthetic capacity in C_3 and C_4 plants, Oecologia, 44:68.

RESPIRATORY METABOLISM IN WHEAT

H. Lambers

University of Groningen
The Netherlands

INTRODUCTION

In many plant species a significant part of the photosynthate produced each day is used in respiration (Table 1). This respiration is not always very efficient in terms of ATP production (Lambers, 1982). Consequently, studies on respiration are of interest both for crop modellers and for others interested in the yield of crops. Respiration in wheat tissues, as in those of many other species, occurs via two electron transport pathways. Respiration via the cytochrome path is coupled to the phosphorylation of three molecules of ADP per molecule of endogenous NADH, whereas respiration via the alternative path produces at most one molecule of ATP per molecule of NADH.

RESPIRATION PATHWAYS IN WHEAT

In intact leaves of wheat, the alternative path contributed about 50% to respiration after several hours of photosynthesis, but none after the night (Azcón-Bieto et al., 1983b). The higher activity of the alternative path after photosynthesis correlated with a higher level of soluble and insoluble carbohydrates in the leaves. When slices of either penultimate and older leaves (Azcón-Bieto et al., 1983b) or flag leaves (Azcón-Bieto et al., 1983a) were used in an investigation of the effect of exogenous sugars on respiration, the same differences in respiration were found as in intact leaves. Addition of sugars to slices of leaves harvested at the end of the night stimulated respiration by about 40%. The stimulation was mainly due to an increased activity of the alternative path. Exogenous sugars did not affect the respiration of slices made from

Table 1. Loss of C in respiratory processes, expressed as a percentage of the C gained in photosynthesis.

Species	Growth conditions	Age (days)	C-loss in respiration	Reference
Hordeum distichum	Nutrient solution	7 24	60 38	Farrar (1980)
Lupinus albus	Sand + nutrients (nitrate-fed)	c.60	31	Pate et al. (1979)
Nicotiana tabaccum	Field grown	22-57	41-47	Peterson and Zelitch (1982
Pterido-phyllum racemosum	Growing in natural habitat (Central Japan)		77	Kimura (1970)
Zea mays	Nutrient solution	56	39	Massimino et al. (1981)

leaves after a period of five or more hours of photosynthesis. Thes results agree with an "energy overflow model". This model predicts that respiration in plant tissues occurs via the non-phosphorylating alternative path only when the input of sugars into such tissues is higher than their carbohydrate demand for the generation of energy, the production of carbon skeletons etc. (Lambers, 1982).

The contribution of the alternative path to the respiration of wheat roots varied with the growth conditions. In roots of wheat plants grown in full nutrient solution, 30% of root respiration occurred via the alternative path (Lambers et al., 1982). In roots of wheat plants grown in soil under nitrogen-limited conditions, the alternative path contributed about 20% to respiration during most of the growth period (Lambers et al., 1982).

In an experiment on effects of drought on two wheat varieties, the alternative path was found to decrease from 20-30% of root respiration to 0% in about four days after the start of the treatmen (M. E. Nicolas et al., personal communication). This decrease in respiration via the non-phosphorylating path correlated with an increase in the concentration of organic solutes in the roots. The results were interpreted similarly to those obtained with *Plantago coronopus*, grown in a non-saline solution and then exposed to 50 mM NaCl in the root environment (Lambers et al., 1981). This concentration does not affect the growth of the plants in long term (Blacquière and Lambers, 1980) or short term experiments (Lambers et al., 1981). Upon addition of NaCl, both leaves and roots accumulated sorbitol, a poly-alcohol with six carbon atoms that is a compatible osmotic solute in this species. During the accumulation of sorbitol, the activity of the alternative path decreased to a very low level, but increased again to the level of control plants after a new steady-state sorbitol level had been reached, about 24 h after addition of NaCl (Lambers et al., 1981). During the adaptation period, the activity of the cytochrome path was virtually unaffected. The amount of carbon "saved", by not using the alternative path during the adaptation period, was the same as that used in the production of sorbitol in the roots. It is hypothesized that the alternative path in roots plays a role in adaptation of a plant to changing water potentials in the root environment.

Much less information is available on the operation and significance of the alternative path in other organs of the wheat plant. Respiration of 0.5 cm sections of wheat stems was inhibited by about 30% on addition of 25 mM salicylhydroxamic acid, SHAM (H. Lambers, unpublished), an inhibitor of the alternative path, used in this concentration to measure the activity of this pathway in intact roots or leaves (Lambers et al., 1983). Similarly, in developing wheat grains, cut into halves, 25 mM SHAM inhibited respiration by about 30% (Fig. 1). This suggests that a non-phosphorylating path contributes significantly to respiration in stems and grains of wheat as well. However, a further kinetic analysis of inhibitor sensitivity is required, prior to any definitive statements on respiration via different pathways in these organs.

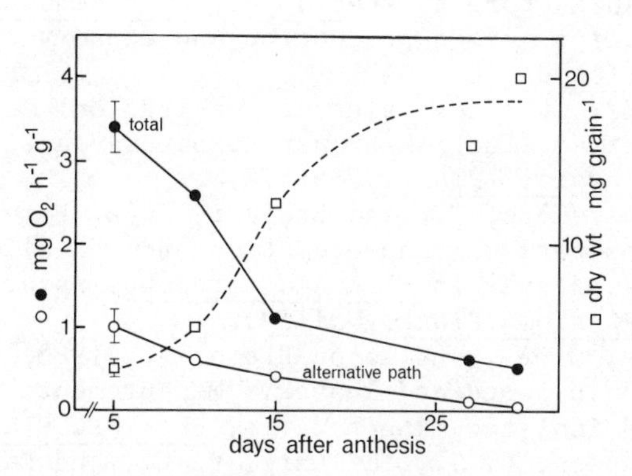

Fig. 1. Respiration and growth in grains of Triticum aestivum L. cv. Condor. Grains were collected from plants grown in the phytotron in Melbourne, Australia. Unpublished data from R. W. Gleadow and H. Lambers.

CONCLUSIONS

It is concluded that respiration in wheat, as in many other species, does not occur with maximum efficiency, i.e. a non-phosphorylating path significantly contributes to the respiration of various organs of the wheat plant. The activity of this path varies diurnally, due to variations in the amount of carbohydrates in the tissues. Environmental conditions, e.g. the supply of nitrogen and the osmotic potential of the root environment, greatly affect the activity of the alternative path. Respiration via the alternative path accounts for a significant portion of the photosynthate. The question whether selection against the alternative path as a wasteful trait reduces the plasticity of a plant, so that it can only marginally cope with environmental fluctuations, remains unresolved. It will be the subject of further investigation using other species.

REFERENCES

Azcón-Bieto, J., Day, D. A., and Lambers, H., 1983a, The regulation of respiration in the dark in wheat leaf slices, Pl. Sci. Lett., 32:313.

Azcón-Bieto, J., Lambers, H., and Day, D. A., 1983b, Effect of photosynthesis and carbohydrate status on respiratory rates and the involvement of the alternative pathway in leaf respiration, Pl. Physiol., 72:598.

Blacquière, T., and Lambers, H., 1981, Growth, photosynhesis and respiration in Plantago coronopus as affected by salinity, Physiol. Plant., 51:265.

Farrar, J. F., 1980, The pattern of respiration rate in the vegetative barley plant, Ann. Bot., 46:71.

Kimura, M., 1970, Analysis of production of an undergrowth of subalpine Abies forest, Pteridophyllum racemosum population. 2. Respiration, gross production and economy of dry matter, Bot. Mag. (Tokyo), 83:304.

Lambers, H., 1982, Cyanide-resistant respiration: a non-phosphorylating electron transport pathway acting as an energy overflow, Physiol. Plant., 55:478.

Lambers, H., Blacquière, T., and Stuiver, C. E. E., 1981, Interactions between osmoregulation and the alternative respiratory pathway in Plantago coronopus as affected by salinity, Physiol. Plant., 51:63.

Lambers, H., Day, D. A., and Azcón-Bieto, J., 1983, Cyanide-resistant respiration in roots and leaves. Measurements with intact tissues and isolated mitochondria, Physiol. Plant., 58:148.

Lambers, H., Simpson, R. J., Beilharz, V. C., and Dalling, M. J., 1982, Translocation and utilization of carbon in wheat (Triticum aestivum), Physiol. Plant., 56:18.

Massimino, D., André, M., Richaud, G., Daguenet, A., Massimino, J., and Vivoli, J., 1981, The effect of a day at low irradiance of maize crop. I. Root respiration and uptake of N, P and K, Physiol. Plant., 51:150.

Pate, J. S., Layzell, D. B., and Atkins, C. A., 1979, Economy of carbon and nitrogen in a nodulated and non nodulated (NO_3-grown) legume, Pl. Physiol., 64:1083.

Peterson, R. B., and Zelitch, I., 1982, Relationship between net CO_2 assimilation and dry weight accumulation in field-grown tobacco, Pl. Physiol., 70:677.

GRAIN GROWTH OF WHEAT AND ITS LIMITATION BY CARBOHYDRATE AND NITROGEN SUPPLY

J. H. J. Spiertz[1] and J. Vos

Centre for Agrobiological Research
Wageningen, The Netherlands

INTRODUCTION

Grain yields of wheat have been increased considerably in north-west Europe. From 1950 onwards, average farm yields of winter wheat in the Netherlands increased from about 3.8 t ha^{-1} to 6.7 t ha^{-1} in 1982. This yield improvement is attributable to significant changes in genotype and in crop management practices. Modern cultivars have greater resistance to lodging, can utilize larger amounts of fertilizer nitrogen more efficiently and show a higher harvest index than their ancestors (Austin et al., 1980; Evans, 1981). The genetic potential for total dry matter production and nitrogen yield has remained comparatively stable; however the fraction of assimilates stored in the grains has increased. The grain protein content has not declined with increasing dry matter yields, because of increases in the rates of nitrogen application. After 1970 a further boost to grain yield followed the introduction of systemic fungicides, aphicides and split dressings of nitrogen (Ellen and Spiertz, 1980). The interactive effects of crop protection and split dressings of nitrogen on the grain yield of winter wheat are shown in Fig. 1. The beneficial effects of these management practices are shown by Dilz et al. (1982), Tinker and Widdowson (1982) and Becker and Aufhammer (1982). In recent multifactorial experiments at Rothamsted Experimental Station, Prew et al. (1983) found that aphicide and fungicide treatments had the greatest effect on yield; the larger amount of nitrogen always

---

[1] Present address: Research Station for Arable Farming and Field Production of Vegetables, P.O. Box 430, 8200 AK Lelystad.

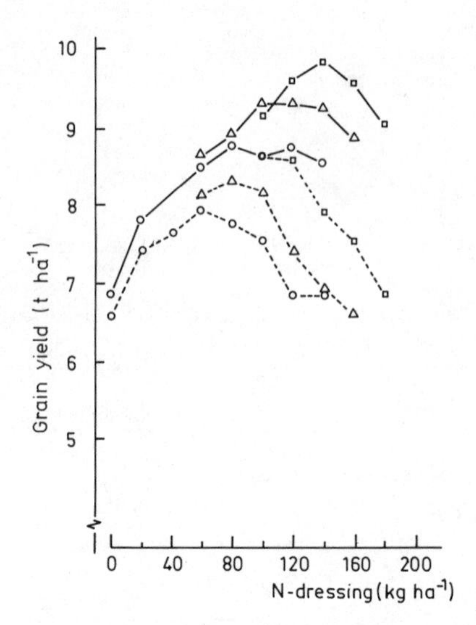

Fig. 1. The response of grain yield in winter wheat to the amount o the major N dressing (x) and to two additional nitrogen dressings, with (-----) and without (———) fungicide. ○---○, 0 + 0 + x dressing; △---△, 60 + 0 + x; □---□, 60 + 40 + x kg ha^{-1}. Timing of N-dressings: early spring, GS32 and GS45. Data derived from Dilz and Schepers (personal communication).

increased nitrogen uptake but decreased yield in the absence of aphicide and fungicide treatments.

The actual difference between mean farm yields and maximum yields in The Netherlands is about 2.5 t ha^{-1}. A better understanding of the physiological processes involved in grain production and of the factors causing the shortfall from maximum yields will lead to a more efficient deployment of existing technology and to the development of new agronomic methods. Modelling wheat growth and production may help to quantify the contributions to yield improvement by genetic traits, nitrogen supply, soil conditions and the control of weeds, pests and diseases. In this paper the various aspects of the post-floral carbohydrate and nitrogen economies are discussed.

POST-FLORAL CARBOHYDRATE AND NITROGEN ECONOMY

A key question in wheat production is whether grain yield can b increased by improving overall photosynthetic production or by enhancing the storage capacity of the grain. Discussions on whether assimilate supply (= source) or storage capacity (= sink) limits yield mostly refer to the grain filling stage, since most grain growth is supported by concurrent photosynthesis rather than by stored stem reserves. However, nitrogen for grain growth is mainly derived from reserves in the vegetative parts and to a smaller extent from post-anthesis uptake. The number of grains is also to a large extent determined by growth and development processes before anthesis. This interrelationship can be expressed as the ratio between grain number and dry matter yield at anthesis, which amounts to about 20 grains per g dry weight under temperate growing conditions (Spiertz, 1982). This ratio will be modified by the rate of net crop photosynthesis per unit thermal time during the pre-floral phase of crop development. Photosynthesis is only slightly influenced by temperature in contrast to the effect on the rate of development. High temperatures shorten the period from ear initiation to anthesis, thus reducing the supply of photosynthate relative to the rate of development; consequently number of grains per ear decreases with increasing temperature.

Generally, grain yield is more closely associated with grain number than with grain weight provided the conditions after anthesis are favourable for grain growth. Under such conditions the number of grains per unit area sets the potential yield of the crop. The demand for assimilates by the grains will be met by current assimilation and remobilization of reserves. This relationship can be represented by a simple equation:

GRAIN YIELD = POST-FLORAL ASSIMILATION + AVAILABLE RESERVES

In general, the proportions of assimilates contributed during grain filling of wheat are:

	Current Assimilation	Available Reserves
- for carbon	80 - 90%	10 - 20%
- for nitrogen	20 - 50%	50 - 80%

Thus the carbon requirement for grain growth is met largely by photosynthesis after anthesis, whereas the nitrogen requirement can be provided to a large extent by nitrogen taken up before anthesis.

RELATIONS BETWEEN RATE AND DURATION OF GRAIN GROWTH

The rate and duration of grain growth are strongly governed by temperature: a rise in temperature enhances the rate of carbohydrate and nitrogen accumulation, but shortens the duration (Sofield et al., 1977; Spiertz, 1977). Nitrogen relocation and protein accumulation show a stronger response to temperature than carbohydrate accumulation (Vos, 1981). Within the temperature range from 16 to 20°C the Q_{10}-values for the rates of protein deposition and grain growth amounted to 2.0 and 1.6, respectively. Nevertheless the effect of temperature on protein yield of the grains is smaller than on dry matter yield, because protein depletion of the leaves leads to decreases in photosynthetic capability and to increased leaf senescence. The effects of temperature and radiation on grain yield were evaluated with a simulation model (Vos, 1981; Chapter 13). It was predicted from the model that grain yield should decrease by 30 to 40 g m^{-2} per °C rise in average daily temperature during grain filling.

The effect of temperature on the duration of grain filling, D, can be described by an inverse exponential relationship between D and the mean temperature or by relating D to accumulated temperature above a base temperature (Vos, 1981). For the purpose of modelling, the latter is the more attractive approach. The reciprocal of duration, 1/D, was plotted against temperature, T, and yielded the following equation for wheat growth under controlled environmental conditions:

$$1/D = 0.0016\ T - 0.0096 \qquad (1)$$

As is shown in Fig. 2 1/D tends to zero (the duration becomes infinite) at a temperature of about 6°C. Taking this minimum temperature as a base, the heat sum for the grain filling period amounts to 625°C d. Davidson and Campbell (1983) also reported a linear relationship between duration and the mean daily temperature over the range from 5 to 25°C. Although the potential duration of grain growth is determined by temperature, the actual duration will depend on supply of assimilates to the grains. The duration of the period in which grain filling is not restricted by shortage of

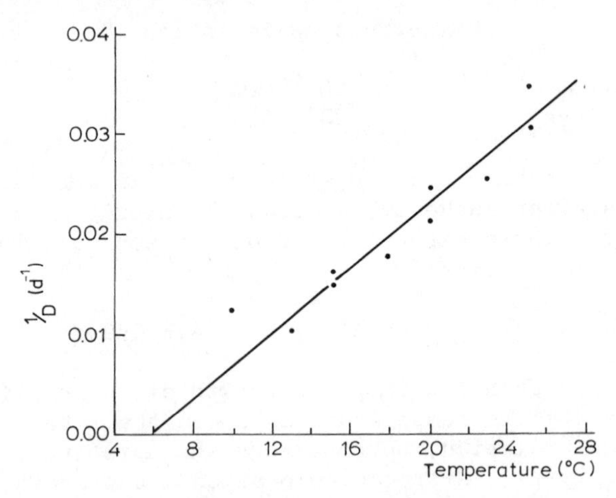

Fig. 2. The reciprocal of the duration (D) of the period between anthesis and maximum grain dry weight plotted against mean temperature, for wheat grown under controlled conditions. Data derived from Vos (1981).

assimilates may be derived as follows:

$$\text{Duration (days)} = \frac{\text{Available crop reserves (kg ha}^{-1})}{\text{Rate of grain growth (kg ha}^{-1}\text{d}^{-1}) - \text{Rate of post-floral assimilation (kg ha}^{-1}\text{d}^{-1})} \quad (2)$$

For example,

(i) Duration based on use and supply of carbohydrates.

$$D = \frac{\text{Available carbohydrate reserves}}{\text{Rate of carbohydrate accumulation} - \text{Rate of net photosynthesis}} = \frac{2000}{200-150} = 40 \text{ days} \quad (3)$$

(ii) Duration based on use and supply of nitrogen.

$$D = \frac{\text{Available N-reserves}}{\text{Rate of nitrogen accumulation} - \text{Rate of post-floral N-uptake}} = \frac{100}{4-1.5} = 40 \text{ days} \quad (4)$$

It follows that a rise in temperature shortens the duration of linear grain growth, if the enhanced rate of carbohydrate and nitrogen use by the grains not compensated for by an increase of post-floral assimilation or remobilization of stem reserves.

The amount of mobile carbohydrate ranges from 1.5 to 3.0 t ha^{-1}, when crop dry weight at anthesis amounts to about 10.0 t ha^{-1} (Austin et al., 1977; Spiertz and Ellen, 1978). If we take respiration and relocation to the roots into account then the net contribution of stored carbohydrate to the grains will be limited to between 0.5 and 2.0 t ha^{-1} (Austin et al., 1977; de Wit et al., 1979).

GROWTH EFFICIENCY AND RESPIRATION

A considerable fraction of assimilated carbon is lost as respiration. A suitable parameter to express the magnitude of respiration is the growth efficiency, GE, defined as the ratio between the increment in plant dry weight over a certain time interval, ΔW_X, and the sum of ΔW_X and R, where R is the total respiration rate, integrated over the same time interval and expressed in glucose units. Thus growth efficiency can be defined as:

$$GE = \frac{\Delta W_x}{\Delta W_x + R} \quad (5)$$

Analyses of data on respiration and growth and on retention of assimilated ^{14}C (e.g. Ryle et al., 1976) suggest that for many species, under various growing conditions, GE is unlikely to exceed 0.65 during the vegetative phase. During the reproductive phase GE is usually somewhat smaller, especially when large amounts of lipid are stored. Relocation of constituents from vegetative parts to storage organs represents an energy expenditure without resulting in a new increment of total dry matter and as a consequence the GE-value will fall.

An example of the change with time of GE during the post-floral period of wheat is given in Fig. 3. Growth efficiencies are calculated for the change in total plant dry weight (GE_t) and for the increment in grain dry weight (GE_g) for periods between successive harvests. Initially GE_g is smaller than GE_t, resulting from a gain in structural dry matter and storage of soluble carbohydrates in vegetative parts. At later stages the relation is reversed because of relocation of stored materials to the grains and non-metabolic losses of vegetative tissue (e.g. loss of senesced leaves). In this example, GE_t varies between 0.5 and 0.6, but declines rapidly during ripening when respiration is high relative to any additional increment of dry matter. Over the whole post-floral period GE_t and GE_g both average to 0.5. GE values derived from

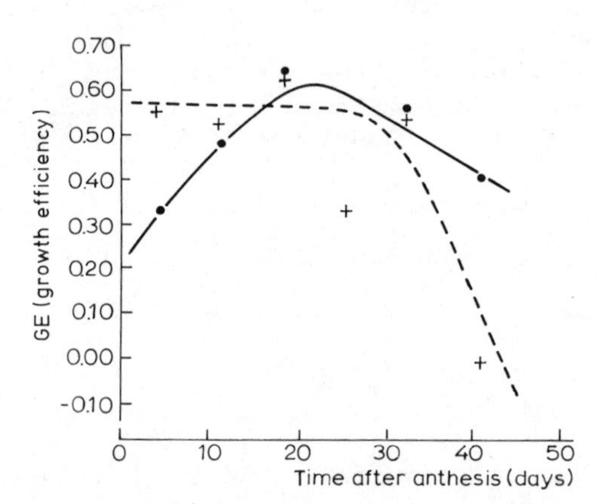

Fig. 3. Change with time after anthesis of the growth efficiency (GE), calculated in terms of the increment in grain dry weight (●, ———) or total plant dry weight (+, -----). Data derived from Vos (1981).

grain yields and estimates of post-floral total plant respiration given by Pearman et al. (1981) range between 0.57 and 0.70: however, these figures are not directly comparable with the present data because Pearman et al. neglected root respiration. Vos (1981) concluded that growth respiration of grains could be estimated quite accurately, using coefficients derived from a consideration of the energy cost of various metabolic and transport processes. However, the ratio between observed and calculated respiration rates was between 1.0 and 1.3 for leaf blades, about 2.0 for stem and leaf sheaths and from 3.0 to 6.0 for roots. After anthesis, respiration of these organs is mainly via maintenance and transport and the derived coefficients were inadequate to calculate these respiratory terms correctly.

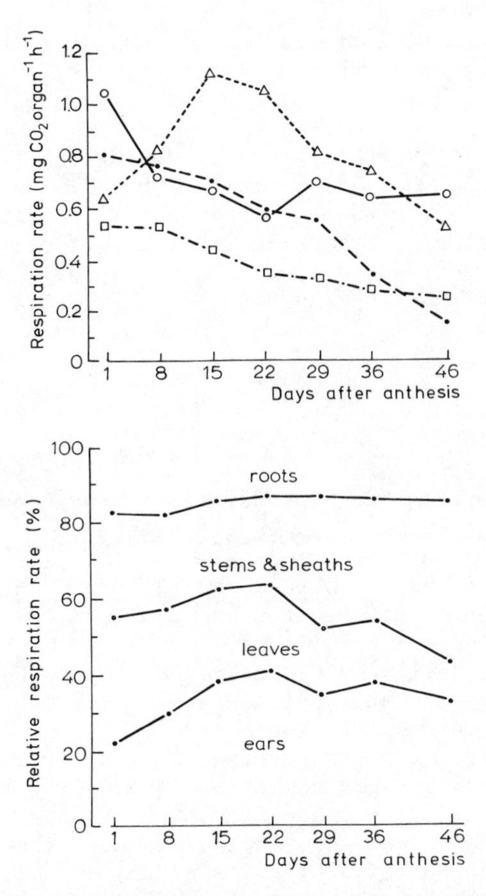

Fig. 4. Change with time after anthesis of respiration rates for component parts of the wheat plants; data after Vos (1981). (a) in absolute units: △---△, ears; •---•, leaves; O---O, stems and sheaths; □-•-•-•□, roots. (b) expressed as cumulative percentages of the total plant respiration rate.

Changes of respiration rates during the grain-filling period are given in Fig. 4a (absolute units) and in Fig. 4b (relative units). Ear respiration represents a major fraction of shoot respiration (Pearman et al., 1981), although in the middle of grain filling the respiration which is directly linked with grain growth represents at its maximum about 0.25 of total plant respiration.

The response of the respiration rate to an instantaneous change in temperature can usually be described satisfactorily by assuming a Q_{10} of 2. The rate of growth processes (conversions) is affected by temperature but their efficiencies remain unaltered. To what extent overall efficiency is affected by temperature depends on the ratio between maintenance respiration and growth respiration before and after the temperature change. For longer periods (weeks) temperature effects will presumably be small. With respect to nitrogen it can be noted that GE varied only slightly with increase in nitrogen concentrations in the plant (Vos, 1981).

If the overall growth efficiency during the grain-filling phase amounts to about 0.50, a grain yield of 8 t ha^{-1} dry matter of which 2 t ha^{-1} is contributed by pre-anthesis assimilates requires a post-anthesis gross photosynthesis of 12.0 t carbohydrate ha^{-1} (or 17.6 t (CO_2) ha^{-1}). Assuming an average duration of 40 days for grain filling this implies average daily gross photosynthesis of 300 kg CH_2O ha^{-1}.

PHOTOSYNTHESIS

Since most carbohydrate stored in the grain is derived from post-anthesis assimilation, the duration of high photosynthetic activity of the green plant parts during the post-floral phase is critical for achieving high grain yields. Usually it is observed that crop photosynthesis declines shortly after anthesis because of advancing senescence, moving from the older to the younger leaves.

During grain filling nitrogen relocation to the grains normally exceeds nitrate uptake and assimilation. High temperature accelerates protein accumulation in the grains and therefore speeds up the depletion of nitrogen reserves in the vegetative parts of the wheat plant. This nitrogen depletion, especially of the leaves, might accelerate leaf senescence and thus reduce photosynthesis rate too quickly by comparison with the potential duration of grain filling (Migus and Hunt, 1980; Evans, 1983; see also Chapter 14). Gregory et al. (1981) found that the maximum photosynthesis rate of flag leaves after anthesis (between 1 and 3 g CO_2 m^{-2} h^{-1}) was related to the fractional loss of nitrogen from these leaves. Therefore, net loss of nitrogen from the leaves will result in a corresponding reduction in the rate of crop photosynthesis. It has been shown that late nitrogen top dressings appreciably increase the nitrogen content of the leaves and delay their senescence via higher post-floral nitrogen uptake (Spiertz and Ellen, 1978). Under conditions of ample water and nutrient supply and moderate temperatures, it was found that the maximum rate of crop photosynthesis could be maintained at about 6 g CO_2 m^{-2} h^{-1} for four successive weeks after anthesis (de Vos, 1977). The duration of a high photosynthetic capacity of the crop is strongly influenced by limitations exerted by temperature, water and nitrogen supply, pests and diseases. Each of these limiting factors may affect photosynthesis either by hastening senescence, which is irreversible, or by adaptive changes in crop behaviour in response to environmental constraints, and such changes may be reversible.

Gallagher and Biscoe (1978) expressed the growth efficiency of a crop as the ratio between net photosynthetic productivity and absorbed radiation. They found an average value of 2.2 g MJ^{-1} for the period from emergence to final harvest under favourable conditions. In our own experiments with winter wheat, the similar parameter relating net photosynthesis to <u>intercepted</u> radiation had a value of about 3 g MJ^{-1} for a closed crop canopy under optimal conditions. Using this value, total net photosynthesis was calculated from intercepted radiation data and compared with crop growth rate. During a fortnight in June, average net photosynthesis amounted to 271 kg CH_2O ha^{-1} d^{-1} and crop growth rate to 193 kg CH_2O ha^{-1} d^{-1}; as a consequence 78 kg CH_2O ha^{-1} d^{-1} was relocated to the roots (J. H. J. Spiertz, unpublished data). The main part (55-70%) of the assimilates relocated to the roots before ear emergence are used for root growth, but a considerable amount is lost by root respiration and rhizosphere deposition (Martin, 1977; Sauerbeck and Johnen, 1977). After anthesis, when the grains become an active sink for assimilates, it is supposed that assimilate supply to the roots becomes less important. However, to ensure an optimal water and nutrient supply during grain-filling, the root system has to maintain a high activity and some root growth for exploring the deeper soil layers is to be expected. Böhm (1978) and Gregory et al. (1978) found that the length and dry weight of the roots decreased in the top soil but increased in the deeper soil layers throughout the post-anthesis period.

NITROGEN UPTAKE AND PARTITIONING

Nitrogenous compounds for grain growth are mainly supplied by the vegetative aerial parts (65-80%); the remainder originate from uptake and relocation by the roots after anthesis (Spiertz and Ellen, 1978; Gregory et al., 1981; Vos, 1981). Nitrogen available for relocation can be estimated from the difference between the nitrogen content of the crop at anthesis (about 150 kg ha^{-1}) and the nitrogen residues in straw and chaff (about 60 kg ha^{-1}).

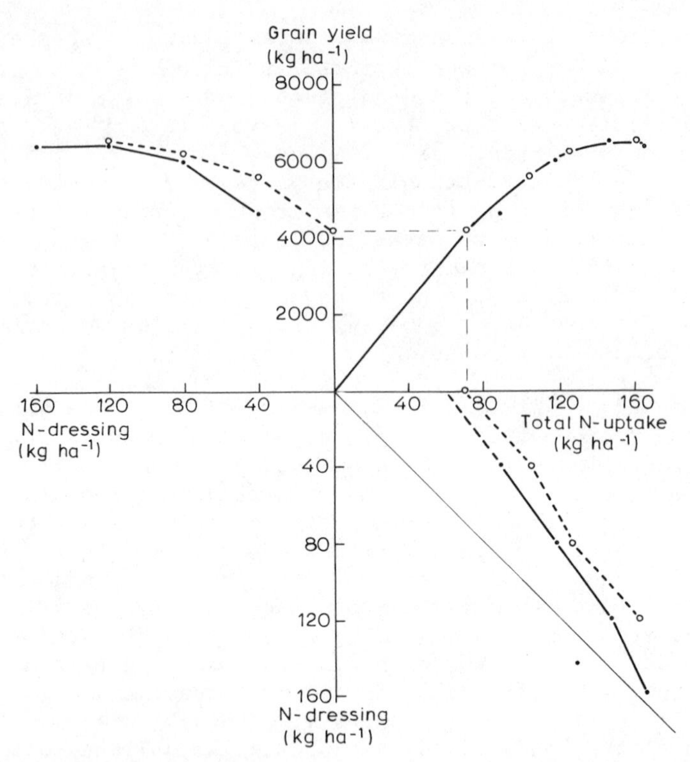

Fig. 5. The relationship between nitrogen dressing, nitrogen uptake and grain yield of wheat, for cv. Donata grown at Wageningen in 1979, with none o---o and 40 kg ha^{-1} •——• N fertilizer in autumn.

The nitrogen harvest-index (NHI), the proportion of the total amount of nitrogen taken up that is in the grains at harvest, usually ranges between 0.74 and 0.82. If water shortage occurs during the grain-filling stage, translocation of nitrogen from the vegetative tissue to the grains may be hampered, resulting in a lower nitrogen harvest-index, and a decrease in the nitrogen use efficiency. The efficiency of utilization of nitrogen by the wheat crop, defined as the amount of grain produced per unit nitrogen absorbed, is determined on the one hand by NHI and on the other hand by the nitrogen concentration in the grain (Fig. 5). At low nitrogen levels the relation between grain yield and total nitrogen uptake is linear with a rate of about 60 kg dry weight per kg nitrogen absorbed by the grains. At higher nitrogen uptake levels the yield response curve deviates from a straight line, reflecting an increase in the nitrogen content of the grains.

Large variations in nitrogen uptake from soil reserves and in recovery of nitrogen fertilizer exist between years and soils. The former variations mainly result from nitrogen residues of the preceding crop, the amount of nitrogen leached and the rate of mineralization of soil organic matter. Nitrogen recovery is strongly related to soil conditions and to crop growth. The factors affecting the uptake of nitrogen by a crop are imperfectly known, but growth rate, rooting depth, soil temperature, soil water content and nitrogen content in the soil solution are all involved.

CONCLUSIONS

Grain production of wheat in temperate regions is strongly associated with the amount of post-floral net photosynthesis and of remobilization of nitrogen reserves in the vegetative parts of the crop. Crop productivity has been increased by maintaining the photosynthetic activity of the canopy for longer after anthesis, through a combination of improved nitrogen supply and optimal control of diseases and pests. An extended post-floral period of canopy photosynthesis may also promote root activity after anthesis and as a consequence the post-floral nitrogen uptake may be enhanced. This late nitrogen uptake is necessary to meet the nitrogen demands of the grains.

During the grain-filling period, the interrelationships between assimilating organs, storage sites and growing organs are complex and depend on environmental conditions. Temperature, in particular, is a main driving force determining rate as well as duration of grain growth. Crop yield limitations are difficult to evaluate because variations in weather, soil conditions and genetic traits are interactive. In recent years, crop modelling has been used extensively as a means of integrating the photosynthetic behaviour of crop plants in response to weather, water and nutrient supply. Further progress in wheat modelling has to be made by quantifying crop development and the various feedback mechanisms between development, assimilation and storage processes. The systems approach attempts to integrate existing knowledge of wheat production, but does not eliminate the need for experimentation that can fill gaps in present knowledge.

REFERENCES

Austin, R. B., Edrich, J. A., Ford, M. A., and Blackwell, R. D., 1977, The fate of the dry matter, carbohydrate and ^{14}C lost from the leaves and stems of wheat during grain-filling, Ann. Bot., 41:1309.

Austin, R. B., Bingham, J., Blackwell, R. D., Evans, L. T., Ford, M. A., Morgan, C. L., and Taylor, M., 1980, Genetic improvements in winter wheat yields since 1900 and associated physiological changes, J. agric. Sci., Camb., 94:675.

Becker, F. A., and Aufhammer, W., 1982, Nitrogen fertilisation and methods of predicting the N-requirements of winter wheat in the Federal Republic of Germany, Proc. Fert. Soc., 211:33.

Böhm, W., 1978, Untersuchungen zur Wurzelentwicklung bei Winterweizen, Z. Acker- und Pflanzenbau, 147:262.

Davidson, H. R., and Campbell, C. A., 1983, The effect of temperature, moisture and nitrogen on the rate of development of spring wheat as measured by degree-days, Can. J. Pl. Sci., 63:833.

de Vos, N. M., 1977, Crop photosynthesis: Wheat, in: "Crop Photosynthesis: Methods and Compilation of Data Obtained with a Mobile Field Equipment", Th. Alberda, ed., Agric. Res. Rep. no. 865, Pudoc, Wageningen.

de Wit, C. T., van Laar, H. H., and van Keulen, H., 1979, Physiological potential of crop production, in: "Plant Breeding Perspectives", J. Sneep and A. J. T. Hendriksen, eds., Pudoc, Wageningen.

Dilz, K., Darwinkel, A., Boon, R., and Verstraeten, L. M. J., 1982, Intensive wheat production as related to nitrogen fertilization crop protection and soil nitrogen experience in the Benelux, Proc. Fert. Soc., 211:93.

Ellen, J., and Spiertz, J. H. J., 1980, Effects of rate and timing of nitrogen dressing on grain yield formation of winter wheat (T. aestivum L.), Fert. Res., 1:177.

Evans, J. R., 1983, Nitrogen and photosynthesis in the flag leaf of wheat (Triticum aestivum L.), Pl. Physiol., 72:297.

Evans, L. T., 1981, Yield improvement in wheat: empirical or analytical? in: "Wheat Science - Today and Tomorrow", L. T. Evans and W. J. Peacock, eds., Cambridge University Press, Cambridge.

Gallagher, J. N., and Biscoe, P. V., 1978, Radiation absorption, growth and yield of cereals, J. agric. Sci., Camb., 91:47.

Gregory, P. J., McGowan, M., Biscoe, P. V., and Hunter, B., 1978, Water relations of winter wheat. 1. Growth of the root system, J. agric. Sci., Camb., 91:91.

Gregory, P. J., Marshall, B., and Biscoe, P. V., 1981, Nutrient relations of winter wheat. 3. Nitrogen uptake, photosynthesis of flag leaves and translocation of nitrogen to grain, J. agric. Sci., Camb., 96:539.

Martin, J. K., 1977, Factors influencing the loss of organic carbon from wheat roots, Soil Biol. Biochem., 9:1.

Migus, W. N., and Hunt, L. A., 1980, Gas exchange rates and nitrogen concentrations in two winter wheat cultivars during the grain-filling period, Can. J. Bot., 58:2110.

Pearman, I., Thomas, S. M., and Thorne, G. N., 1981, Dark respiratio of several varieties of winter wheat given different amounts of nitrogen fertiliser, Ann. Bot., 47:535.

Prew, R. D., Church, B. M., Dewar, A. M., Lacey, J., Penny, A., Plumb, R. T., Thorne, G. N., Todd, A. D., and Wiliams, T. D., 1983, Effects of eight factors on the growth and nutrient uptak of winter wheat and on the incidence of pests and diseases, J. agric. Sci., Camb., 100:363.

Ryle, G. J. A., Cobby, J. M., and Powell, C. E., 1976, Synthetic and maintenance respiratory losses of $^{14}CO_2$ in uniculm barley and maize, Ann. Bot., 40:571.

Sauerbeck, D. R., and Johnen, B. G., 1977, Root formation and decomposition during plant growth, in: "Soil Organic Matter Studies", I.I.A.E.A., Vienna.

Sofield, I., Evans, L. T., Cook, M. G., and Wardlaw, I. F., 1977, Factors influencing the rate and duration of grain-filling in wheat, Aust. J. Pl. Physiol., 4:785.

Spiertz, J. H. J., 1977, The influence of temperature and light intensity on grain in relation to the carbohydrate and nitrogen economy of the wheat plant, Neth. J. agric. Sci., 25:182.

Spiertz, J. H. J., and Ellen, J., 1978, Effects of nitrogen on crop development and grain growth of winter wheat in relation to assimilation and utilisation of assimilates and nutrients, Neth. J. agric. Sci., 26:210.

Spiertz, J. H. J., 1982, Grain formation and assimilate partitioning in wheat. 1. Ear development, assimilate supply and grain growth of wheat, in: "Simulation of Plant Growth and Crop Production", F. W. T. Penning de Vries and H. H. van Laar, eds., Pudoc, Wageningen.

Tinker, P. B., and Widdowson, F. V., 1982, Maximising wheat yields, and some causes of yield variation, Proc. Fert. Soc., 211:149.

Vos, J., 1981, Effects of temperature and nitrogen supply on post-floral growth of wheat: measurement and simulations, Agric. Res. Rep. no. 911, Pudoc, Wageningen.

ASPECTS OF MODELLING POST-FLORAL GROWTH OF WHEAT AND CALCULATIONS OF THE EFFECTS OF TEMPERATURE AND RADIATION

J. Vos

Centre of Agrobiological Research
Wageningen
The Netherlands

INTRODUCTION

This paper summarizes some aspects of modelling post-floral growth of wheat. A more extensive account of the experimental work and of the simulation model was given by Vos (1981; see also Chapter 12). Effects of ambient temperature and/or nitrogen supply on post-floral growth of wheat were studied in experiments both under controlled conditions and outdoors; measurements included photosynthesis, respiration and analyses of water-soluble carbohydrates and of nitrogen content.

The simulation model, mainly based on data and relations determined under controlled conditions, was developed to estimate the effects of ambient temperature and radiation after anthesis on grain yield in the field. In the model, quantitative descriptions are given of production and utilization of carbohydrates, N-redistribution, and senescence as modified by temperature; some of these details are described below. Time steps of one day were used in the model.

PARTITIONING OF ASSIMILATES

There is usually negligible structural growth of vegetative organs after anthesis. Carbohydrate partitioning thus largely amounts to defining the respiratory terms and the potential growth rate of the grains at a known reference temperature. The growth rate during the linear phase seems to be an intrinsic property of the grain, primarily determined by the genotype and possibly modified by growth conditions till the endosperm cell number is fixed. Similar

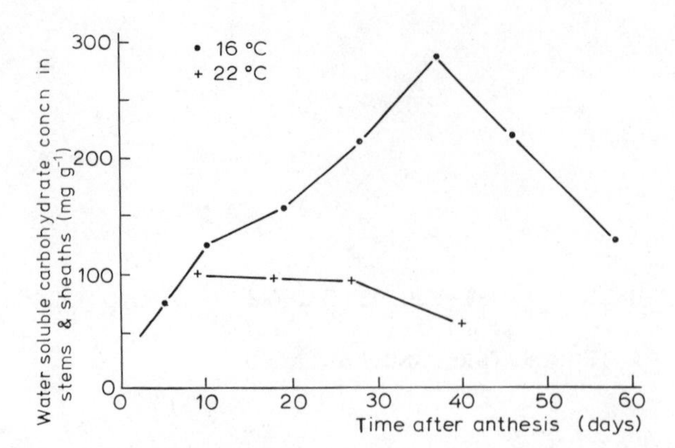

Fig. 1. Changes with time after anthesis in the concentrations of water-soluble carbohydrates in the dry matter (mg g^{-1}) of stems and sheaths for two temperature treatments. The plants were grown under controlled conditions. At six days after anthesis some of the plants were transferred from 16°C to 22°C. The data are derived from Vos (1981; his Expt. IV).

remarks apply to the rate of nitrogen deposition in the grains. Carbohydrates not used for growth and respiration are stored, mainly in the stem (and leaf sheaths). The amount of reserves changes with fluctuations in production and utilization of carbohydrates as illustrated in Fig. 1. The amount of carbohydrate reserves and its change with time should therefore not be modelled as following a predetermined pattern.

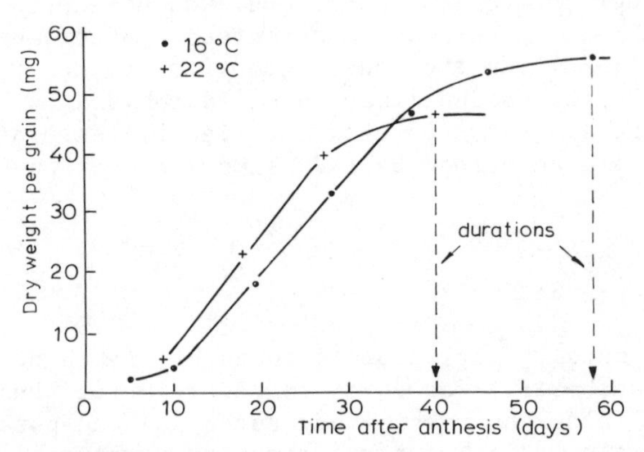

Fig. 2. Changes with time after anthesis in dry weight per grain (mg) for two temperature treatments (see legend to Fig. 1 for details).

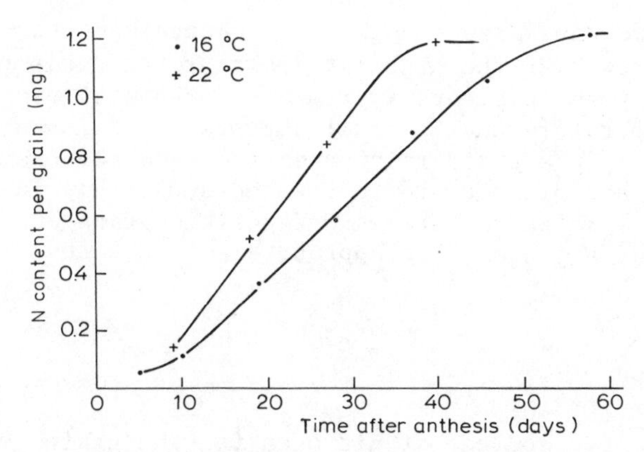

Fig. 3. Changes with time after anthesis in nitrogen content per grain (mg) for two temperature treatments (see legend to Fig. 1 for details).

EFFECTS OF TEMPERATURE

Effect on the Duration of Grain Filling

The duration is shorter at higher temperatures, as shown in Figs 2 and 3. It is well established that duration is mainly controlled by temperature and can be modelled in relation to thermal time, the sum of daily temperature above a minimum temperature (see Chapter 12). Differences among authors in duration may arise from genotypic variation and from different ways of assessing anthesis and maturity.

Effect of Temperature on the Rate of Grain Filling

The rate of grain growth during the linear phase increases with temperature (Fig. 2), though, when expressed as Q_{10}, the rate enhancement declines from about 3 in the temperature range of 10°C to 14°C to 1 at temperatures around 25°C and higher. It was found that a Q_{10} of 1.5 was appropriate for the rate of accumulation of "non-protein" constituents in the range of 16°C to 22°C, and this temperature response was incorporated in the model. The increase in rate with temperature was not sufficient to counterbalance the decrease in duration. Of course when evaluating temperature responses, only data from experiments where carbohydrate supply was not likely to have limited grain growth should be included.

Effect of Temperature on Nitrogen Deposition in Grains

The amount of nitrogen present in the grains at maturity was similar at 16°C and 22°C (Fig. 3). Apparently, the increase in deposition rate was large enough to counterbalance the shorter duration of growth at the higher temperature and this phenomenon has also been observed in other experiments. The difference between the effect of temperature on the final amounts of nitrogen and of carbohydrates in the grain reinforces the need for the economies of carbohydrates and nitrogen to be treated separately in models. Data to evaluate the relation between N-deposition rate and temperature are scarce, but a Q_{10} of 2 is appropriate in the temperature range of 16°C to 22°C.

THE NITROGEN REGIME

There are two sources of nitrogen for the grains, additional uptake and relocation from vegetative parts. The latter is usually the larger (Fig. 4; see also Chapter 12). Analyses revealed that the fraction of the daily N demand of the grains that was met by additional uptake plus relocation from roots (NUF) diminished with increase in the concentration of nitrogen in the shoot at anthesis, and a relationship of this form was incorporated into the model

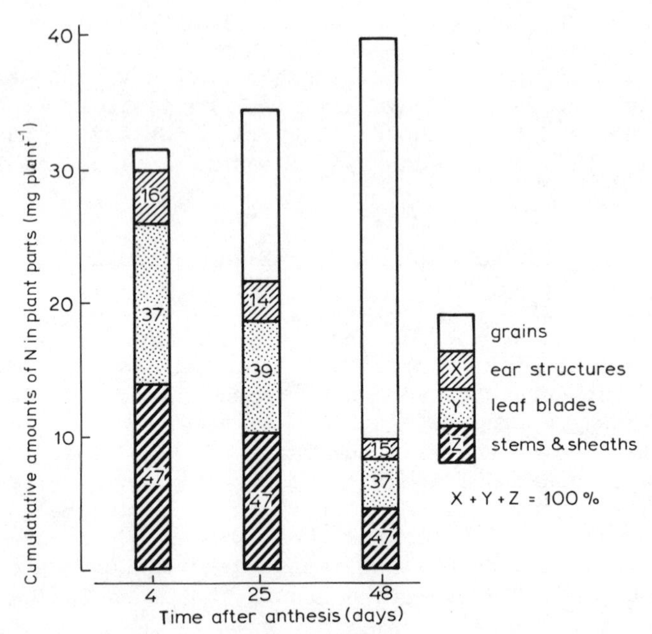

Fig. 4. Changes with time after anthesis in the amounts of nitrogen (mg) in component aerial parts. The figures in the graph represent the distribution (%) of nitrogen in the shoot (except the grains) divided into stems and sheaths, leaf blades and ear structures. Data derived from Vos (1981).

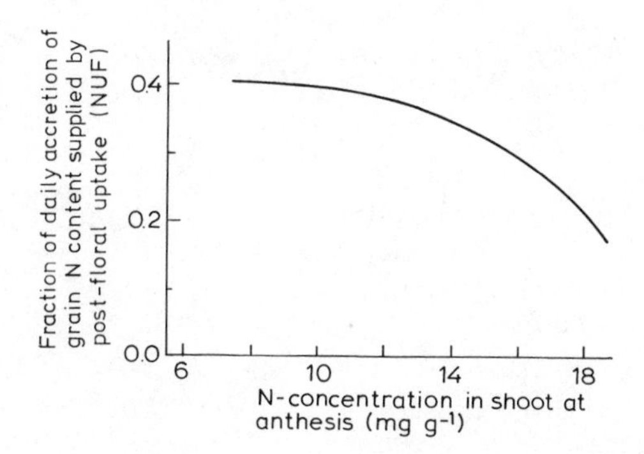

Fig. 5. The fraction of the daily accretion of the amount of nitrogen in the grains that is supplied by uptake from the soil and relocation from the roots as a function of the nitrogen concentration (mg g^{-1}) in aerial parts at anthesis (after Vos, 1981).

(Fig. 5). This certainly is a weak point in the model since there are other factors that control N uptake.

The relative distribution of shoot N between the different vegetative organs (including ear structures, but not the grains) remains remarkably constant throughout grain filling (Fig. 4). Thus, each class of organs contributes to the total amount of N translocated to the grains daily in proportion to the relative amount of shoot-N present in that organ. This feature was not influenced by temperature or nitrogen fertilization. This means that redistribution of N can be described quite satisfactorily, while the treatment of its complement, uptake, still presents some problems.

CALCULATION OF EFFECTS OF TEMPERATURE AND RADIATION ON GRAIN YIELD

The model was tested against data from field experiments using observed daily mean temperatures and radiation totals. Good agreement was obtained between measured and simulated courses of events, including nitrogen uptake. Given this agreement, it was considered that the effects of temperature and radiation after anthesis could be evaluated by changing weather parameters by a fixed quantity, e.g. adding 2°C to each measured input value for daily mean temperature or diminishing each value for the daily radiation total by 100 J cm^{-2}. This maintains the original day-to-day variation in weather variables and only alters the overall value. An example of such an exercise is shown in Fig. 6. On average over a number of simulations, grain yield dropped by between 30 and 40 g m^{-2} per °C rise in daily mean temperature after anthesis, and increased by about 20 g m^{-2} for each 100 J cm^{-2} (PAR) increase in daily radiation total. It follows that mean daily radiation totals would have to

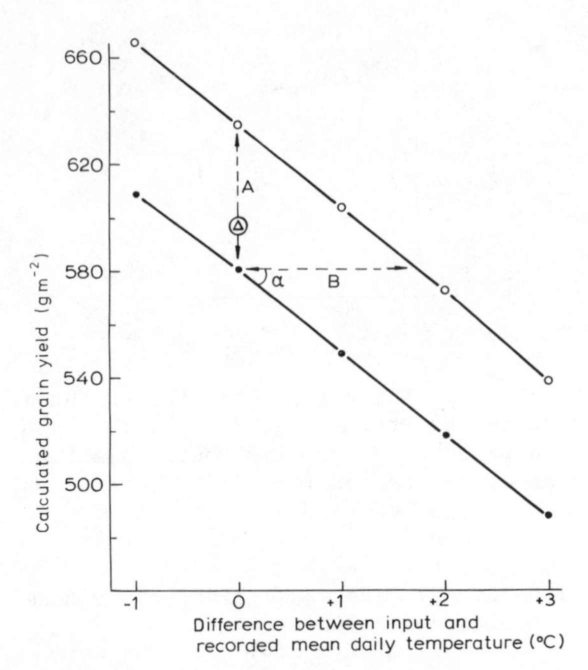

Fig. 6. Model calculations of grain dry-matter yield for different modifications of input weather data. The encircled triangl represents the yield measured and simulated with recorded weather data as input. The X-axis gives the temperature increment added to or subtracted from recorded mean daily temperatures in the simulation runs. Closed symbols; 100 J cm^{-2} subtracted from each value for the daily radiation total: open symbols; 200 J cm^{-2} added to recorded values. See text for further explanation (after Vos, 1981).

increase by between 130 and 180 J cm^{-2} to compensate for the adverse effect on grain yield of the increase in daily mean temperature by 1°C. These effects of temperature and radiation only apply to the yield levels examined, viz. 650 to 800 g m^{-2}. This analysis depends upon the relative uptake of nitrogen after anthesis as well as the pattern of redistribution, being only marginally affected by changes in temperature and radiation; this can be shown to be largely true.

REFERENCE

Vos, J., 1981, Effects of temperature and nitrogen supply on post-floral growth of wheat; measurements and simulations, Agric. Res. Rep., 911, Pudoc, Wageningen.

RELATIONSHIPS BETWEEN PHOTOSYNTHESIS, TRANSPIRATION AND NITROGEN IN THE FLAG AND PENULTIMATE LEAVES OF WHEAT

L. A. Hunt

University of Guelph
Guelph, Ontario, Canada

INTRODUCTION

Most of the carbohydrates used for grain growth in wheat are produced after anthesis (for reviews see Thorne, 1974; Austin and Jones, 1975; Evans et al., 1975). The amount of carbon assimilated during grain filling thus constitutes the ultimate upper limit to yield. It has been shown with some genotypes, however, that the amount of carbon assimilation after anthesis can be more than adequate to ensure that all grains are well filled (Fischer and Hille RisLambers, 1978; Aguilar-M and Hunt, 1984), and thus does not impose a limit to yield. Such a contradiction may reflect that fact that most wheat breeding programmes include good grain filling as an explicit objective of selection - a selection criterion that may only be satisfied when carbon assimilation substantially exceeds the demands of grain growth (Fischer et al., 1977). Thus, carbon assimilation during the grain-filling period may limit yield even though physiological analyses show that grain requirements for carbohydrates are less than the post-anthesis supply. Clarification of this point is essential for good deployment of effort in a wheat improvement programme and it would be aided by a comprehensive model of carbon assimilation during the grain-filling period. This paper reviews and reports work on factors that must be accounted for in such a model.

PHOTOSYNTHESIS DURING GRAIN FILLING

A decline in the photosynthetic rate of the flag and penultimate leaves during the grain-filling period has often been reported (Dantuma, 1973; Stoy, 1975; Spiertz, 1977), as has a decline in canopy photosynthesis over the same period (Puckridge, 1971; Puckridge and Ratkowsky, 1971; de Vos, 1975; Spiertz and van Laar, 1978). Some studies have shown a generally linear decline in photosynthesis from anthesis for individual leaves (Patterson and Moss, 1979) and for canopies (Puckridge, 1971), whereas others have found little decline for several weeks after anthesis (Wittenbach, 1979 for individual leaves; de Vos, 1975 for a canopy), or a rapid decline around anthesis to reach a minimum about one week later, followed by an increase (Evans and Rawson, 1970; Rawson and Evans, 1971). Where several genotypes have been included in one study, differences between genotypes in the pattern of decline in photosynthesis have been reported (Aslam and Hunt, 1978; Winzeler an Nosberger, 1980).

Some of the changes with time in the rate of flag leaf photosynthesis have been attributed to feedback effects dependent or the demand for carbohydrates within the plant (for a review see Evar et al., 1975). However, in other research (e.g. Austin and Edrich, 1975) photosynthesis of the flag leaf has been found to be independent of the assimilate requirement of the ear. Such differences have been explained in terms of either the influence on leaf photosynthesis of the activity of sources and sinks other than the ear-flag leaf system, or the control of photosynthesis through stomatal responses rather than changes in the photosynthetic system _per se_ (Rawson et al., 1976).

TRANSPIRATION DURING GRAIN FILLING

Recent research has indicated that stomata may remain more open and consequently the transpiration rate may be greater after anthesi than before (Morgan, 1977; Clarke and McCaig, 1982). Measurements made in controlled conditions, where water supply was adequate, have shown a decrease in transpiration after full flag-leaf expansion until anthesis, followed by an increase to a peak late in ontogeny, and then a final decrease as senescence became advanced (Aslam and Hunt, 1978; Migus and Hunt, 1980).

NITROGEN DURING GRAIN FILLING

In contrast to carbon, much nitrogen for grain growth in wheat is assimilated before anthesis and translocated from leaves and othe structures to the grain (Neales et al., 1963; Austin et al., 1977). However, some high grain-protein cultivars appear to take up some of their nitrogen in the grain-filling period (Mikesell and Paulsen, 1971), and some high yielding wheat crops assimilate much of their grain nitrogen after anthesis (Spiertz and Ellen, 1978).

The pattern of nitrogen remobilization from leaves has been examined in a number of studies (Spiertz and Ellen, 1978; Gregory et al., 1981). Comparing genotypes, McNeal et al. (1966) showed that there was a slight increase in the nitrogen content of the leaves of Mindum (a durum wheat) between flowering and the middle of grain filling, whereas for four other genotypes (all common wheats) the nitrogen content either remained constant or decreased. More recently, Dalling et al. (1976) found that the percentage of nitrogen extracted from the leaves between anthesis and maturity was the same for the two cultivars studied, but that remobilization of the nitrogen in the glumes (accounting for 15% of the total plant nitrogen at anthesis) varied from 76% for one cultivar to 57% for the other. Migus and Hunt (1980) showed that the pre-anthesis pattern of nitrogen remobilization from the flag and penultimate leaves differed between cultivars when well supplied with nitrogen from solution culture, but did not differ when the nitrogen supply was lower. After anthesis the patterns did not differ between cultivars.

Whether differences in the pattern of decline of leaf nitrogen are reflected in the final distribution of nitrogen between grain and other above-ground structures is not known. However, several workers have shown that the final distribution varies among genotypes (Johnson et al., 1967; Lal et al., 1978; Austin et al., 1977; Löffler and Busch, 1982), and also that the differences among genotypes depend on the nitrogen supply. Canvin (1976) used data presented by Hucklesby et al. (1971) to show that the nitrogen harvest index decreased with the amount of nitrogen supplied for one cultivar, but

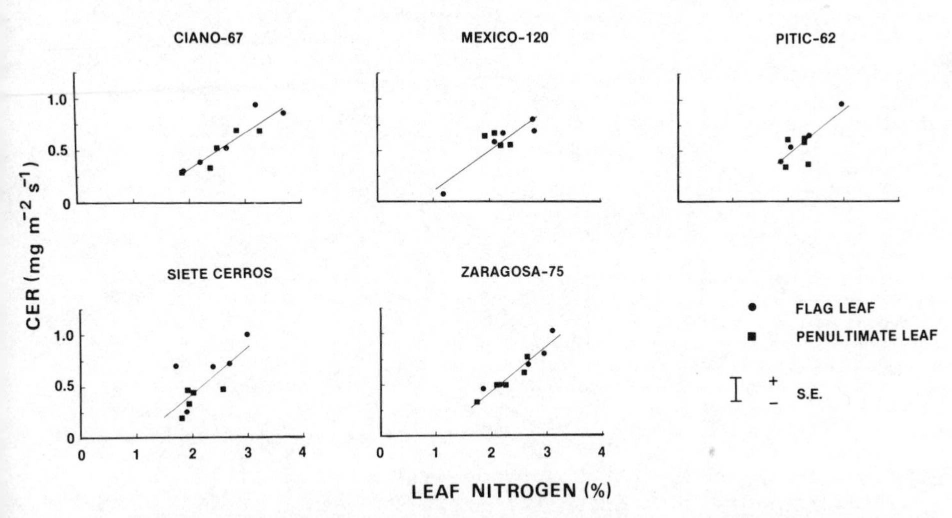

Fig. 1. Relationship between CO_2 exchange rate (CER) and nitrogen content for the main-stem flag and penultimate leaves of five spring wheats. Data from van der Poorten (1979).

remained constant for two others. In other words, at high nitrogen, the leaves and vegetative structures of one genotype retained more nitrogen late into the grain-filling period than did the other two genotypes.

PHOTOSYNTHESIS, TRANSPIRATION AND NITROGEN

There is considerable correlation between photosynthesis and nitrogen content during the life of an individual leaf. Khan and Tsunoda (1970a,b) reported that leaf photosynthetic rate was correlated with the nitrogen content per unit of leaf surface area i both wild and cultivated wheats. Subsequently, a similar relationshi has been found between photosynthetic rate and nitrogen concentratic during leaf ageing (Osman and Milthorpe, 1971; Dantuma, 1973). Further, Migus and Hunt (1980) and Gregory et al. (1981) showed that photosynthesis was little affected by differences in leaf nitrogen induced by the nutrient regime. Such findings led Gregory et al. (1981) to suggest that photosynthesis was affected by the change in nitrogen status rather than the absolute nitrogen content.

Van der Poorten (1979) examined the relationships between photosynthesis, transpiration and nitrogen content in five wheat genotypes. He showed that the time course of photosynthesis of both

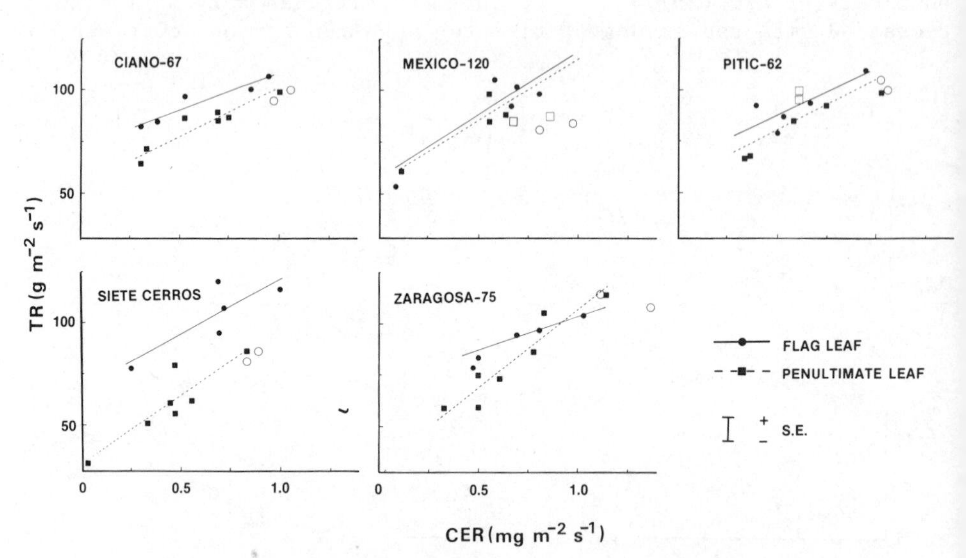

Fig. 2. Relationship between transpiration rate (TR) and CO_2 exchange rate (CER) for the main-stem flag and penultimate leaves of five spring wheats. The unfilled symbols were data not used for regression. Data from van der Poorten (1979).

the flag and penultimate leaves varied among cultivars, but that the changes in photosynthetic rates were closely related to alterations in the nitrogen contents of the leaves (Fig. 1). Transpiration generally changed in concert with photosynthesis (Fig. 1), but for most flag leaves the first measurements of transpiration after anthesis were less than expected on the basis of the remaining post-anthesis data, and on some occasions in the middle of the flag leaf's life transpiration was greater than expected. The stomata thus seemed more sensitive to internal influences than was the photosynthetic system per se, as suggested by Fischer et al. (1981). Stomatal changes had little influence on photosynthesis, suggesting that the photosynthetic system was not responsive to small changes in intercellular-space CO_2 concentration (Farquhar and Sharkey, 1982).

CONCLUSIONS

The work discussed here emphasizes the importance of nitrogen metabolism in determining the amount of carbon assimilated during the grain-filling period (Sinclair and de Wit, 1975; Spiertz and Ellen, 1978; Gregory et al., 1981; Swank et al., 1982). It also suggests, however, that the pattern of nitrogen mobilization from leaves varies among genotypes. The extent of such genotypic variation and its physiological basis need to be examined further. In particular, it seems important to determine if the observed differences merely reflect variations in the array of tissues from which the ear can draw nitrogen, or dissimilarities in the timing of the response to some senescence signal (cf. Lindoo and Nooden, 1976) originating in the ear. Any comprehensive model needs to account for both genotypic differences in the nitrogen content of the straw before grain filling (Kramer, 1979), and the mechanism of senescence. Thomas and Stoddart (1980) pointed out that "it is as necessary to understand senescence as it is photosynthesis and translocation when establishing physiological and biochemical criteria for crop improvement". The same obviously applies when considering the development of a comprehensive model for wheat.

REFERENCES

Austin, R. B., and Edrich, J., 1975, Effects of ear removal on photosynthesis, carbohydrate accumulation and on the distribution of assimilated ^{14}C in wheat, Ann. Bot., 39:141.

Austin, R. B., and Jones, H. G., 1975, The physiology of wheat, A. Rep. Pl. Breed. Inst., 1975, 20.

Austin, R. B., Ford, M. A., Edrich, J. A., and Blackwell, R. D., 1977, The nitrogen economy of winter wheat, J. agric. Sci., Camb., 88:159.

Aguilar-M, I., and Hunt, L. A., 1984, Genotypic variation in some physiological traits in winter wheat grown in the humid continental climate of Ontario, Can. J. Pl. Sci., 64:113.
Aslam, M., and Hunt, L. A., 1978, Photosynthesis and transpiration c the flag leaf in four spring-wheat cultivars, Planta, 141:23.
Canvin, D. T., 1976, Interrelationship between carbohydrate and nitrogen metabolism, in: "Genetic Improvement of Seed Proteins", National Academy of Sciences, Washington, D.C.
Clarke, J. M., and McCaig, T. N., 1982, Evolution of techniques for screening for drought resistance in wheat, Crop Sci., 22:503.
Dalling, M. J., Boland, G., and Wilson, J. H., 1976, Relation betwee acid proteinase activity and redistribution of nitrogen during grain development in wheat, Aust. J. Pl. Physiol., 3:721.
Dantuma, G., 1973, Rates of photosynthesis in leaves of wheat and barley varieties. Neth. J. agric. Sci., 21:188.
de Vos, N. M., 1975, Field photosynthesis of winter wheat during the grain-filling phase under highly fertile conditions, in: "Proceedings of the 2nd International Winter Wheat Conference", Zagreb, Yugoslavia.
Evans, L. T., and Rawson, H. M., 1970, Photosynthesis and respiration by the flag leaf and components of the ear during grain development in wheat, Aust. J. Biol. Sci., 23:245.
Evans, L. T., Wardlaw, I. F., and Fischer, R. A., 1975, Wheat, in: "Crop Physiology; Some Case Histories", L. T. Evans, ed., Cambridge University Press, Cambridge.
Fischer, R. A., and Hille RisLambers, D., 1978, Effect of environmen and cultivar on source limitation in wheat, Aust. J. agric. Res., 29:443.
Fischer, R. A., Aguilar-M, I., and Laing, D. R., 1977, Post-anthesis sink size in a high yielding dwarf wheat: yield response to grain number, Aust. J. agric. Res., 28:165.
Fischer, R. A., Bidinger, F., Syme, J. R., and Wall, P. C., 1981, Leaf photosynthesis, leaf permeability, crop growth and yield o short spring wheat genotypes under irrigation, Crop Sci., 21:367.
Farquhar, G. D., and Sharkey, T. D., 1982, Stomatal conductance and photosynthesis, A. Rev. Pl. Physiol., 33:317.
Gregory, P. J., Marshall, B., and Biscoe, P. V., 1981, Nutrient relations of winter wheat. 3. Nitrogen uptake, photosynthesis o flag leaves and translocation of nitrogen to grain, J. agric. Sci., 96:539.
Hucklesby, D. P., Brown, C. M., Howell, S. E., and Hageman, R. H., 1971, Late spring applications of nitrogen for efficient utilization and enhanced production of grain and grain protein of wheat, Agron. J., 63:274.
Johnson, V. A., Mattern, P. J., and Schmidt, J. W., 1967, Nitrogen relations during spring growth in varieties of Triticum aestivum L. differing in grain protein content, Crop Sci., 7:664.
Khan, M. A., and Tsunoda, S., 1970a, Evolutionary trends in leaf

photosynthesis and related leaf character among cultivated wheat species and its wild relatives, Jap. J. Breed., 20:133.

Khan, M. A., and Tsunoda, S., 1970b, Differences in leaf photosynthesis and leaf transpiration rates among six commercial wheat varieties, Jap. J. Breed., 20:344.

Kramer, T. L., 1979, Environmental and genetic variation for protein content in winter wheat (Triticum aestivum L.), Euphytica, 28:209.

Lal, P., Reddy, G. G., and Modi, M. S., 1978, Accumulation and redistribution pattern of dry matter and N in triticale and wheat varieties under water stress condition, Agron. J., 70:623.

Lindoo, S. J., and Nooden, L. D., 1976, The interrelation of fruit development and leaf senescence in 'Anoka' soybeans, Bot. Gaz., 137:218.

Löffler, C. M., and Busch, R. H., 1982, Selection for grain protein, grain yield and nitrogen partitioning efficiency in hard red spring wheat, Crop Sci., 22:591.

McNeal, F. H., Berg, M. A., and Watson, C. A., 1966, Nitrogen and dry matter in five spring wheat varieties at successive stages of development. Agron J., 58:605.

Migus, W. N., and Hunt, L. A., 1980, Gas exchange rates and nitrogen concentrations in two winter wheat cultivars during the grain filling period, Can. J. Bot., 58:2110.

Mikesell, M. E., and Paulsen, G. M., 1971, Nitrogen translocation and the role of individual leaves in protein accumulation in wheat grain, Crop Sci., 11:919.

Morgan, J. M., 1977, Changes in diffusive conductance and water potential of wheat plants before and after anthesis, Aust. J. Pl. Physiol., 4:75.

Neales, T. F., Anderson, M. J., and Wardlaw, I. F., 1963, The role of the leaves in the accumulation of nitrogen by wheat during ear development, Aust. J. agric. Res., 14:725.

Osman, A. M., and Milthorpe, F. L., 1971, Photosynthesis of wheat leaves in relation to age, illuminance and nutrient supply. II. Results, Photosynthetica, 5:61.

Patterson, T. G., and Moss, D. N., 1979, Senescence in field grown wheat, Crop Sci., 19:635.

Puckridge, D. W., 1971, Photosynthesis of wheat under field conditions. III. Seasonal trends in carbon dioxide uptake of crop communities, Aust. J. agric. Res., 22:1.

Puckridge, D. W., and Ratkowsky, D. A., 1971, Photosynthesis of wheat under field conditions. IV. The influence of density and leaf area index on the response to radiation, Aust. J. agric. Res., 22:11.

Rawson, H. M., and Evans, L. T., 1971, The contribution of stem reserves to grain development in a range of wheat cultivars of different height, Aust. J. agric. Res., 22:851.

Rawson, H. M., Gifford, R. M., and Bremner, P. M., 1976, Carbon dioxide exchange in relation to sink demand in wheat, Planta, 132:19.

Sinclair, T. R., and de Wit, C. T., 1975, Photosynthate and nitrogen requirements for seed production by various crops, Science., 189:565.

Spiertz, J. H. J., 1977, The influence of temperature and light intensity on grain growth in relation to the carbohydrate and nitrogen economy of the wheat plant, Neth. J. agric. Sci., 25:182.

Spiertz, J. H. J., and Ellen, J., 1978, Effects of nitrogen on crop and grain growth of winter wheat in relation to assimilation an utilization of assimilates and nutrients, Neth. J. agric. Sci., 26:210.

Spiertz, J. H. J., and van Laar, H. H., 1978, Differences in grain growth, crop photosynthesis and distribution of assimilates between a semi-dwarf and a standard cultivar of winter wheat, Neth. J. agric. Sci., 26:233.

Stoy, V., 1975, Use of tracer techniques to study yield components i seed crops, in: "Tracer Techniques for Plant Breeding", International Atomic Energy Agency, Vienna.

Swank, J. C., Below, F. E., Lambert, R. J., and Hageman, R. H., 1982 Interaction of carbon and nitrogen metabolism in the productivity of maize, Pl. Physiol., 70:1185.

Thomas, H., and Stoddart, J. L., 1980, Leaf senescence, A. Rev. Pl. Physiol., 31:83.

Thorne, G. N., 1974, Physiology of grain yield of wheat and barley, Rothamsted Exp. Stn Rep., 1973, Pt. II, 5.

van der Poorten, G., 1979, Characteristics of gas exchange, grain growth and leaf nitrogen in some selected wheat cultivars of different yield potentials, M.Sc. thesis, University of Guelph.

Winzeler, H., and Nosberger, J., 1980, Carbon dioxide exchange of spring-wheat in relation to age and photon-flux density at different growth temperatures, Ann. Bot., 46:685.

Wittenbach, V. A., 1979, Ribulose bisphosphate carboxylase and proteolytic activity in wheat leaves from anthesis through senescence, Pl. Physiol., 64:884.

MODELLING THE EFFECTS ON GRAIN YIELD OF GENETIC VARIATION IN SOME CROP CHARACTERISTICS

R. B. Austin

Plant Breeding Institute
Cambridge, U.K.

INTRODUCTION

Except in adverse environments, the great majority of the carbohydrate in harvested grain originates from photosynthesis during the grain filling period, from anthesis to maturity. However, though the carbohydrate yield of a wheat crop is realised during grain filling, the potential for carbohydrate production is determined before anthesis. Protein, constituting about 10% of the grain dry matter, is synthesized in the grain mainly from organic nitrogen (amino acids and amides) that is translocated from the vegetative organs as they senesce and the proteins they contain are hydrolysed. This knowledge forms the basis of the simulation model used in this paper.

THE MODEL

The model calculates daily carbohydrate production throughout the grain filling period. Summing over the period gives the total carbohydrate available for grain filling. To this total, protein and water are added to give grain of average composition so providing an estimate of potential grain yield.

Daily carbohydrate production is calculated as a function of solar radiation and daylength and of those plant characteristics which determine the light interception by the canopy and the rates of photosynthesis of individual leaves, as described by Monteith (1965). Details of other features of the model have been described elsewhere (Austin, 1982). The plant characteristics (symbols or acronyms in brackets) considered are:-

(1) Leaf area at anthesis (ALAI).

(2) The duration of the grain filling period in days from anthesis to the time when no green tissue is present (DUR).

(3) The proportion of incident photosynthetically active radiation not intercepted by unit leaf area index (S) - the sun-flecked area fraction.

(4) The fractional transmission of photosynthetically active radiation by a leaf (T).

(5) The light saturated rate of photosynthesis of single leaves (P_{max}), expressed here as mg CO_2 dm^{-2} h^{-1}.

(6) The efficiency of photosynthesis at limiting irradiance (1/B) expressed as g carbohydrate produced per Joule of incident radiation.

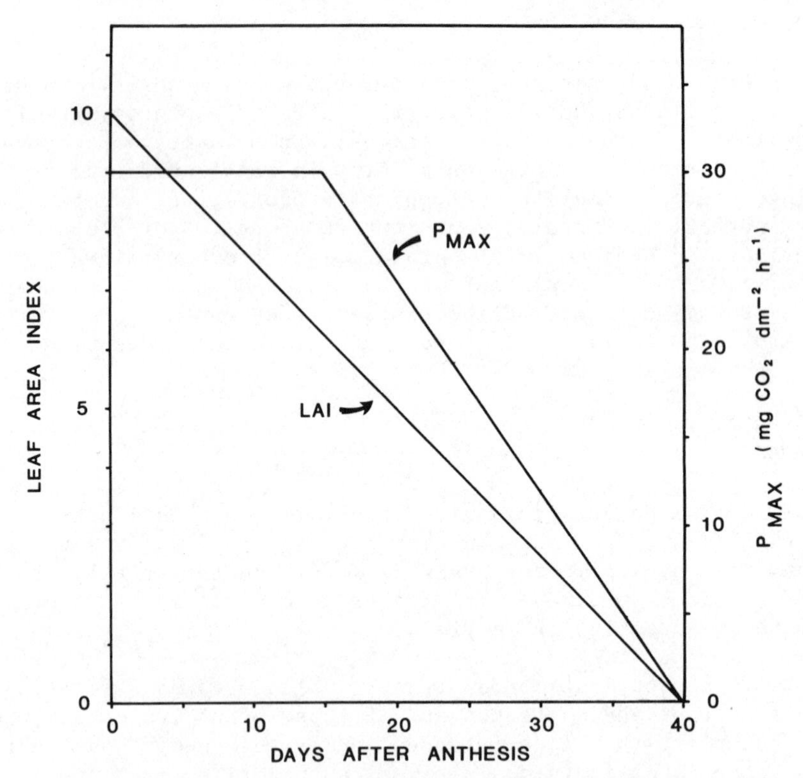

Fig. 1. Changes in leaf area index and P_{max} during the grain filling period. Values used in this figure are from the 'standard case' (Austin, 1982).

(7) The time of onset of decline in P_{max}, in days after anthesis (SENO).

(8) Loss of carbohydrate in dark respiration, as a percentage of net photosynthesis (RESP).

The interception of photosynthetically active radiation by the ears is taken as 33%, and the ears are taken to contribute 15% of the total canopy photosynthesis. Figure 1 shows how leaf area index and P_{max} are taken to vary during grain filling. Parameter values for the standard case are given in Austin (1982).

RESULTS

Taking values of the above parameters established from measurements on well grown crops where nutrients and water supply were considered to be non-limiting, and taking the daily radiation as 8.4 MJ m^{-2} d^{-1} (total solar and sky, 400-700 nm) and the daylength as 16 h, the calculated 'potential' yield is 12 t ha^{-1} (15% moisture, 10% protein on a fresh weight basis).

Assuming the reflection of photosynthetically active radiation as 10% of that incident on the crop and using Monteith's (1965) expression to calculate radiation interception by the leaves, the efficiency of utilisation of solar energy for carbohydrate production can be calculated. In the model this is done on a daily basis. Figure 2 shows, for the standard case parameter values (which predict a potential yield of 12 t ha^{-1}), net carbohydrate production per day, and the corresponding efficiencies. The average efficiency, expressed as net carbohydrate produced during the entire grain filling period divided by the total photsynthetically active radiation absorbed by the canopy, is 3.47 g MJ^{-1}, a value close to that measured over weekly periods on well-grown crops at anthesis, when the absorption of radiation by senescent plant parts is small (P. V. Biscoe, personal communication). However, the model will overestimate the efficiency that would be measured from radiation absorption by the entire crop during the grain filling period because it takes no account of dead tissue, which, as the crop ages, becomes an increasing proportion of the biomass. In the model, the main reason for the decline in efficiency with age is that, as leaf area index declines, the average intensity to which unit leaf area is exposed increases, and efficiency decreases as leaves approach light saturation.

The sensitivity of potential yield to changes in several plant characteristics is illustrated in Fig. 3, which is adapted from Austin (1982). The extent to which the characteristics could be varied independently of one another by breeding is, in the main, not known. There is evidence (Austin et al., 1980) that modern cultivars

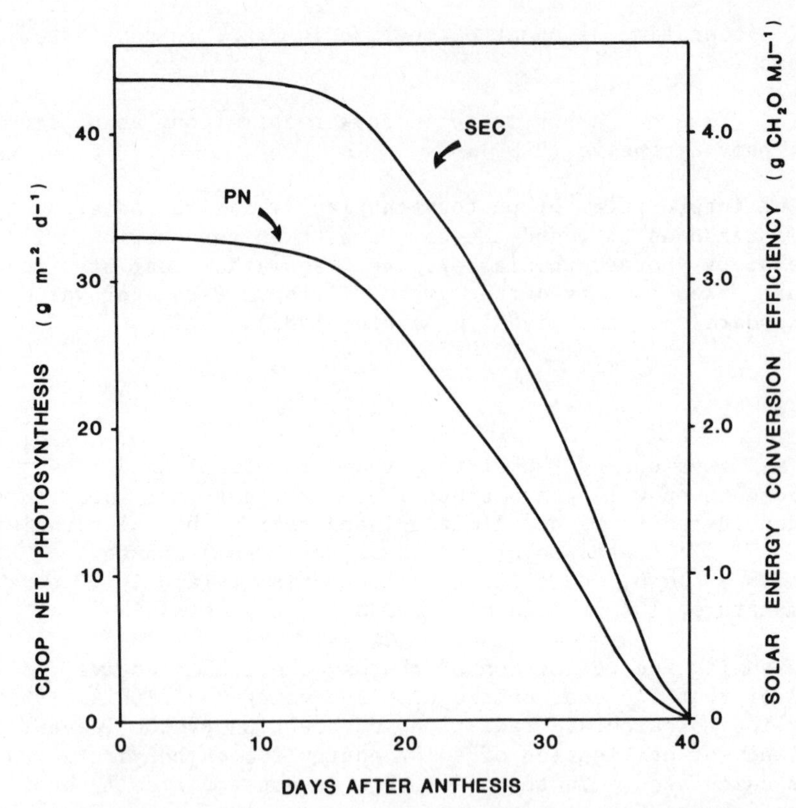

Fig. 2. Changes in the calculated daily net photosynthesis, PN, and efficiency of conversion of absorbed photosynthetically active solar radiation, SEC, during the grain filling period. The values shown in this figure have been calculated for the 'standard case' parameters.

have a longer grain-filling period because they flower sooner but mature at the same time as old cultivars. The leaf lamina area inde at anthesis appears to differ with cultivar, but there is likely to be compensating variation in the amount of other photosynthetic tissue (stems and leaf sheaths), modern cultivars having less than older ones by virtue of their reduced height. Considerable variatio exists in leaf posture, which determines the pattern of light interception by wheat canopies, as measured by the parameter S. However, experimental evidence (Innes and Blackwell, 1983) supportin the results of the simulations, shows that over the relevant range

→

Fig. 3. Calculated effects on potential grain yield of variations i some crop characteristics. Closed circles represent the 'standard case' parameter values and yield.

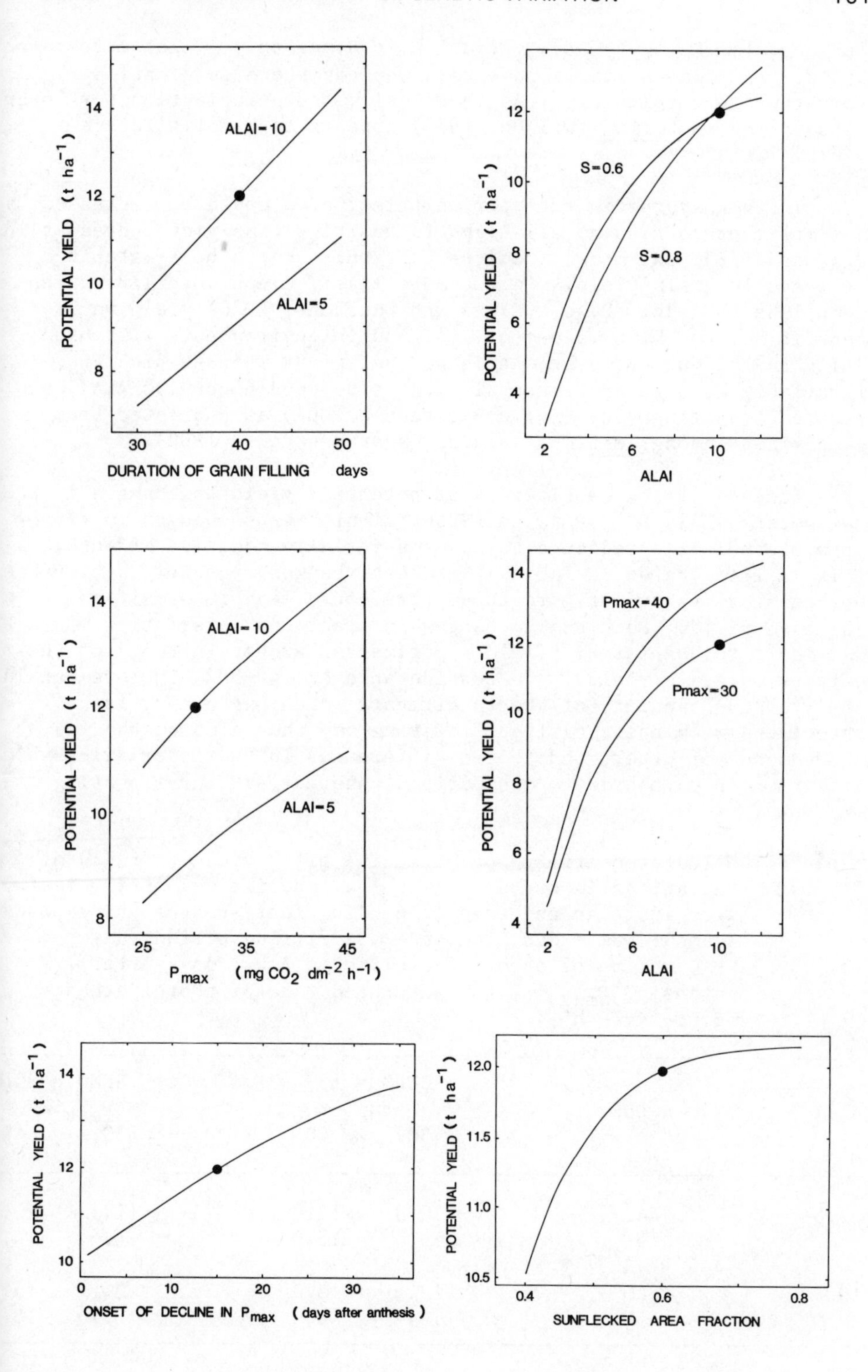
POTENTIAL YIELD (t ha^{-1})
14
12
10
8
ALAI=10
ALAI=5
30
40
50
DURATION OF GRAIN FILLING days
POTENTIAL YIELD (t ha^{-1})
12
10
8
6
4
S=0.6
S=0.8
2
6
10
ALAI
POTENTIAL YIELD (t ha^{-1})
14
12
10
8
ALAI=10
ALAI=5
25
35
45
P_{max} (mg CO_2 dm^{-2} h^{-1})
POTENTIAL YIELD (t ha^{-1})
14
12
10
8
6
4
Pmax=40
Pmax=30
2
6
10
ALAI
POTENTIAL YIELD (t ha^{-1})
14
12
10
0
10
20
30
ONSET OF DECLINE IN P_{max} (days after anthesis)
POTENTIAL YIELD (t ha^{-1})
12.0
11.5
11.0
10.5
0.4
0.6
0.8
SUNFLECKED AREA FRACTION

for cereals, (c. 0.6-0.8), variation in S has only slight effects on yield. At present, it is uncertain whether there is genetic variation in respiratory loss in cereals, though selection for lower respiration in *Lolium* (Wilson, 1975) appears to benefit forage yield.

If photosynthetic rate per unit leaf area could be increased by a similar proportion at all light intensities (implying increases in P_{max} and 1/B), the model predicts that there would be a matching increase in grain yield. In the more likely event that 1/B varies much less than does P_{max}, (Evans and Dunstone, 1970) yield would benefit most at high values of ALAI and in environments with high daily radiation. However, it is not yet clear whether high P_{max} is invariably associated with small leaf size (and hence low ALAI) and reduced leaf longevity (and hence reduced DUR) as indicated from comparisons among *Triticum* species (Austin et al., 1982).

Because of the sensitivity of potential yield to changes in the parameters ALAI, DUR, P_{max} and SENO, simultaneous changes in two or more of them are predicted to have essentially additive effects. This is illustrated in Table 1. Modest changes, possibly achievable by breeding, in two or more parameters could lead to significant gains in yield. Such small changes in plant characteristics would b difficult to quantify in field experiments, except in terms of their effects on grain yield. The considerable (c. 40-50%) improvement in the yielding capacity of wheat cultivars which, as noted, is associated with their earlier flowering may thus also be the consequence of other, undetected differences in characteristics whic affect grain carbohydrate production. Whether earlier flowering is

Table 1. Calculated effects on potential grain yield (t ha^{-1}) of variations in certain plant characteristics: ALAI = leaf area index at anthesis (including leaf sheaths and exposed stems); DUR = duration of grain filling period, days; SENO = time of onset of decline in P_{max}, days after anthesis; P_{max} = light saturated rate of photosynthesis, mg CO_2 dm^{-2} h^{-1}.

ALAI	DUR	SENO = 15 P_{max}		SENO = 30 P_{max}	
		30	40	30	40
7.5	40	10.9	12.6	12.1	13.
	50	13.2	15.2	14.6	16.
10	40	12.0	13.8	13.5	15.
	50	14.4	16.7	16.1	18.

consequence of the earlier cessation of vegetative growth and is responsible for the reduced straw (mainly stem) yield is also not known.

LIMITATIONS OF THE MODEL

The model is very simple and, except for daily radiation and daylength, does not take explicit account of environmental factors, such as temperature, water and nutrient status which can affect growth. These environmental factors are likely to act through their effects on some of the model parameters particularly P_{max}, SENO and DUR. To the extent that such effects can be specified, environmental effects can be allowed for.

The model takes no account of environmental or genetic effects on the size and condition of the crop up to anthesis, except insofar as these affect ALAI. Further, it is assumed that the capacity of the grain to accumulate carbohydrate never limits grain growth or yield. To the extent that grain number and possibly potential grain size is proportional to plant size at anthesis, and hence ALAI, this assumption is probably satisfactory, except for extreme values of some of the parameters.

Despite these limitations, the model provides predictions of the effects of certain plant characteristics on grain yield which can be tested by experiment.

ACKNOWLEDGEMENTS

I am grateful to Miss M. A. Ford for help in running the simulation model.

REFERENCES

Austin, R. B., Bingham, J., Blackwell, R. D., Evans, L. T., Ford, M. A., Morgan, C. L., and Taylor, M., 1980, Genetic improvements in winter wheat yields since 1900 and associated physiological changes, J. agric. Sci., Camb., 94:675.

Austin, R. B., Morgan, C. L., Ford, M. A., and Bhagwat, S. G., 1982, Flag leaf photosynthesis of *Triticum aestivum* and related diploid and tetraploid species, Ann. Bot., 49:177.

Austin, R. B., 1982, Crop characteristics and the potential yield of wheat, J. agric. Sci., Camb., 98:447.

Evans, L. T., and Dunstone, R. L., 1970, Some physiological aspects of evolution in wheat, Aust. J. Biol. Sci., 23:725.

Innes, P., and Blackwell, R. D., 1983, Some effects of leaf posture on the yield and water economy of winter wheat, J. agric. Sci., Camb., 101:367.

Monteith, J. L., 1965, Light distribution and photosynthesis in fiel
crops, Ann. Bot., 29:17.
Wilson, D., 1975, Variation in leaf respiration in relation to growt
and photosynthesis of Lolium, Ann. appl. Biol., 80:323.

A DESCRIPTION OF PARTITIONING IN PLANTS

M. R. Thorpe and Alexander Lang

Physics and Engineering Laboratory
DSIR, Lower Hutt, New Zealand

INTRODUCTION

The importance of partitioning patterns in plants as a factor that determines growth and yield has long been recognised, and is a question which arises in a number of ways when crop growth is considered. For instance, the growth of crops is closely related to the amount of radiation absorbed, over a wide range of conditions and genotypes (Monteith, 1977; Legg et al., 1979; Fischer, 1983) and it follows that many factors have their major influence on growth in an indirect way through their effects on the allocation of new assimilate into leaf area rather than by their direct effects on metabolism. Again, the useful part of many plants is, as for wheat, only some fraction of total plant mass, and improvement in the yield of modern crops is attributed to higher harvest index (Evans and Dunstone, 1970), as well as to increased leaf area duration.

Crop growth and yield are determined by factors which can be divided approximately into those affecting light interception, assimilation from atmosphere and soil, allocation, and losses by respiration, pests or mechanical means. Research covers various levels of the biological hierarchy: from the highest level of crop stands, through the levels of individual plants, organs, tissues, cells, organelles, and then on to biochemistry and molecular biology. Thornley (1980) has discussed how biologists are mostly reductionist by nature and "feel that, if only they could understand the level below the one at which they are working, then all their problems would be solved". Thornley pointed out that in all these processes we are not so well served at the whole plant level, and especially in the area of translocation and partitioning, where the mechanistic details are contentious and we lack a quantitative organ-level description of translocation that can be used in understanding partitioning. He argues, as we have (Lang and Thorpe, 1983), that to seek a mechanistic understanding prematurely may be misguided; the facts must first be represented in a good description of the process at its own level. This description is then available both for synthesis to the level above, and for explanation at the level below. Holistic modelling of phloem tracer transport has been used in this way, for example, to describe the dominant processes at work and then give a mechanistic interpretation (Minchin and Thorpe, 1982). It is also plain that our description of the process at its level should be dynamic, enabling the evolution of the system to be explained.

In this paper we reproduce the descriptive scheme of Lang and Thorpe (1983) for analysing partitioning in an attempt to fulfil those two roles. We start with a simple definition of a general plant, and outline a simple descriptive scheme for its dynamics. We then describe some experimental observations in these terms, showing some ways in which the analysis provides specific questions to be answered, and answers to questions posed at the level above.

THE DESCRIPTIVE SCHEME

A plant is usually thought of as an assembly of sources, paths and sinks for assimilate, with the regulation of growth coming about through the properties of these elements. We have proposed (Lang and Thorpe, 1983) a rather simpler concept, in which the plant consists of a number of pools of specified materials between which exchange takes place by means of numerous processes. The term pool refers to a specified material within a specified boundary, and the term process to all avenues of exchange between pools. The concept of pool overlaps to some extent with source or sink, and that of process with path. However by using the term pool we avoid the connotation of function as implied by 'source' and 'sink', and the term process includes not only translocation, but all aspects of exchange between pools, including chemical interconversion. By describing the system in this way, we focus on the net flows in or out of pools, and not on the component flows which assist our interpretation of results but are experimentally more difficult to measure. By contrast, measurements of pool mass are straightforward and from these the time rate of change of pool mass may be inferred. This is termed the pool rate with dimensions mass per time ($M\ T^{-1}$).

We can choose to specify pool content and extent at any level in the hierarchy of biological processes. For example, pool content may be a particular chemical compound, a class of compounds, or even dry matter. Pool extent may be independently specified too at any level, for example cell, organ, or crop community. Broadening the definition of content emphasises the processes of transport, and broadening the definition of extent emphasises transformation. In fact we are concerned in this paper only with transport of carbohydrate, quantified as dry matter.

Having defined pool rates, we need to establish a way to describe them. Flows are commonly related to potentials - a standard procedure in physics, especially useful where the flow is a vector field and yet the potential is a scalar. This is a purely empirical description - as are Ohm's, Fick's and Darcy's laws. A similarly empirical relationship is used in micrometeorology where vertical fluxes are related to gradients, using the same coefficient for momentum and mass - sometimes with poor results if the empiricism is forgotten. In the same way we associate a potential P_i with each pool i. The plant system (Fig. 1) consists of a connected set of pools, each of which perceives a common plant potential V, and the pool rate I_i we describe as

$$I_i = S_i \, (V - P_i) \tag{1}$$

where S_i is a parameter which we term the permeance, and V is a measure of availability of assimilate. It is a property of the plant that is sensed by each pool, and is analogous to a voltage; it is not a rate of supply. Each pool has two properties: its permeance, S, which is analogous to a conductance; and its potential, P, which indicates the demand of the pool for assimilate. The net flow to the pool is proportional to the pool permeance, and to the difference between the plant potential and the pool potential.

Since we have no measure of V, it might appear that we have begun a fruitless exercise, especially as a measure of V seems to be a prerequisite for measuring P and S. However the essence of the model is that the pool rates are all described by only one plant parameter, V, that is common to all pools (Equation 1). The change in partitioning patterns that arises from a perturbation in V can be used to evaluate P and S for several pools in relative terms. A slightly different derivation of this model (Lang and Thorpe, 1983) assumes that the pool rate I_i is some unknown function of a plant parameter V, and that in our model we have the tangent to that function, in terms of a pseudo-potential P_i and a slope S_i. V is dimensionless, but we have chosen to name a unit (pel) for its scale to emphasize that it is implicit in the dimensions of the pool parameters S_i (mass time^{-1} pel^{-1}) and P_i (pel). P_i and S_i describe different aspects of the competition between pools: a pool with a highly negative potential will suffer a proportionately smaller change in pool rate than a pool with a less negative one - it would have priority for assimilate, maintaining its growth to a larger extent than others would when assimilate is short. We would expect shoot apices and reproductive structures to have highly negative potentials. Pools with large permeances (S_i) will tend to have high pool rates; the ability of many storage organs to quickly gain or lose dry weight implies that they have large pool permeances and not very negative pool potentials. Potential is an intensive property, it is independent of size; permeance is extensive and depends on pool size.

It is not necessary to have observations for all parts of the plant, as we may describe the competition for assimilate amongst only some of the pools. Referring to Fig. 1, the relative rates of two pools depend on the ratio of their permeances and the difference between their potentials. On the other hand it may be useful to include an environmental pool (i.e. the rest of the world; see Fig. 1) to describe net exchanges with the plant's surroundings in terms of plant state. A description of such phenomena as sink effects on photosynthetic rate could then be made.

The procedure for estimating S, P and V can take two courses. It is straightforward to estimate them if the experiment has the plant in only two states (i.e. control and perturbed), although this is unsatisfactory because it does not show whether the model can describe partitioning throughout that range of plant states - the analysis merely draws a straight line between the two points. We define the two states as V = 0 and 1 pel, and denote the changes in pool rate as ΔI_i, then from Equation (1)

$$S_i = \Delta I_i \ (M\ T^{-1}\ pel^{-1}); \qquad P_i = -I_i(V = 0)/\Delta I_i \ (pel). \tag{2}$$

It is statistically more satisfactory to have a lot more than two treatments, when there will be a set of pool rates I_{ij} for pool i in treatment j, and the task is to find the set of V_j, P_i and S_i values which best describe the data. The precision of measured pool rates will vary from pool to pool (according to their specific growth rates for example), making it advisable to normalize the rates for each pool by some parameter z_i so that the data for each pool

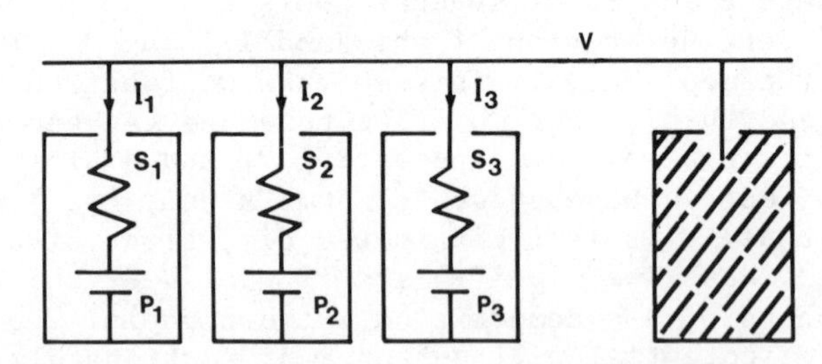

Fig. 1. Schematic representation of three competing pools in a plant. Each pool is represented by a permeance, S, and a pool potential, P, and perceives a plant assimilate potential, V. The net of all exchanges between these pools and their environment (i.e. the remainder of the plant and the rest of the world) are represented by the shaded box.

have the appropriate statistical weight in determining the model parameter values. The normalized pool rates X_{ij} can then be written:

$$X_{ij} = I_{ij}/z_i = a_i + b_{ij} \quad (3)$$

i.e. the mean over all treatments, a_i, plus the perturbation of that mean in treatment j, b_{ij}. The last term can in general be written as the finite series

$$b_{ij} = q_1 U_{i1} V_{j1} + q_2 U_{i2} V_{j2} + q_3 U_{i3} V_{j3} + \ldots . \quad (4)$$

where the q, U and V terms form matrices that can be determined by the singular value decomposition of the matrix b_{ij} (Mandel, 1971). The q_k values are positive and in order of decreasing magnitude, and the number of terms (maximum value of k) is equal to the number of pools. The algorithm has U_{ik} and V_{jk} of order unity (the sum over j of V_{jk} is zero, and of V_{jk}^2 is unity) so that the terms in Equation (4) contribute according to the magnitudes of the q_k. Our use of a model of the form of Equation (1) is equivalent to assuming that the first term in Equation (4) is dominant, and our model, with only the one parameter for plant state, will then describe the data satisfactorily. The procedure is very similar to principal components analysis. If we take only the first term in Equation (4) then, by comparison with Equation (1), the pool parameters are:

$$S_i = z_i q_1 U_i; \quad P_i = -a_i/(q_1 U_i). \quad (5)$$

The name pel is retained for the scale of V_j for the reasons given earlier. This procedure provides both the scaling V_j of treatments and a pair of P and S values to describe the competitive properties of each pool.

SOME APPLICATIONS OF THE MODEL

The response of partitioning between 5 pools in young cowpea plants to perturbations in the availability of assimilate can be described by this scheme. The perturbations were induced by a range of defoliation treatments that removed varying amounts of source and sink tissue. Pool rates were inferred from harvests at the time of treatment (day 0), and on two successive days. This gave dry weights at days 0, 1 and 2. Plotting these against time the slopes of the lines are the pool rates, and the standard errors of the slopes were used for the statistical weights, z_i (Equation 3). The plant potential values V_j corresponding to the 12 treatments were determined by the analysis described above. The ranking of the 12 treatments by the estimates of V_j is in close agreement with what we would expect from their severity (see Fig. 2, where the details of treatment are indicated).

The pool parameter values are given in Table 1. The pools fall into two distinct groups according to their potentials. Pools 4 and 5, which are the younger parts of the plant, have far lower potentials than the others, reflecting the much smaller proportional changes in their pool rates brought about by the experimental treatments. Pool 3 was a source leaf, and the treatments have had no effect on it that we can detect - these leaves were almost fully expanded at the time of the experiment. Pools 1 and 2 were older tissue and have much higher potentials than pools 4 and 5, reflecting their much bigger proportional changes, and their permeances are also higher, indicating bigger actual changes in pool rates.

It was not possible to demonstrate, by this analysis, that the values of the plant parameters were affected by treatments. Treatment effects would have appeared as systematic variation of the data from the model. The changing pattern of assimilate distribution, including apparent interactions, is adequately described by the model in terms of just one plant parameter, without any need to suppose that the plant parts responded to the treatments in any active way - despite the very wide range of treatments that were used. This result needs to be further tested. By using defoliation, these treatments affected the availability of assimilate and thus plant potential without deliberately altering the nature of the remaining pools. Some treatments (e.g. nutrition) would have their effects through altering the parameters of all pools, and it would not be possible to analyse such data as the whole plant would have been changed in nature by the treatment. However, if the treatments were confined to one pool (warming, or metabolic inhibitors, say), one could test whether allocation to the other pools was predicted by their pool parameter values. This would show whether plant potential could be perturbed in fundamentally different ways and the same model still apply. This consistent behaviour may be quite common since, for example, roots often suffer more (i.e. have higher pool potential) than other organs when a plant grows under conditions unfavourable for a variety of reasons whereas the

Table 1. Parameter values for the cowpea data presented in Fig. 2.

Pool i	Pool mass mg	Permeance S_i (mg d^{-1} pel^{-1})	Potential P_i (pel)
1	73	6.5	-0.06
2	31	7.3	-0.16
3	60	-0.9	0.10
4	4.5	3.2	-0.73
5	4.1	3.5	-0.70

shoot apex continues to receive a stable supply of assimilate (i.e. has a lower pool potential) in many unfavourable growth environments (Evans, 1975).

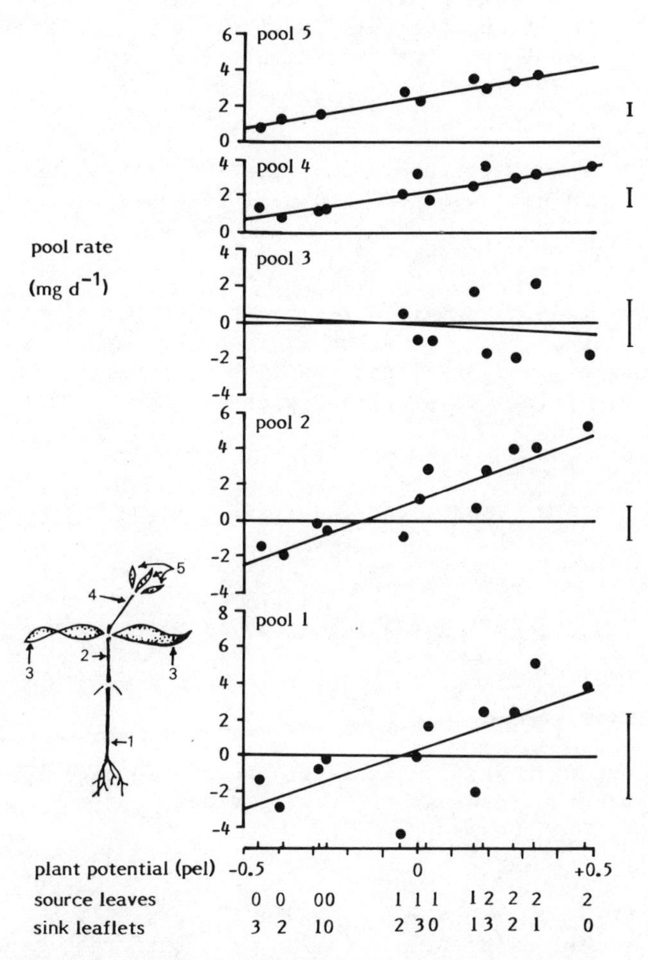

Fig. 2. Partitioning in ten-day-old cowpea seedlings. The measured pool rate (I_{ij}) is plotted against plant potential (V_j) which was calculated from the data using the analysis described in the text. The 5 pools are indicated on the inset: (1) root and hypocotyl; (2) first internode; (3) a unifoliate (source) leaf; (4) shoot apex and petiole of the first trifoliate leaf; (5) a (sink) leaflet of the first leaf. Error bars show 2 s.e. of the estimates of pool rate. The complement of source leaves and sink leaflets present for each plant state are shown below the abscissa.

A second result of modelling the cowpea data was the evidence of differences in parameter values between pools; the difference in potentials indicates that the allocation ratios changed with assimilate potential in the plant. It follows that we can predict the allocation that would occur in other conditions, because the assimilate potential V can be inferred, given the pool parameter values, from the amount of assimilate available, or from the rate of any one pool. To illustrate this possibility, consider just two pools (r, s) of a plant. From Equation (1) we have:

$$I_r = S_r(V - P_r); \qquad I_s = S_s(V - P_s).$$

Eliminating V between the above two equations gives the following linear relationship between the two pool rates:

$$I_s = (S_s/S_r)\, I_r - S_s(P_s - P_r). \tag{6}$$

If the plant tissue is physiologically stable then specific permeance g, rather than permeance S, is a more appropriate property for describing dry weight allocation, where we define specific permeances g_s, g_r for pools of mass s and r such that:

$$S_s = s\ g_s; \qquad S_r = r\ g_r. \tag{7}$$

Noting that I_s is defined as net pool rate, i.e.:

$$I_s = \frac{ds}{dt} \qquad I_r = \frac{dr}{dt}$$

we may substitute equation (7) into (6) and write:

$$\frac{1}{s}\frac{ds}{dt} = \frac{g_s}{g_r}\frac{1}{r}\frac{dr}{dt} - g_s(P_s - P_r). \tag{8}$$

This growth equation with conveniently separated variables may be integrated over the interval t_o to t, giving:

$$\log s = A + B \log r - C\ t \tag{9}$$

where $C = g_s(P_s - P_r)$, $B = g_s/g_r$, and $A = \log s_o - B \log r_o - C\ t_o$. Equation (9), derived by Barnes (1979) from different assumptions, predicts a linear relationship at any particular time between the logarithms of plant part masses, but with an intercept that varies linearly with time. Barnes (see Fig. 3) produced clear evidence of the steady shift in intercept with time in several sets of data from the literature for crops with storage roots (carrot, beet), which is a further demonstration that allocation of net assimilate can be described with the two parameters permeance and potential (with potentials distinct) but now over a period when plants grew in size by two to three orders of magnitude. In our derivation of Equation (9) we have integrated net flows, and the coefficient C arises from

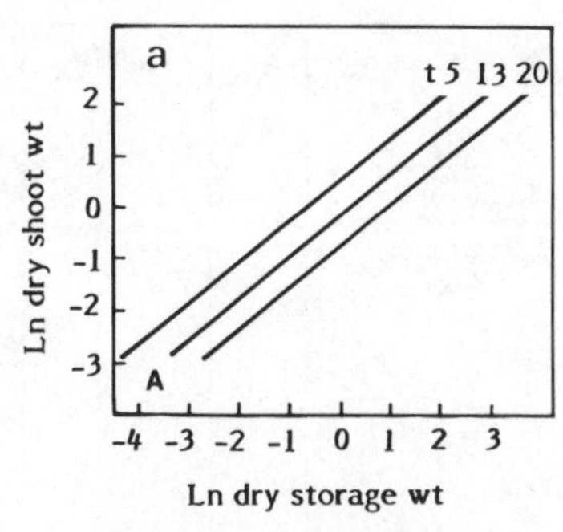

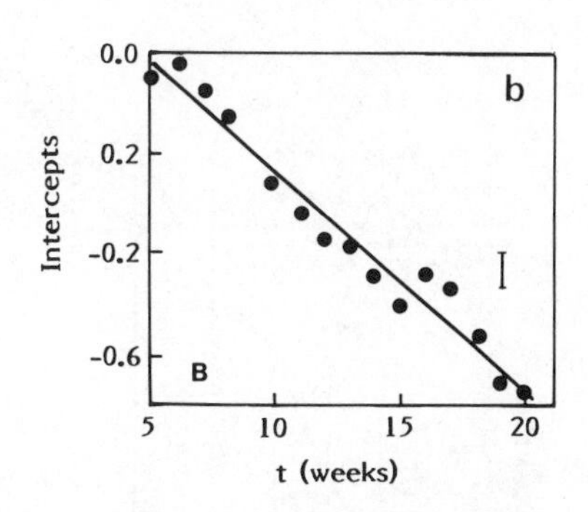

Fig. 3. (a) The relationship observed at time t since emergence, between the logarithms of dry matter in shoot and storage root of carrot. (b) The variation with time of the intercept obtained as in (a). Redrawn from Barnes (1979) with permission.

differences in pool potential. Barnes considered gross flows of assimilate to sinks, being the product of sink size and specific sink activity; assimilate was converted to dry matter with a conversion efficiency for each organ, and losses to respiration and death resulted in the time-dependent term. In our treatment we are explicitly concerned with net flows, and the pool parameters represent those processes and all others responsible for plant behaviour. The data assembled by Barnes (e.g. Fig. 3) show that, if we assume constant pool parameters, reasonable growth patterns are predicted; known changes in parameter values could be incorporated in Equation (9). For example, Barnes noted that root:shoot ratio would be strongly influenced by C; a result of respiration rates in his interpretation, in ours of a difference in pool potentials.

This scheme for describing the competition between pools by their associated potentials and permeances can also be useful in describing the interactions between flows of assimilate that have been observed on a shorter time scale. Compensatory behaviour within plants following manipulations of sources and sinks has frequently been observed with radio-isotopes (e.g. Pickard et al., 1978; Fondy and Geiger, 1980). The responses fall readily within the same descriptive framework, with some extensions because of the shorter time scale. For example (Fig. 4), stopping translocation from a source (A) can bring about an immediate compensatory change in translocation from a second source (B). A succession of demands made on the second source were satisfied only to a decreasing extent, and when demand was relaxed, the flow reduced to below the original rate, even to the extent of changing direction in some cases. A set of assimilate pools (A, B and sink C), connected as in Fig. 5, is suggested by this behaviour, and the dynamic adjustment in the plant

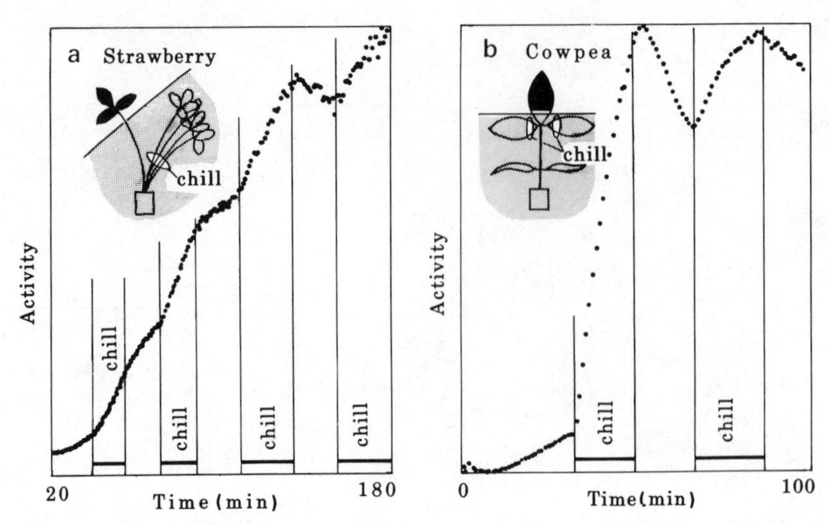

Fig. 4. Accumulation of ^{11}C-labelled assimilate which has been exported from a juvenile leaf of (a) strawberry, and (b) cowpea. The insets show the experimental arrangement with the fed leaf black and monitored region shaded. Flow rate (proportional to slope) increased when sink demand increased due to chilling of other sources, and decreased when these were rewarmed. Note reversal of flow direction (import by the fed leaf) after the third rewarming (in a) and the first rewarming (in b). Reproduced from Thorpe and Lang (1983) with permission.

suggests storage pools (D, E) of limited capacity alongside the path from each source. The compensatory increase in translocation to C comes from storage pool E adjacent to the second source; when this pool (E) empties, its potential falls (since the extra demand cannot be satisfied indefinitely), and if demand is relaxed the potential of pool E can be sufficiently low for flow in this path to reverse. Such buffering flows have been thought (Moorby, 1977; Thorpe and Lang, 1983) to prevent detection of product-inhibition of photosynthesis.

With our present knowledge of phloem and cell physiology, the physical analogue is more readily interpreted at short time scales (minutes to days), than when describing the longer term behaviour (weeks or more). The longer term responses include the processes evident in short-term behaviour, as well as other less well understood processes of sink and source metabolism. However a conceptual framework does seem to be arising in the literature, as reviewed for example by Gifford and Evans (1981). Longitudinal transport in the phloem is commonly believed to arise from pressure

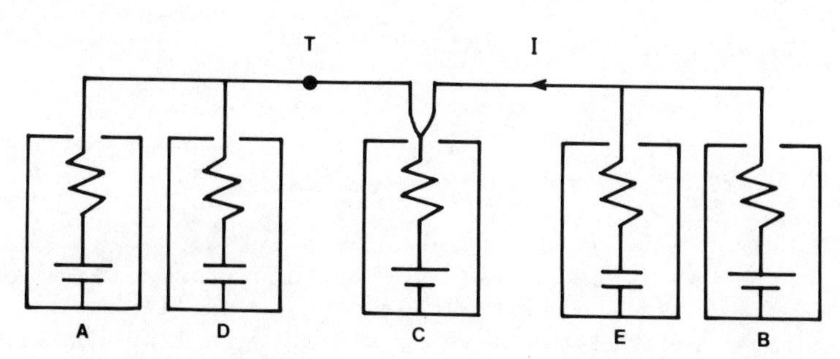

Fig. 5. Schematic representation of the assimilate pools proposed to explain the dynamic interactions (Fig. 4) between flows of assimilate supplying a common sink. Flow I from source B to sink C represents the observations in Fig. 4. Flow at T is stopped intermittently.

gradients generated osmotically by loading and unloading of the transported sugar (usually sucrose). This osmotic flow of water into sieve tubes has been shown to accompany phloem loading in wheat leaves (Minchin and Thorpe, 1982). Potassium may also have an important osmotic function (Lang, 1983). Our pool potential for this process probably represents sieve tube pressure potential.

Flows of assimilate between the phloem pathway and the sink or source tissue seem commonly to involve the apoplast, e.g. in source leaves (Giaquinta, 1983), reproductive storage tissue (Ho and Gifford, 1984; Thorne and Rainbird, 1983) and secondary thickening (Minchin and Thorpe, 1984), though a symplastic route in some sinks seems likely, e.g. the cortex of the root (Dick and ap Rees, 1975). The potentials that describe loading and unloading flows seem likely to be related to (i) sieve tube sucrose, or perhaps turgor, and (ii) the concentration of apoplast sucrose: and not the values for bulk tissue. For example, bean stem (Minchin and Thorpe, 1984) showed no systematic longitudinal variation of total sucrose, whereas sucrose in the apoplast decreased steadily from source to sink, suggesting that phloem loading and unloading conserved the ratio of sucrose concentrations across the phloem-apoplast barrier. Unloading into the apoplast increased when transport of assimilate to the sink was prevented; i.e. the flow into pool D (Fig. 5) was detected in the apoplast. A control of unloading via apoplast sucrose was suggested since no increase in apoplastic flow was detected when the concentration of sucrose there had been previously boosted by exogenous application. It appears then that the apoplast may be a site where these buffering processes occur. Less well understood are those processes involved in the uptake of assimilate by the sink tissue itself. Apoplastic sucrose does seem to be a factor, but processes within the sink symplast are also implicated (Gifford and

Evans, 1981) and the interpretation of sink pool parameters awaits further work.

CONCLUSION

With this simple view of plants as a set of pools competing for material to be partitioned, we have been able to describe allocation with the use of only two pool parameters, potential and permeance. The analysis is effectively a physical analogue like that advocated some time ago (Monteith, 1972) as 'a framework to discuss plant-environment relationships and partitioning that might help to prevent woolly thinking and tautologous statements'. Many of the tautologies are still with us, but now framed in a more useful way. Being set at a level between plant physiology and crop science, the analysis should form a useful link between the two disciplines. For example, relationships between the growth of different plant parts have been explained in terms of our model.

We have also shown that a single physical analogue of the same type can be used to describe allocation phenomena on several time scales. On the shortest scale it is possible to begin interpreting the description from knowledge of phloem physiology. An apoplastic sucrose pool was suggested as a link between the transport system and the source, sink and storage pools. The instantaneous flows were related to the distributions of assimilate there and in the phloem. Longer term responses, exemplified by the cowpea data (Fig. 2), depend on the properties of the sources and sinks, which we can describe, but interpretation is less straightforward, especially in the case of the sinks. Integration leads to an equation identical to that used by Barnes (1979) to describe the growth of plants over several months.

REFERENCES

Barnes, A., 1979, Vegetable plant part relationships. II. A quantitative hypothesis for shoot/storage root development, Ann. Bot., 43:487.

Dick, P. S., and ap Rees, T., 1975, The pathway of sugar transport in roots of *Pisum sativum*, J. exp. Bot., 26:305.

Evans, L. T., 1975, Beyond photosynthesis - the role of respiration, translocation and growth potential in determining productivity, in: "Photosynthesis and Productivity in Different Environments", J. P. Cooper, ed., Cambridge University Press, Cambridge.

Evans, L. T., and Dunstone, R. L., 1970, Some physiological aspects of evolution in wheat, Aust. J. Biol. Sci., 23:725.

Fischer, R. A., 1983, Wheat, in: "Potential Productivity of Field

Crops under Different Environments", International Rice Research Institute, Manila.
Fondy, B., and Geiger, D. R., 1980, Effect of rapid changes in sink-source ratio on export and distribution of products of photosynthesis in leaves of Beta vulgaris L. and Phaseolus vulgaris L., Pl. Physiol., 66:945.
Giaquinta, R. T., 1983, Phloem loading of sucrose, A. Rev. Pl. Physiol., 34:347.
Gifford, R. M., and Evans, L. T., 1981, Photosynthesis, carbon partitioning and yield, A. Rev. Pl. Physiol., 32:485.
Ho, L. C., and Gifford, R. M., 1984, Accumulation and conversion of sugars by developing wheat grains. V. The endosperm apoplast and apoplastic transport, J. exp. Bot., 35:58.
Lang, A., 1983, Turgor-regulated translocation, Pl. Cell Environ., 6:683.
Lang, A., and Thorpe, M. R., 1983, Analysing partitioning in plants, Pl. Cell Environ., 6:267.
Legg, B. J., Day, W., Lawlor, D. W., and Parkinson, K. J., 1979, The effects of drought on barley growth: models and measurements showing the relative importance of leaf area and photosynthetic rate, J. agric. Sci., Camb., 93:703.
Mandel, J., 1971, A new analysis of variance model for non-additive data, Technometrics, 13:1.
Minchin, P. E. H., and Thorpe, M. R., 1982, Evidence for a flow of water into sieve tubes associated with phloem loading, J. exp. Bot., 33:233.
Minchin, P. E. H., and Thorpe, M. R., 1984, Apoplastic phloem unloading in the stem of bean, J. exp. Bot., 35:538.
Monteith, J. L., 1972, Introduction: Some partisan remarks on the virtues of field experiments and physical analogues, in: "Crop Processes in Controlled Environments", A. R. Rees, K. E. Cockshull, D. W. Hand and R. G. Hurd, eds., Academic Press, London.
Monteith, J. L., 1977, Climate and efficiency of crop production in Britain, Phil. Trans. R. Soc. Ser. B, 281:277.
Moorby, J., 1977, Integration and regulation of translocation within the whole plant, in: "Integration of Activity in the Whole Plant", D. H. Jennings, ed., Cambridge University Press, Cambridge.
Pickard, W. F., Minchin, P. E. H., and Troughton, J. H., 1978, Real time studies of carbon-11 translocation in moonflower. I. The effects of cold blocks, J. exp. Bot., 29:993.
Thorne, J. H., and Rainbird, R. M., 1983, An in vivo technique for the study of phloem unloading in seed coats of developing soybean seeds, Pl. Physiol., 72:268.
Thornley, J. H. M., 1980, Research strategy in the plant sciences, Pl. Cell Environ., 3:233.
Thorpe, M. R., and Lang, A., 1983, Control of import and export of photosynthate in leaves, J. exp. Bot., 34:231.

ASSIMILATE PARTITIONING AND UTILIZATION DURING VEGETATIVE GROWTH

G. K. Hansen and H. Svendsen

Royal Veterinary and Agricultural University
Taastrup, Denmark

INTRODUCTION

In whole plant or crop models, plants are often divided into organs e.g. shoots and roots. It is often convenient to divide these organs into two further compartments - structural and storage material (Warren Wilson, 1972; Thornley, 1977). Little is known about the kinetic parameters for partitioning assimilate between the compartments. A within-organ model by Thornley (1977) involved the division of structural materials into degradable and non-degradable components though experimentally-derived values of rate parameters were not available. Barnes and Hole (1978) showed that, based on Thornley's (1977) model, respiration could still be described as consisting of two processes - growth and maintenance - and estimates of the parameters were made. Farrar (1980) and Prosser and Farrar (1981) estimated parameters of a three compartment model consisting of soluble, storage and structural carbon.

In this paper we describe the basis and application of a model of assimilate partitioning and utilization. The model has been developed for Lolium multiflorum, and the examples quoted are for this species, but the principles are appropriate for the vegetative phase of growth of wheat.

BASIS OF THE MODEL

Based on separate measurements of CO_2-exchange rates and Thornley's (1977) model, we have tried to determine the kinetic parameters for allocation of carbon, not only within shoots and roots but also from shoots to roots. The model for carbon (Fig. 1)

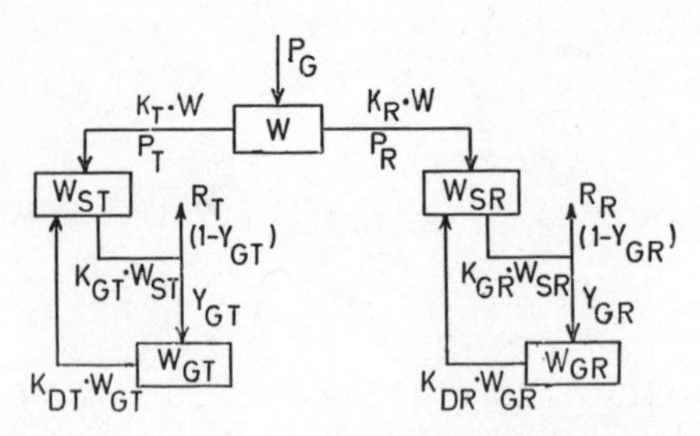

Fig. 1. The model. The five state variables are: W, the primary pool of assimilates; W_{ST} and W_{SR}, the storage pools of assimilates in shoots and roots; W_{GT} and W_{GR}, the structural materials. The six first-order rate constants of shoots (T) and roots (R) are: K_T, K_R, K_{GT}, K_{GR}, K_{DT} and K_{DR}. The efficiencies of conversion of storage materials into structural materials are Y_{GT} and Y_{GR}. P_G is gross photosynthesis, the driving variable, and R_T and R_R are shoot and root respiration rates.

operates with first-order kinetic constants in a single direction. Previously determined parameters for growth and maintenance respiration (Hansen, 1978) were used and structural materials were not divided into degradable and non-degradable compartments. The parameters (Table 1) were estimated at 20°C and a temperature coefficient, Q_{10}, of 2 was assumed. In shoots as well as in roots the values of K_G (K_{GR} and K_{GT}) were much higher than K_D (K_{DR} and K_{DT}), which gives stability to the model. The ratio, K_T/K_R, of the parameters that govern the partitioning of carbon from the primary storage pool W_P to shoots and to roots had a value of about 3. In young barley plants, Prosser and Farrar (1981) found K_{GT} = 1.98 and K_{DT} = 0.26 at 20°C, in near agreement with those estimated here.

Table 1. Values of rate constants (d^{-1}) at 20°C and the respiratory parameters (Y_G) (Hansen, 1978).

K_T	K_R	K_{GT}	K_{GR}	K_{DT}	K_{DR}	Y_{GT}	Y_{GR}
0.70	0.24	2.20	3.20	0.15	0.079	0.83	0.54

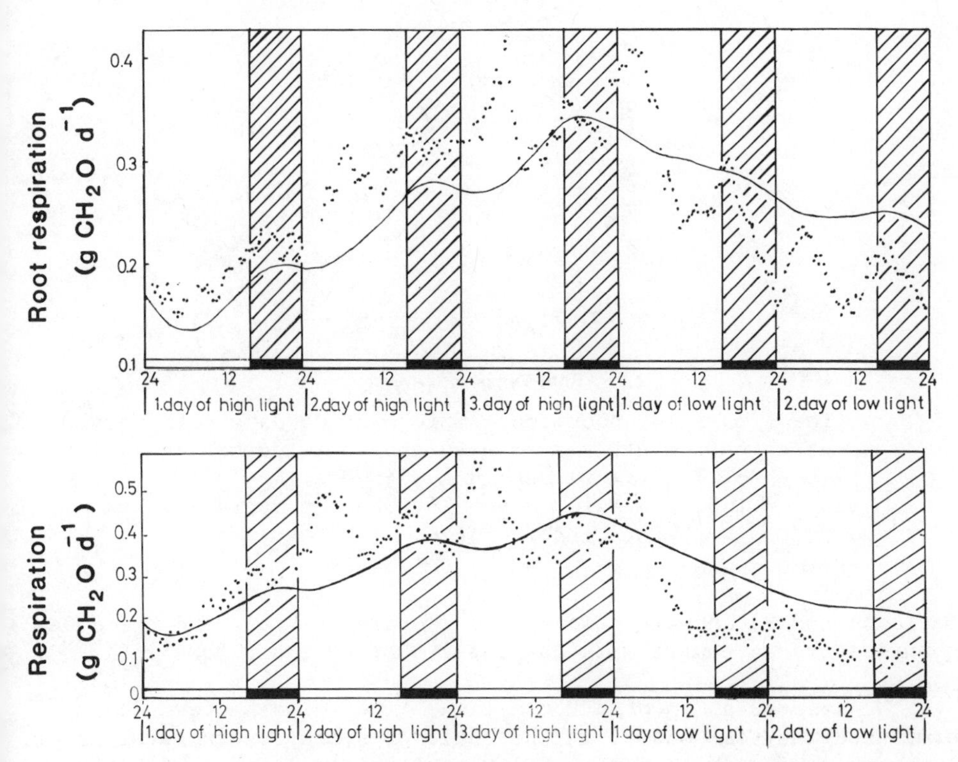

Fig. 2. The time course of simulated (———) and measured (·····) values of root respiration rates, before (upper) and after (lower) cutting the tops.

VALIDATION OF THE MODEL

The model can be tested by its predictions of the behaviour of either the state variables or the rate variables. The assumption of two carbohydrate pools in the shoots is difficult to test by chemical analysis. The total material in the two pools together was in near agreement with available results from soluble carbohydrate assays. Root respiration rates from Hansen (1977) were compared with simulated values (Fig. 2). The simulated results lag behind the measurements when the irradiance was changed, but after a day at the new conditions, the agreement between observed and simulated results was quite close. The model was unable to mimic the two diurnal peaks found in root respiration rates; these may result from an alternative respiration pathway as suggested by Lambers (Chapter 11).

The modelling of the partition of assimilates between shoots and roots was validated by modifying either shoot activity (P_G) or root activity through changing root temperature (Fig. 3). The results were in qualitative agreement with the semi-empirical formulation of

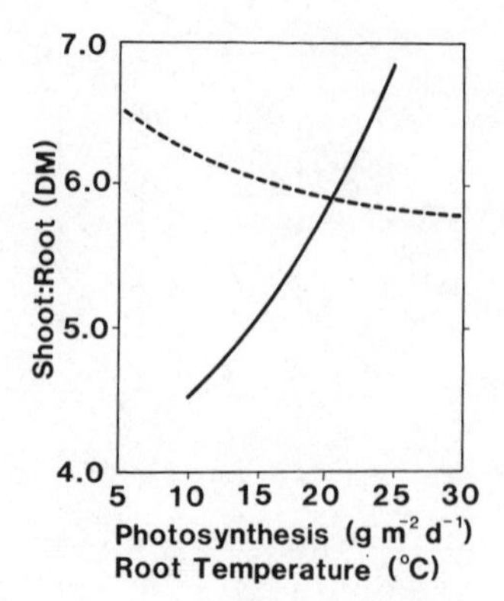

Fig. 3. The simulated shoot:root ratio over 25 days with varying gross photosynthesis (-----) and root temperatures (———). Initial values in g m^{-2} were: W_{ST} = 20, W_{GT} = 160, W_{SR} = 10, W_{GR} = 26. Shoot temperature was 20°C and photoperiod was 16 h.

Davidson (1969) and with Cooper and Thornley (1976), and there was quantitative agreement with the results of Ryle and Powell (1976).

We assumed a Q_{10} of 2 for all the kinetic parameters and simulated the time course of the ratio of storage:structural materia in the tops (W_{ST}/W_{GT}) for a 25-day period using different but constant values of P_G during the period (Fig. 4). The simulated results showed a decrease in W_{ST}/W_{GT} because of the increase in W_{GT}. According to the model, the relative growth rate of W_G, μ, is given by:

$$\mu = (W_S/W_G)\ K_G\ Y_G - K_D$$

Assuming the same Q_{10} value for K_{GT} and K_{DT} then, at $\mu = 0$, $W_{ST}/W_{GT} = 0.082$. The rate at which W_{ST}/W_{GT} approaches this value will depend on the plant size and on P_G, as shown in Fig. 4.

DISCUSSION

The estimation of the parameter values in the present model was based on experiments in which a prolonged dark period of two days wa followed by a prolonged photoperiod of two days. This may have caused changes in the enzymic system from that under normal day/nigh periods (Penning de Vries, 1975). However, as the model gave reasonable values of state and rate variables compared to those available from experiments, it is believed that the parameters have nearly the right values.

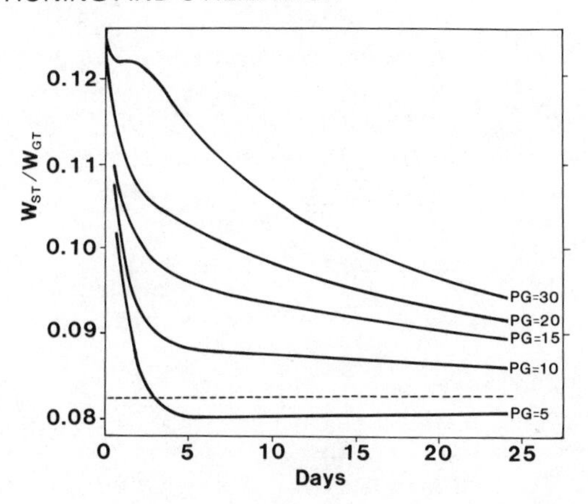

Fig. 4. The time course of relative amount of shoot storage material, W_{ST}/W_{GT}, at varying values of gross photosynthesis, P_G (g m^{-2} d^{-1}). The dotted line represents zero growth. Initial conditions are given in Fig. 3, with root temperature 20°C.

The model is based on the assumption that partitioning and utilization of assimilates are related to the amount of assimilate available. This is only true to a certain extent. If nitrogen is limiting the utilization parameters will change (Hansen, 1977) and this will influence the kinetic parameters in the model. As root respiration decreases when nitrate is not available, lack of nitrogen will decrease the shoot:root ratio.

Although the model is simple, it is convenient to use in connection with crop growth models and can be expanded to deal with the reproductive phase of growth.

REFERENCES

Barnes, A., and Hole, C. C., 1978, A theoretical basis of growth and maintenance respiration, Ann. Bot., 42:1217.

Cooper, A. J., and Thornley, J. H. M., 1976, Response of dry matter partitioning, growth and carbon and nitrogen levels in the tomato plant to changes in root temperature: Experiment and theory, Ann. Bot., 40:1139.

Davidson, R. L., 1969, Effect of root/leaf temperature differentials on root shoot ratios in some pasture grasses and clover, Ann. Bot., 33:561.

Farrar, J. F., 1980, Allocation of carbon to growth, storage and respiration in the vegetative barley plant, Pl. Cell Environ., 3:97.

Hansen, G. K., 1977, Adaption to photosynthesis and diurnal

oscillation of root respiration rates for *Lolium multiflorum*, *Physiol. Plant*., 39:275.

Hansen, G. K., 1978, Utilisation of photosynthesis for growth, respiration and storage in tops and roots of *Lolium multiflorum*, *Physiol. Plant*., 42:5.

Penning de Vries, F. W. T., 1975, The cost of maintenance processes in plant cells, *Ann. Bot*., 39:77.

Prosser, J., and Farrar, J. F., 1981, A compartmental model of carb allocation in the vegetative barley plant, *Pl. Cell Environ*., 4:303.

Ryle, G. J. A., and Powell, C. E., 1976, Effect of rate of photosynthesis on the pattern of assimilate distribution in th graminaceous plant, *J. exp. Bot*., 27:189.

Thornley, J. H. M., 1977, Growth, maintenance and respiration: a reinterpretation, *Ann. Bot*., 41:1191.

Warren Wilson, J., 1972, Control of crop processes, *in*: "Crop Processes in Controlled Environments", A. R. Rees, K. E. Cockshull, D. W. Hand and R. G. Hurd, eds., Academic Press, London.

ROOTS AND WATER ECONOMY OF WHEAT

J. B. Passioura

CSIRO, Division of Plant Industry
Canberra, Australia

INTRODUCTION

A wheat plant, from one important point of view, is an evaporating and assimilating surface. There is a nexus between the carbon assimilated and the water evaporated by this surface that results, in a given environment, in the dry weight of the plant being proportional to the water transpired. Where the water available for transpiration is limited, the dry weight of the plant will be correspondingly limited. My aim is to explore the influence of the roots on the productive performance of this surface when water is limiting, while keeping in mind the needs of those who wish to model the behaviour of a wheat crop at incremental time scales ranging from a few minutes to a day.

Drought occurs in many different ways, but there are two extreme, though not unusual, patterns that are useful for organizing our thoughts around. The first typifies moderately humid conditions such as those that occur through much of Europe, and results in the upper parts of the soil profile becoming dry, but not the lower, so that there is little danger that a crop would ever completely run out of available water. The second typifies semi-arid conditions, such as those of the Middle East or of Australia, in which the water supply is so small that there is a substantial danger that a crop could completely exhaust the soil of available water, and hence die well before maturity, with consequently a very low grain yield; in these conditions the pattern of water use through the season becomes as important as the size of the water supply (Passioura, 1983). It seems appropriate for this meeting to concentrate more on semi-humid rather than on semi-arid conditions, and so I will largely ignore complications arising from the pattern of water use through a season.

While there is a nexus between transpiration and production, there is also one between a crop's production and the light that it intercepts (Legg et al., 1979). The connection between all three, i.e. transpiration, production, and light interception, is through leaf area. Accordingly, I plan to explore the effects of the activity of roots on the production of leaf area, not only through the roots' influence on the water relations of the shoot, but also through other puzzling influences that appear to be unrelated to the water relations of the shoot and may be mediated through growth regulators.

MODELLING UPTAKE OF WATER BY ROOTS

There have been two main approaches to modelling the uptake of water by roots. One treats the root system, quite empirically, as a distributed sink (e.g. Allmaras et al., 1975), and while it may be useful for summarizing data, this approach seems to offer little prospect of improving either our understanding or our ability to predict the behaviour of a crop. The other treats the root system, with much greater focus, as a collection of essentially individual cylindrical sinks to which water is presumed to move radially throug the adjacent soil. This "single-root" model is very popular, both because of its apparent geometric fidelity and because it is open-ended in that it seems to provide a sound base on which to build increasingly more realistic models. Unfortunately, in practice its predictive power is often low, and it is particularly poor in estimating the extraction of the influential water that lies deep in the subsoil. Let us examine its features in some detail.

The Single-Root Model

The assumptions that this model embodies have been most thoroughly articulated by Tinker (1976). In essence, these are that the roots in a given volume of soil are assumed to be a collection o cylinders of radius a, each of which occupies the centre of and has effectively sole access to a cylinder of soil whose outer radius, b, is given by $(\pi L)^{-\frac{1}{2}}$ where L, the rooting density, is the total length of root per unit volume of soil. Flow of water through the soil to a root is assumed to obey the cylindrical diffusion equation

$$\frac{\partial\theta}{\partial t} = \frac{1}{r}\frac{\partial}{\partial r}\left(r\ D(\theta)\ \frac{\partial\theta}{\partial r}\right) \tag{1}$$

where θ is the volumetric water content, r is the radial distance from the centre of the root, and D(θ) is the diffusivity of soil water. Equation (1) is troublesome to solve, so it is usual to simplify it by assuming that $\partial\theta/\partial t$ is either zero or constant. Both assumptions give adequate solutions except in extreme circumstances, although that of constant $\partial\theta/\partial t$ is much the better. Various boundary conditions have been used. That at b is straightforward, for it is either set by the condition at a if $\partial\theta/\partial t$ is assumed to be zero (i.e. the flow is steady state), or, more realistically, the flow at b can be set at zero (i.e. $\partial\theta/\partial r = 0$), because b marks the watershed between adjacent roots.

The boundary at a is also troublesome, for it is the integral of a host of complicated processes occurring within the plant and its atmospheric environment. It is our inability to specify this boundary accurately that probably accounts for many of the failures of the single-root model. Early users (e.g. Gardner, 1960; Cowan, 1965) specified a quasi-steady flux at a, which was a useful stratagem for exploring the behaviour of the model when flow through the soil was clearly limiting, but gave insufficient weight to resistance within the plant to be realistic.

One of the most useful and accurate of the approximate solutions of Equation (1) is the following, adapted from Passioura (1980):

$$Q = \frac{2}{b^2(\ln (b/a)-\frac{1}{2})} \int_{\theta_a}^{\theta_b} D(\theta)\, d\theta \qquad (2)$$

where Q is a quasi-steady rate of withdrawal of water by the roots (and can be expressed as $-d\bar{\theta}/dt$, the fall in mean water content, $\bar{\theta}$, with time, t) and θ_a, θ_b are θ at $r = a$, and $r = b$, respectively. We can simplify this equation further by assuming that D is constant (which it effectively is in many soils during the extraction of about the last 25% of available water - see, for example, Rose (1968) and Passioura (1980)) and that $(\ln (b/a)-\frac{1}{2}) \simeq 4$ (which is a common value when roots are sparse enough for the equation to be of interest), whereupon, $Q \simeq D(\theta_b-\theta_a)/2b^2$. Given the cylindrical geometry, which results in much steeper gradients in θ at a than at b, $r_b \simeq \bar{\theta}$, so we have $Q \simeq D(\bar{\theta}-\theta_a)/2b^2$. If we now assume that the roots can hold θ_a steady at the value to which they can ultimately lower $\bar{\theta}$ (equivalent to a suction of, say, 2.0 MPa), we can derive an expression for a time constant, t*, that would describe the rate of withdrawal of water when flow through the soil was entirely limiting. This time constant is $2b^2/D$, which takes the value b^2 days when b is expressed in cm and $D = 2\ cm^2\ d^{-1}$, a common value when soil-water suction is within the range 0.5 to 2.0 MPa. Even with an effective rooting density as low as $0.1\ cm\ cm^{-3}$, t* is only 3 d, which implies that $(\theta - \theta_a)$ would be reduced to 1/e (i.e. 0.37) of its initial value in 3 days. Admittedly, the roots may hold θ_a low only during daylight, so that t* may be, say, 6 days, but this is still quick and leaves us with the puzzle of why some crops fail to extract quite substantial amounts of water from the subsoil even though they seem to have plenty of time and plenty of roots. This issue is discussed further by Passioura (1983).

Equation (2) is useful not only for constant θ_a, but also for variable θ_a, provided that $d\theta_a/dt$ is not too large (e.g. it takes about 0.1 b^2/D for the equation to be useful after a substantial ste change in θ_a, cf. Crank (1975, p.85). θ_a depends on Q and the hydraulic resistances within both plant and soil. Newman (1969) and Reicosky and Ritchie (1976) were among the first to emphasise the influence of the hydraulic resistance of the root system on the boundary condition at a, and Landsberg and Fowkes (1978) dissected this resistance into radial and longitudinal components, and explore some interactions between the two, especially the fact that the root are joined together to form a network. When we realise that both L and $D(\theta)$ vary with depth, z, in the soil profile; that both radial and longitudinal resistances vary along a root and among roots; that the radial resistance may be a function of flow rate (Weatherley, 1982); and that $\tau(\theta,z)$ (the soil water suction) is needed to convert θ at a (derived from Equation (1)) into a water potential, it is perhaps not surprising that the model often fails: we can never hope to measure all the parameters that we need to make it realistic.

But there is another, more fundamental, reason that may account for many of the failures: the model itself may be wrong, and if so i is a waste of time to continue using it. The following are some of the doubts that have been raised about it.

Inapplicability of Fick's First Law

This law asserts (for our purposes) that the flux density of water is proportional to the gradient in θ; Equation (1) is derived from it. It may be inapplicable where the path length for flow traverses only a small number of soil particles. Certainly it works well for flow over large distances, and certainly it means little fo flow past a single particle, but little is known about its applicability in between. The law probably works well when we are integrating flow over a large volume of soil rather than over a large distance through soil, and for practical purposes, the use of Equation (1) usually relates to such a large volume of soil that its applicability on Fickian grounds is not in doubt. Detailed laboratory experiments are needed to give a definitive answer. Over attempts to provide an answer have been, collectively, inconclusive. Both Herkelrath et al. (1977) and Faiz and Weatherley (1978) found that the single-root model greatly underestimated the hydraulic resistance of the root:soil system, and they invoked a large interfacial resistance, rather than the breakdown of Equation (1), t explain their results. However, I have found (Passioura, 1980) that Equation (1) does provide a good first-order explanation for the flo of water to the root surface (Fig. 1). The differing outcomes of

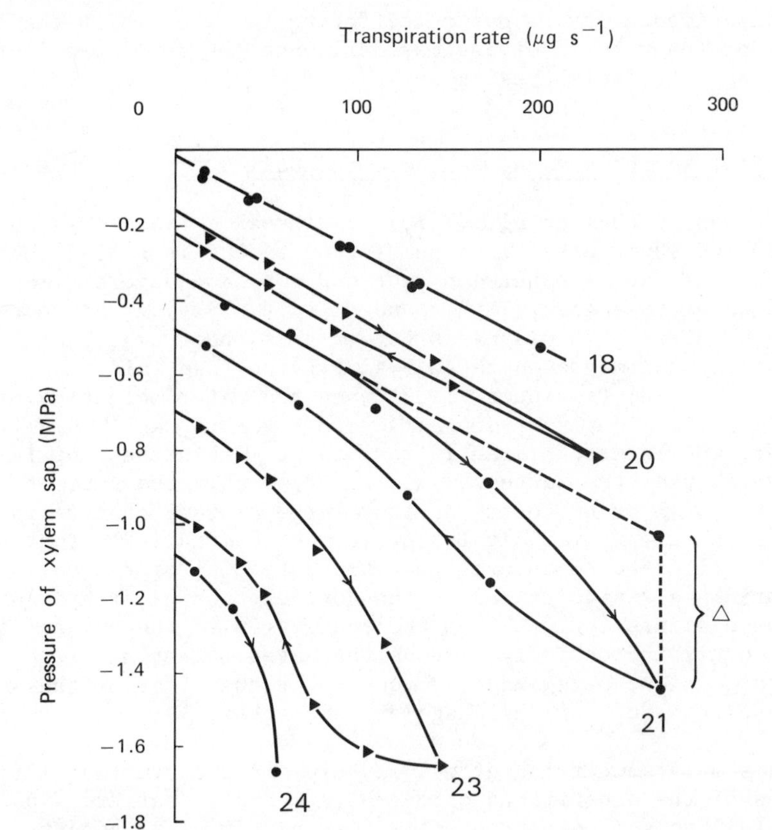

Fig. 1. Pressure of the xylem sap of a leaf as a function of total transpiration rate for a wheat plant that is monotonically drying the soil occupied by its roots (redrawn from Passioura, 1980). The soil was not watered after sowing. Numbers on lines denote days from sowing. Bulk soil water potential at the start of each of the successive runs shown was -40, -190, -330, -580 and -950 kPa. The broken line for the third run depicts the component of the pressure that arises from flow through the plant, added to the bulk soil water potential; the deviation (△) between the solid and broken lines depicts the fall in soil water potential from its bulk value to its value at the surface of the roots. The agreement betwen this deviation and Equation (2) is very good if it is assumed that one third of the total root length is taking up water. This length is equal to that of the morphologically distinct roots that had apparently grown after the soil water potential was less than about -100 kPa. The hydraulic resistance of the plant was essentially constant. For further discussion, especially on the hysteresis, see Passioura (1980).

these experiments may be due to differing rates at which the soil dried, for the rate of drying may influence the development of a dominating interfacial resistance.

Interfacial Resistance Between Root and Soil

A paper by Huck et al. (1970) on diurnal shrinkage in cotton roots rekindled interest in a suggestion by Philip (1957) that a vapour gap may develop between root and soil and severely impede the flow of water (see also, Taylor and Willatt, 1984). For roots growing in pores with diameters larger than their own, such a gap seems likely and may even be self-amplifying (see Faiz and Weatherley, 1978; Passioura, 1982) when the water potential of the plant is falling during a diurnal cycle. For if, as is widely believed, the largest hydraulic resistance within the root is at the endodermis, then the cortex, which is presumably the site of shrinkage, will be protected against large diurnal changes in plant-water potential provided it is in intimate hydraulic contact with the soil; it will tend to assume the water potential of the soil. But once shrinkage starts to reduce the contact, the major hydraulic resistance starts to move from the endodermis to the epidermis, and so the cortex increasingly assumes the water potential of the rest of the plant, which, being usually much lower than that of the soil, will result in further shrinkage of the cortex.

This self-amplifying effect may explain the greatly differing outcomes of the experiments cited above. Roots that are exposed to a slowly falling soil-water potential may have time to adjust, physiologically or morphologically, or both, so that no contraction of the cortex takes place. Osmotic adjustment could ensure that the cortical cells remain turgid despite having a low water potential, and the way that roots mould themselves to adjacent soil particles when they are growing into fairly dry soil (Passioura, 1980) suggests that they would have a much better chance of maintaining hydraulic continuity with the soil as it continues to dry, than would the smooth cylindrical roots that tend to grow in soil of high water potential. Roots exposed to a rapid fall in soil-water potential may have no time to adjust so that they experience the self-amplifying contraction of the cortex that leads to a high overall hydraulic resistance. Laboratory experiments usually involve much greater rates of change of soil-water potential than are common in the field; those of Faiz and Weatherley (1978) seemed to have been especially rapid.

It is a modeller's nightmare to be faced with a sudden change in an important variable, such as the development of a vapour gap implies. Fortunately, if the above arguments about rates of change are right, the sudden development of a vapour gap in field conditions is unlikely and it is probably safe to continue ignoring the interfacial resistance, except where a large proportion of the roots are growing in pores larger than themselves.

The Effect of Clumping of Roots

The single-root model assumes that roots are evenly distributed, so that it is possible to define the outer radius, b, of a cylinder of soil to which each root is assumed to have exclusive access. This assumption makes sense for the disturbed soil of the plough layer, but in deeper undisturbed soil it is common for roots to be constrained to extensive macropores such as fissures and pores resulting from biological activity (Ehlers et al., 1983). The resultant clumping of the roots may greatly diminish their ability to extract water. The following discussion explores some of the consequences of clumping.

Let us assume, first, that the roots are constrained to biologically-produced pores such as wormholes. This creates conditions that are mathematically similar to those of the single-root model, with the distinction being that instead of each root having exclusive access to a cylinder of soil of radius b, we assume that each group of roots has exclusive access to a cylinder of soil of radius B, where $B = (\pi L^*)^{-\frac{1}{2}}$ and L^* is the length of occupied pore per unit volume of soil. With b = B Equations (1) and (2) remain applicable. B is a very influential parameter whose importance can be gauged from the time constant for this system, which by analogy with the single-root model is $2B^2/D$. Putting $D = 2\ cm^2\ d^{-1}$ as before, we see immediately that if B exceeds about 4 cm, which it can easily do (Goss and Fischer, personal communication), the time constant becomes a large proportion of the life of a crop. Furthermore, roots growing in holes much larger than themselves may have poor hydraulic continuity with the soil, which would further decrease their ability to extract the water. However, little is known about this. Possibly root hairs may be important in facilitating contact.

In some soils, roots are often constrained to grow in planar voids associated with fracture planes. We can explore this type of constraint by assuming a few different idealized arrangements of these planes, namely: (i) the planes are parallel, at a spacing of 2B, (ii) the planes are arranged two-dimensionally to give hexagonal prisms, which for our purposes we can take to be cylinders of radius B, and (iii) the planes are arranged three-dimensionally to delineate blocks of soil that we may take to be spheres of radius B. The early stages of the flow of water to roots in these voids will presumably be radial, but the cylinders of influence of the roots will soon interact so that eventually the collective behaviour of the roots will approach that of a distributed sink spread evenly through the voids. The transition from one pattern of flow to the other is difficult to analyse, but for the sake of illustration, let us assume that the roots are behaving as an evenly distributed sink. Making similar assumptions to those used in deriving Equation (2), we get the following:

$$Q = \frac{2n}{B^2} \int_{\theta_B}^{\theta_o} D(\theta)\, d\theta \qquad (3)$$

where θ_o is θ at the centre of the slab, cylinder, or sphere, and $n = 1$ for the slab, 2 for the cylinder, and 3 for the sphere. The time constant, derived using the same argument as before, is $B^2/2nD$. Putting $D = 2$ cm^2 day^{-1}, as before, we see that the time constant becomes large (say, ≥15 day) when B exceeds about 8, 12, or 15 cm for the slab, cylinder and sphere, respectively. The main point to note about these time constants is that they are independent of overall rooting density, L, unless the single-root model has a time constant as large as them, when it becomes the appropriate model to use, for the roots in the planar voids are then so sparse that they can no longer be collectively considered as a distributed sink within the voids.

If a clumped-root model is an appropriate one to use, there are considerable consequences for the data we need to collect. Values of L determined on soil samples obtained by auger are irrelevant for describing flow through the soil, unless they are so low that the clumped model is not appropriate; what is relevant is the geometry of the extensive macropores, which unfortunately is much more difficult to determine. However, the geometry of these macropores will not be so ephemeral as L (unless they arise from current faunal activity), so any measurements of these would be useful over many seasons. It is still necessary, however, to determine $L(z,t)$ in some detail whenever hydraulic resistance within the roots is comparable to, or greater than, the effective hydraulic resistance in the soil, and when we wish to describe uptake as a function of depth.

Because of the differing geometries, Equation (3) is not as good an approximation as Equation (2), and it gets worse the larger is n. This is because these equations are derived by assuming that $\partial\theta/\partial t$ is essentially independent of position at any given time, although it changes with time. This assumption is better the more convergent are the lines of flow. It is examined in detail for one-dimensional flow by Passioura (1977). But the approximations are good enough to show the essence of the differing geometries, which is that the length scale (i.e. B) that is required to give similar time constants for the extraction of water increases sharply as the flow lines change from convergent (e.g. to a wormhole) through parallel, to divergent (e.g. out of a sphere). Note that if we call the area of occupied planar void per unit volume of soil the void-area density, A (cf. L, the root-length density), then $B \simeq n/2A$ and $t^* = n/8A^2D$, i.e. t^* increases with n at constant A, owing to the increase in B with increased n at constant A.

Data on the shape and distribution of occupied macropores are rare, but from anecdotal information it looks as though roots being restricted to cylindrical macropores such as wormholes may in practice be rather more significant than their being restricted to fracture planes, which have to be quite a long way apart to be influential.

Hydraulic Resistance of the Root System

Flow through the roots accounts for most of the fall in water potential between bulk soil and leaf except in quite dry soil, or possibly when the roots are strongly clumped, or there is a large interfacial resistance between root and soil. It is convenient to divide the hydraulic resistance of the roots into two components: radial, from the epidermis to the xylem; and longitudinal, along the xylem. The two interact in interesting ways (Landsberg and Fowkes, 1978).

The linearity of the overall resistance has long been in doubt in dicotyledons (Weatherley, 1982), but in monocotyledons, especially wheat, the resistance is to a good approximation independent of flow rate in both field (Wallace and Biscoe, 1983) and laboratory experiments (see Fig. 1). Non-linearities appear to be associated with plants grown in nutrient solution or very wet soil (Meyer and Ritchie, 1980; Passioura and Munns, 1984) although the reason for the association is obscure.

Because of the way that a root system is arranged as a network, with the flow from lateral roots being channelled into a few major axes before reaching the shoot, longitudinal resistance can be very influential in some circumstances (Passioura, 1972; Greacen et al., 1976; Landsberg and Fowkes, 1978), especially when most of the water supply to the shoot is being collected deep in the subsoil and being transported through a dry topsoil by only a few axes. Data on the distribution and longitudinal resistance of the deeply penetrating axes are sparse, despite their occasionally great importance; they are, of course, very difficult to collect. The work of Busscher and Fritton (1978) is often invoked as a reason for ignoring the longitudinal resistance (e.g. Steinhardt et al., 1981), but their calculations seem to be about 3 orders of magnitude in error and should be ignored.

The Extraction of Recent Rain from Previously Dry Topsoil

Losses of water by direct evaporation from the soil may be as high as 50% of current rainfall in some circumstances (Fischer and Turner, 1978). These losses must be particularly severe when light rain falls on previously dry topsoil under a sparsely-leafed crop. Some attempts have been made to explore this loss (Tanner and Jury, 1976; Shawcroft and Gardner, 1983) but our ignorance of the processes involved, especially the recovery of the root system from quiescence is very great. The speed of recovery may depend on the presence of specialised roots as reported by Vartanian (1981).

General Remarks

Much of the above discussion has concerned the detailed physics and physiology of the uptake of water by roots. But how much detail do we need? If our primary aim is to predict or improve the yield of wheat, it may be unnecessary to have a detailed picture of the extraction of water as a function of depth and time throughout the soil profile. All that is essential to know is the extent to which the ability of roots to extract water limits the growth of the shoot. Using a few robust approximations (for example, that growth is not limited until 80% of the available water supply is used - Ritchie, 1974), and a small set of data, we may get answers that are more useful than those provided by elaborate models containing dozens of disposable parameters that are beyond our means to measure.

ROOTS AND THE DEVELOPMENT OF LEAF AREA

One of the aims of modelling the uptake of water is to predict the relation between transpiration rate and leaf water potential in the hope that this will enable us to predict stomatal conductance, or more generally, the rate of photosynthesis. But given the nexus between water-use and growth (most thoroughly elaborated on recently by Tanner and Sinclair, 1983), it may often be worthwile to bypass thoughts of photosynthesis, and to concentrate primarily on predicting transpiration where water is likely to be limiting growth. Now transpiration depends partly on stomatal conductance, but largely on leaf area index when this is less than about 2.5 (Ritchie, 1974; Legg et al., 1979), so it is the development of leaf area that assumes greatest importance. How do roots affect this development? They presumably do so through their effects on the water relations of the shoot. The water supplied by them today presumably influences the increment in leaf area accrued by tomorrow although the mechanisms involved are elaborate and poorly understood. Availability of photosynthate may play less part than we commonly assume, if the concentrations of both soluble and insoluble carbohydrates increase in water-limited wheat as they do in cotton

(Ackerson, 1981). And the role of turgor in the growing cells is much in doubt (Michelena and Boyer, 1982), although techniques for measuring the turgor of growing cells are very prone to producing artifacts: for example, the water potential of such cells must change rapidly after they have been excised and placed in a psychrometer chamber, i.e. the attempt to measure the water potential changes it.

But there is increasing circumstantial evidence that the roots may influence the behaviour of the shoot, both in stomatal conductance and growth, by means other than water relations:

(i) Eavis and Taylor (1979) showed that transpiration rates of soybean are influenced by $\bar{\theta}$ but not by L, as though it were the environment of the roots, rather than their ability to take up water, that was important.

(ii) Richards and Rowe (1977) and Carmi and Shalhevet (1983) showed, for peaches and cotton respectively, that plants whose roots were growing in a restricted volume grew poorly despite the roots (and, apparently, the shoots) being supplied with ample water and nutrients.

(iii) Bates and Hall (1981) showed that stomatal conductance of cowpea was related to soil water status rather than to leaf water potential.

Each of these examples suggests that there are messages emanating from the roots that at least partly control the behaviour of the shoot. Cytokinins may well be implicated (Itai and Benzioni, 1976; Carmi and van Staden, 1983) but the production of inhibitors may also be involved. Current models of water relations of roots (e.g. Rowse et al., 1978; Steinhardt et al., 1981; Huck and Hillel, 1983) emphasise the uptake of water by the roots and, sometimes, the effect of soil water status on the growth of the roots. But if the roots are somehow directly sensing soil water status (possibly through their tips, which, having a poor hydraulic connection with the rest of the plant, will have a water potential strongly buffered by that of the adjacent soil), and transmitting that information to the shoot, then we may have to change the emphasis of such models.

CONCLUSION

Current models (state of the art?) of the influence of roots on the water economy and thence the production of wheat lie firmly within the tradition set by Cowan (1965) twenty years ago when he combined, for the first time, the beguilingly simple single-root model of uptake with a function relating stomatal conductance to leaf water potential. Implicit in this tradition is the idea that carbon-fixation is the main driver of growth. But if it is generally true that the carbohydrate status of water-limited plants is higher than that of well-watered ones, it behooves us to explore the possibility that root-controlled leaf growth is an important determinant of dry matter accumulation wherever water is short but not sharply limited. Where the water supply is sharply limited, and the prospects are goo for the roots extracting all the available water, then the productio of dry matter is set by the efficiency with which the leaves can trade water for carbon dioxide (Tanner and Sinclair, 1983).

REFERENCES

Ackerson, R. C., 1981, Osmoregulation in cotton in response to water stress. II. Leaf carbohydrate status in relation to osmotic adjustment, Pl. Physiol., 67:489.

Allmaras, R. R., Nelson, W. W., and Voorhees, W. B., 1975, Soybean and corn rooting in southwestern Minnesota. I. Water-uptake sink, Soil Sci. Soc. Am. Proc., 39:764.

Bates, L. M., and Hall, A. E., 1981, Stomatal closure with soil wate depletion not associated with changes in bulk leaf water status Oecologia (Berl.), 50:62.

Busscher, W. J., and Fritton, D. D., 1978, Simulated flow through th root xylem, Soil Sci., 125:1.

Carmi, A., and Shalhevet, J., 1983, Root effects on cotton growth an yield, Crop Sci., 23:875.

Carmi, A., and van Staden, J., 1983, Role of roots in regulating the growth rate and cytokinin content in leaves, Pl. Physiol., 73:76.

Cowan, I. R., 1965, Transport of water in the soil-plant atmosphere system, J. appl. Ecol., 2:221.

Crank, J., 1975, "The Mathematics of Diffusion", (2nd Edn), Oxford University Press, Oxford.

Eavis, B. W., and Taylor, H. M., 1979, Transpiration of soybeans as related to leaf area, root length, and soil water content, Agron. J., 71:441.

Ehlers, W., Höpke, V., Hesse, F., and Böhm, W., 1983, Penetration resistance and root growth of oats in tilled and untilled loess soil, Soil Till. Res., 3:261.

Faiz, S. M. A., and Weatherley, P. E., 1978, Further investigations into the location and magnitude of the hydraulic resistances in the soil:plant system, New Phytol., 81:19.

Fischer, R. A., and Turner, N. C., 1978, Plant productivity in the arid and semi-arid zones, A. Rev. Pl. Physiol., 29:277.

Gardner, W. R., 1960, Dynamic aspects of water availability to plants, Soil Sci., 89:63.

Greacen, E. L., Ponsana, P., and Barley, K. P., 1976, Resistance to water flow in the roots of cereals, in: "Water and Plant Life", O. L. Lange, L. Kappen, E. D. Schulze, eds., Springer-Verlag, Berlin.

Herkelrath, W. N., Miller, E. E., and Gardner, W. R., 1977, Water

uptake by plants. II. The root contact model, Soil Sci. Soc. Am. J., 41:1039.
Huck, M. G., and Hillel, D., 1983, A model of root growth and water uptake accounting for photosynthesis, respiration, transpiration and soil hydraulics, Adv. Irrig., 2:273.
Huck, M. G., Klepper, B., and Taylor, H. M., 1970, Diurnal variations in root diameter, Pl. Physiol., 45:529.
Itai, C., and Benzioni, A., 1976, Water stress and hormonal response, in: "Water and Plant Life", O. L. Lange, L. Kappen, and E. D. Schulze, eds., Springer-Verlag, Berlin.
Landsberg, J. J., and Fowkes, N., 1978, Water movement through plant roots, Ann. Bot., 42:493.
Legg, B. J., Day, W., Lawlor, D. W., and Parkinson, K. J., 1979, The effect of drought on barley growth: models and measurements showing the relative importance of leaf area and photosynthetic rate, J. agric. Sci., Camb., 92:703.
Meyer, W. S., and Ritchie, J. T., 1980, Resistance to water flow in the sorghum plant, Pl. Physiol., 65:33.
Michelena, V. A., and Boyer, J. S., 1982, Complete turgor maintenance at low water potentials in the elongating region of maize leaves, Pl. Physiol., 69:1145.
Newman, E. I., 1969, Resistance to water flow in soil and plant. I. Soil resistance in relation to amounts of root: theoretical estimates, J. appl. Ecol., 6:1.
Passioura, J. B., 1972, The effect of root geometry on the yield of wheat growing on stored water, Aust. J. agric. Res., 23:745.
Passioura, J. B., 1977, Determining soil water diffusivities from one-step outflow experiments, Aust. J. Soil Res., 15:1.
Passioura, J. B., 1980, The transport of water from soil to shoot in wheat seedlings, J. exp. Bot., 31:333.
Passioura, J. B., 1982, Water in the soil-plant-atmosphere continuum, in: "Encyclopedia of Plant Physiology New Series Vol. 12B. Physiological Plant Ecology II", O. L. Lange, P. S. Nobel, C. B. Osmond, H. Ziegler, eds., Springer-Verlag, Berlin.
Passioura, J. B., and Munns, R., 1984, Hydraulic resistance of plants. II. Effects of rooting medium and time of day, in barley and lupin, Aust. J. Pl. Physiol., (in press).
Passioura, J. B., 1983, Roots and drought resistance, Agric. Water Manag., 7:265.
Philip, J. R., 1957, The physical principles of soil water movement during the irrigation cycle, Proc. Int. Congr. Irrig. Drain., 8:125.
Reicosky, D. C., and Ritchie, J. T., 1976, Relative importance of soil resistance and plant resistance in root water absorption, Soil Sci. Soc. Am. J., 40:293.
Richards, D., and Rowe, R. N., 1977, Root-shoot interactions in peach: the function of the root, Ann. Bot., 41:1211.
Ritchie, J. T., 1974, Atmospheric and soil water influences on the plant water balance, Agric. Meteorol., 14:183.
Rose, D. A., 1968, Water movement in porous materials. III.

Evaporation of water from soil, Brit. J. appl. Phys. (J. Phys. D.) Ser. 2, 1:1779.
Rowse, H. R., Stone, D. A., and Gerwitz, A., 1978, Simulation of the water distribution in soil. II. The model for cropped soil and its comparison with experiment, Pl. Soil, 49:533.
Shawcroft, R. W., and Gardner, H. R., 1983, Direct evaporation from soil under a row crop canopy, Agric. Meteorol., 28:229.
Steinhardt, R., Ehlers, W., and van der Ploeg, R. R., 1981, Analysis of soil-water uptake from a drying loess soil by an oat crop using a simulation model, Irrig. Sci., 2:237.
Tanner, C. B., and Jury, W. A., 1976, Estimating evaporation and transpiration from a row crop during incomplete cover, Agron. J., 68:239.
Tanner, C. B., and Sinclair, T. R., 1983, Efficient water use in crop production: research or re-search? in: "Limitations to Efficient Water Use in Crop Production", H. M. Taylor, W. R. Jordan and T. R. Sinclair, eds., ASA-CSSA-SSSA, Madison.
Taylor, H. M., and Willatt, S. T., 1984, Shrinkage of soybean roots, Agron. J., 75:818.
Tinker, P. B., 1976, Transport of water to plant roots in soil, Phil. Trans. R. Soc., Ser. B, 273:445.
Vartanian, N., 1981, Some aspects of structural and functional modifications induced by drought in root systems, Pl. Soil, 63:83.
Wallace, J. S., and Biscoe, P. V., 1983, Water relations of winter wheat. 4. Hydraulic resistance and capacitance in the soil-plant system, J. agric. Sci., Camb., 100:591.
Weatherley, P. E., 1982, Water uptake and flow in roots, in: "Encyclopedia of Plant Physiology New Series Vol. 12B. Physiology Plant Ecology II", O. L. Lange, P. S. Nobel, C. B. Osmond and H. Ziegler, eds., Springer-Verlag, Berlin.

WATERLOGGING AS A LIMITATION TO WHEAT YIELD IN AN IRRIGATED CLAY SOIL

W. S. Meyer, H. D. Barrs, and N. S. Jayawardane

CSIRO, Centre for Irrigation Research
Griffith, New South Wales
Australia

INTRODUCTION

Wheat production currently occupies a small proportion of the total irrigated area of south-east Australia. At average yield levels of about 2.6 t ha^{-1} it is not an economic alternative to rice or intensive vegetable crops also grown in the area. However as the availability of irrigation water becomes more constrained, wheat, at yield levels near those obtained under the best soil and water management conditions (about 8 t ha^{-1}), would be economically competitive. This possibility, together with the greater yield per unit of water applied, compared with rice, indicates that increasing areas of wheat may be sown. However it is not yet completely understood why average yields of wheat are low. Circumstantial evidence suggests that it is associated with the poor internal drainage properties of the irrigated clay soils. This paper describes an experiment in which some of the processes occurring within the root zone were measured during an irrigation cycle. Having quantified root zone processes, it is intended to include them in a wheat growth and yield model that can be used as an aid in making decisions about optimum soil amelioration, water management and nutrient management of irrigated wheat.

MATERIALS AND METHODS

Details of the experiment are described elsewhere (Meyer et al., 1984) but the outline of the treatments is as follows. A clay soil (Marah clay loam) was collected in large steel cylinders (0.75 m outside diameter by 1.4 m deep). A drainage base was attached to the bottom of each cylinder. The completed assembly was housed in a lysimeter facility. One set of cylinders contained undisturbed soil (bulk density 1.6 mg mm^{-3}) while another set contained the same soil repacked to a bulk density of 1.2 mg mm^{-3}.

Wheat, cv. Egret, was sown in the cylinders on 27 May 1982 at a density of about 180 plants m^{-2}. Both nitrogen (50 kg N ha^{-1}) and phosphate (88 kg P ha^{-1}) fertilizer were added to the top 100 mm of soil prior to sowing. An additional 40 kg N ha^{-1} was watered in 77 days after sowing. The two watering treatments tested were controlled watering, in which small amounts of water were applied every 4 to 7 days to keep the soil well watered but avoid surface ponding, and flooding, in which sufficient water was applied to ensure that the surface soil was completely inundated. This flooding was applied on three separate occasions with the time of flooding ranging from 4 h to 72 h.

Leaf growth was measured throughout the season. Final grain yield was determined at maturity. Root growth was followed within the soil by counting the number of roots intercepting horizontal, clear acrylic observation tubes. Counts were made twice weekly using a fibre optic device and were subsequently converted to root length density values using a previously determined calibration equation.

Changes in soil water content and its distribution were monitored using a neutron soil-water probe down a central, vertical access tube in each cylinder. Values of air-filled porosity (ε_a) were calculated directly from these soil water measurements (Jayawardane and Meyer, 1984). Soil-air or water samples, obtained from samplers inserted through the side of the cylinder at five depths, were analysed for oxygen concentration. These values combined with ε_a values gave the concentration of O_2 (mg) per unit volume of soil.

RESULTS AND DISCUSSION

Soil Water and Oxygen

The total amount of oxygen within the profile was calculated as the sum of the amounts within each of five layers. This total value changed with treatment (Fig. 1). Total soil oxygen in the repacked soil, where ε_a is high while the soil is dry, is much higher than for the undisturbed soil. Frequent water additions applied in the controlled watering treatment kept the water content at all depths fairly constant in both the undisturbed and repacked soil profiles. Larger fluctuations in the soil water content occurred in the flooding treatments due to the longer periods between water additions and to the larger amounts of water applied during flooding.

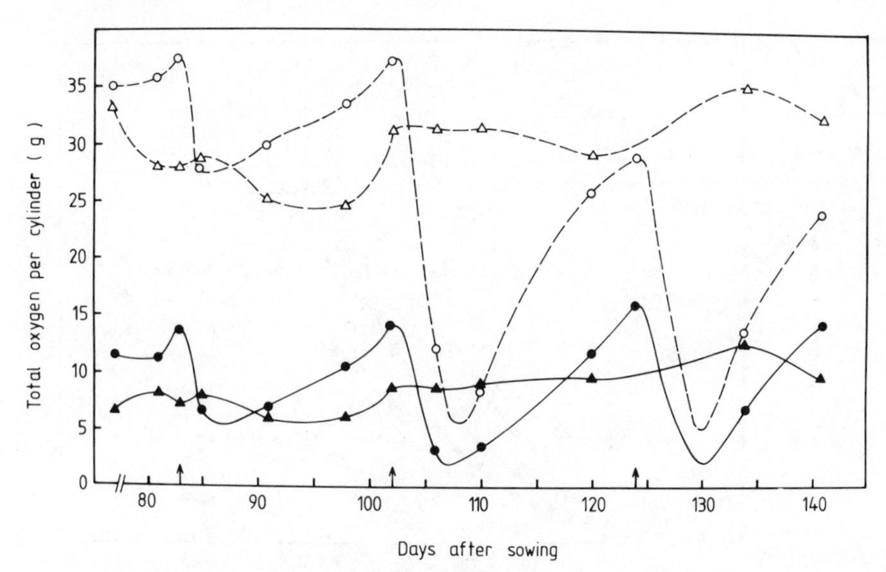

Fig. 1. Total mass of oxygen per cylinder as affected by soil and water addition treatment. (△) repacked, controlled watering; (○) repacked, flooded; (●) undisturbed, flooded; (▲) undisturbed, controlled. Arrows indicate the start of the flooding treatments.

The response of oxygen partial pressure to water additions measured in the cylinders was similar to that previously found in the field (Mason et al., 1983). These responses clearly demonstrate that in slowly draining clay soils, both improved porosity and water management to avoid prolonged surface ponding are needed to ensure that the profile remains well aerated.

Plant Responses

The response of root growth to the root zone treatment is shown in Fig. 2. When O_2 levels are continually low, such as in the undisturbed soil with controlled water additions, root growth is slow. Surface flooding, which forces O_2 concentration down, causes root growth to cease. Examination of the O_2 data suggests that root growth may begin to slow in these soils once soil O_2 is less than 15% of what it would be in a dry aerated soil. Values of soil O_2 less than 10% are very likely to stop root growth.

The effects of the flooding treatments on root growth were not reflected in the measurements of leaf and stem growth. However yield differences between the treatments were large (Table 1). The lack of an observed effect on leaf and stem extension may have been due to

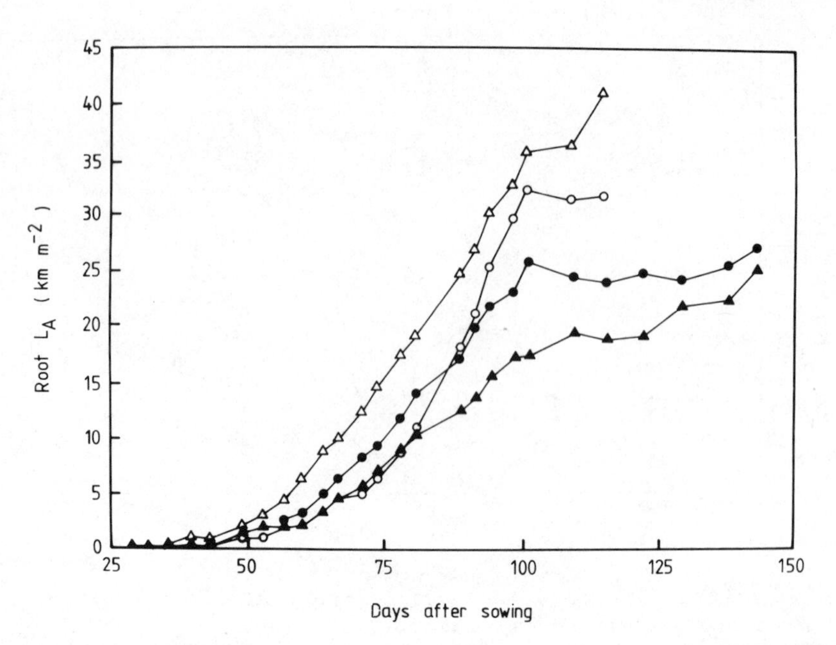

Fig. 2. Change in root length per unit ground area (L_A), symbols as in Fig. 1.

the low nitrogen status of all plants and/or to the advanced stage o development of the plants at the time of the treatments. Subsequent experiments on cotton and maize have shown that leaf growth is reduced by waterlogging but that it takes several days for the effec to be pronounced. Other evidence indicates that this may be the result of reduced root function since a reduction in the rate of uptake of nitrate also follows a similar time trend.

Table 1. Soil bulk density and mean oxygen availability associated with the treatments, and the effects on final grain yield. The mean oxygen value was calculated from the values measured between days 77 and 140 after sowing.

Treatment	Bulk density ($mg\ mm^{-3}$)	Mean O_2 ($mgO_2\ l^{-1}$ soil)	Grain yield ($g\ m^{-2}$)
Undisturbed Control	1.6	8.6	192
Undisturbed Flood	1.6	10.1	250
Repacked Flood	1.2	26.4	312
Repacked Control	1.2	30.1	340

The best yield obtained in these treatments was not large and probably reflects the poor nitrogen status of the plants. The yield from the undisturbed, flooded treatment is similar to the average yield obtained by farmers on similar soils where flood irrigation techniques are used.

Implications for Modelling

The results presented above and the known performance of wheat crops in the irrigated areas indicate that waterlogging associated with surface irrigation of clay soils limits yield. An understanding of the root zone processes causing this limitation can be used to devise crop management strategies which avoid or overcome the limitation. Thus the prediction of soil oxygen conditions and its consequences on plant yield become important.

A wheat growth and yield model is presently being adapted for the area (M. Stapper, personal communication). It is envisaged that this model will form the basis for a computer-based management advice service called SIRAGCROP. This model does not presently account for the effects associated with waterlogging during an irrigation.

The inclusion of a sub-model which describes the effects of transient waterlogging on crop growth and yield will need to be developed. There appear to be two approaches which differ in their degree of complexity. In the simpler approach, air filled porosity (ε_a) of the soil profile can be used as an indicator of O_2 availability. From a simulation of soil water content and its distribution, values of ε_a can be calculated for particular soil layers. To help assess whether a value of ε_a is likely to affect plant yield we are experimenting with semi-empirical limits which indicate when ε_a is either not affecting, restricting or inhibiting yield. The setting of these limits enables integration of the effects of ε_a over both time and soil layers within a profile. The end point of these calculations is an aeration-stress day, a concept originally outlined by Hiler (1969). The effect of aeration-stress days on final yield will depend on crop species, growth stage and root distribution within the profile. In addition, a given relation between aeration-stress days and final yield will occur only within certain temperature limits since ε_a takes no account of the biological demand for oxygen within the root zone. This demand is highly temperature dependent (Currie, 1970). Ultimately then, successful modelling of the effects of soil aeration on crop yield will need to take a more complex approach, simulating both soil water and soil oxygen entering and within the soil. The uniformity of response of soil oxygen concentrations to soil water changes in these soils suggests that modelling the bulk soil rather than at the soil ped level may be possible.

CONCLUSIONS

Surface applications of water to clay soils during an irrigation will result in decreased levels of soil oxygen. In many of the slowly draining soils of south-east Australia the described levels of soil O_2 will restrict and in some cases inhibit root growth, probably leading to substantial yield losses. Modelling of crop yield in this situation will need to consider the possible detrimental effects of an irrigation event on yield. The semi-empirical approach of using air filled porosity as an indicator of the soil aeration status is being investigated. However accurate simulation will only be possible if the effects of soil oxygen concentration on crop growth and yield can be understood and predicted.

REFERENCES

Currie, J. A., 1970, Movement of gases in soil respiration, Soc. Chem. Industry, Monograph 37:152.

Hiler, E. A., 1969, Quantitative evaluation of crop-drainage requirements, Trans. Am. Soc. Agric. Eng., 12:499.

Jayawardane, N. S., and Meyer, W. S., 1984, Measuring air filled porosity changes in an irrigated swelling clay soil, Submitted to Aust. J. Soil Res.,

Mason, W. K., Meyer, W. S., Barrs, H. D., and Smith, R. C. G., 1984, The effects of flood irrigation on air-filled porosity, oxygen and bulk density of a cracking clay soil in relation to the production of summer row crops, in: "Properties and Utilization of Cracking Clay Soils", J. W. McGarity, E. H. Hoult, and H. B. So, eds., University of New England Publishing Unit, Armidale, N.S.W.

Meyer, W. S., Barrs, H. D., Smith, R. C. G., White, N. S., Heritage, A. D., and Short, D. L., 1984, The effect of irrigation on soil oxygen status and root and shoot growth of wheat in a clay soil. Submitted to Aust. J. Agric. Res.,

HARVEST INDEX OF SPRING WHEAT AS INFLUENCED BY WATER STRESS

R. J. Hanks and R. B. Sorensen

Utah State University
Logan, Utah, U.S.A.

INTRODUCTION

Simple models to predict spring wheat yield have usually used a different procedure to predict grain yield than that of above-ground dry matter (Hanks, 1974). The prediction of grain yield has been related to predictions of water stress at different growth stages whereas the prediction of total dry matter yield has been related to total seasonal stress. The use of more complicated models to predict grain yield is justified only if there is a difference in the harvest index (ratio of grain to total dry matter yield) resulting from plant water stress.

There are no clear indications about the factors that influence the harvest index (HI) of a crop (Kanemasu, 1983). Many investigators have indicated that moisture stress, especially during critical phases of growth, has generally decreased harvest index (Passioura, 1977; Hodges, 1978; Rasmussen, 1979; DeLoughery and Crookston, 1979) but, when water was limiting, the harvest index varied from one season to the next in an unexplained way. Our preliminary investigations have generally shown that models to predict seasonal dry matter production, based on seasonal relative water stress, worked equally well for grain production in all but a few cases, and there was little need to take into account growth stage effects. This result points to the consistency of HI. Consequently the present investigation was undertaken to determine whether water stress affected HI in a consistent way under field conditions.

PROCEDURE, RESULTS AND DISCUSSION

The data were collected over several years at Logan, Utah on a Millville silt loam. The studies evaluated the influence of irrigation and other factors on spring wheat yield (dry matter and grain). Irrigation was varied from zero (dry-land) to full irrigation using the line source sprinkler system described by Hanks et al. (1976). Evapotranspiration (ET) was determined from measurements of soil water depletion and estimates of drainage, thus allowing for some of the variability between fields. In an attempt to minimise the year-to-year variation in response to treatment, we use actual ET for a treatment expressed as a fraction of ET for the

Table 1. The harvest index for various cultivars and planting dates (P, see text) with the corresponding relative ET (RET) for experiments in 1975-1979.

Treatment	Year	Irrigation Level					Max ET (mm)
		1	2	3	4	5	
Fremont	1975	0.41	0.50	0.52	0.54		
Bannock	1975	0.47	0.46	0.44	0.47		
Peak '72	1975	0.48	0.45	0.47	0.48		
Lemhi	1975	0.42	0.45	0.47	0.48		
Twin	1975	0.46	0.49	0.52	0.53		
RET	1975	0.71	0.84	0.94	1.00		440
Fremont P1	1979	0.33	0.37	0.48	0.48	0.48	
Borah P1	1979	0.51	0.49	0.55	0.56	0.56	
Twin P1	1979	0.51	0.54	0.60	0.60	0.58	
RET P1	1979	0.43	0.50	0.76	0.85	1.00	550
Fremont P2	1979	0.48	0.48	0.55	0.58	0.56	
Borah P2	1979	0.55	0.50	0.55	0.57	0.58	
Twin P2	1979	0.44	0.46	0.48	0.50	0.50	
RET P2	1979	0.30	0.42	0.71	0.83	1.00	580
Fremont P3	1979	0.43	0.43	0.55	0.52	0.53	
Borah P3	1979	0.46	0.47	0.45	0.43	0.45	
Twin P3	1979	0.44	0.46	0.48	0.50	0.50	
RET P3	1979	0.35	0.40	0.69	0.83	1.00	580
Fremont P4	1979	0.38	0.46	0.46	0.46	0.45	
Borah P4	1979	0.42	0.46	0.46	0.46	0.47	
Twin P4	1979	0.38	0.44	0.46	0.48	0.49	
RET P4	1979	0.25	0.31	0.66	0.82	1.00	520

fully irrigated treatment, as a measure of the stress. The 30 year average precipitation for Logan, Utah is 52, 43, 39, 11, 24 and 27 mm for the months April to September respectively. Harvest index was measured from hand samples of the above-ground plant leaving only very short stubble.

In 1975 four cultivars were studied (Rasmussen and Hanks, 1978). Plots were planted on 1 May and harvested on 17 August when the crops were all mature. Rainfall during this period was 125 mm, 20 mm above normal. Irrigation ranged from zero to 164 mm. The data (Table 1) show that HI tended to increase as relative ET increased for the cultivars Fremont and Twin, but not for Bannock or Peak '72.

In 1979 three cultivars were sown at four planting dates ranging at approximately 15 day intervals from 2 April to 21 May. Plants were harvested on 21 August, when all were mature. Rainfall from 2 April to 21 August was 135 mm, about 30 mm below normal. The maximum amount of irrigation (444 mm) was the same for all planting

Table 2. The harvest index for various cultivars and planting dates (P, see text) with the corresponding relative ET (RET) for experiments in 1981-1983.

Treatment		Year	Irrigation Level					Max ET (mm)
			1	2	3	4	5	
Fremont	P1	1981	0.46	0.48	0.52	0.44	0.45	
RET	P1	1981	0.65	0.72	0.88	0.93	1.00	570
Fremont	P2	1981	0.49	0.49	0.53	0.51	0.46	
RET	P2	1981	0.60	0.71	0.85	0.95	1.00	480
Fremont	P3	1981	0.50	0.50	0.53	0.45	0.43	
RET	P3	1981	0.58	0.65	0.84	0.94	1.00	460
Fremont		1982	0.47	0.45	0.48	0.49	0.49	
Fielder		1982	0.52	0.50	0.51	0.49	0.48	
RET		1982	0.59	0.69	0.79	0.91	1.00	490
Fremont	P1	1983	0.45	0.47	0.49	0.51	0.51	
RET	P1	1983	0.75	0.81	0.89	0.90	1.00	520
Fremont	P2	1983	0.43	0.46	0.48	0.48	0.48	
RET	P2	1983	0.74	0.75	0.86	0.99	1.00	500
Fremont	P3	1983	0.41	0.44	0.43	0.43	0.42	
RET	P3	1983	0.62	0.68	0.84	0.92	1.00	410

dates. The harvest index (Table 1) for the cultivars Fremont and Twin was generally lower for the two lowest irrigation levels than for the three highest irrigation levels but there was no noticeable difference in harvest index for the top three irrigation levels. The cultivar Borah did not show a consistent influence of irrigation level on harvest index. Overall, irrigation level did not influence harvest index for any cultivar provided the relative ET was above 0.50 in this drier than normal year.

In 1981 Fremont was studied at five planting densities, but there were no differences in HI and this aspect of the experiment is not reported in Table 2. There were three planting dates of 15 April, 30 April and 14 May. The plots were harvested in late August. Rainfall from April until harvest was 188 mm, 50 mm above normal, and irrigation ranged up to 250 mm. The lowest value of HI was generally found at irrigation levels 4 or 5 followed by level 1 or 2 and the highest at level 3.

In 1982 the cultivars Fremont and Fielder were studied with five nitrogen levels. The crop was planted on 21 April and harvested on 22 August. Rainfall for the period was 140 mm, about 10 mm above normal, and irrigation ranged up to 310 mm. The results (Table 2) show little influence of irrigation level on HI for either cultivar, little consistent influence of nitrogen level on HI (not shown) and little difference between cultivars.

In 1983 the cultivar Fremont was planted at three dates ranging from early April until late May. Harvest was on 15 August for the first two plantings and 23 August for the third. Rainfall from April to August was about 300 mm, 130 mm above normal. Irrigation ranged up to 250 mm. There was an indication of a smaller HI at the lowest irrigation level (Table 2), and HI decreased generally for later sowing dates.

CONCLUSION

The data from the several years of studies in Utah, USA, showed a slight trend for the harvest index to be lower at the lowest water levels (highest stress) for some years and for some cultivars. However, looking at all data from five years, there was no clear indication of any influence of water stress on HI. Thus, for many field situations like those studied, models that predict dry matter can also predict grain yield with reasonable accuracy.

REFERENCES

DeLoughery, R. L., and Crookston, R. K., 1979, Harvest index of corn affected by population density, maturity rating, and environment, Agron. J., 71:577.

Hanks, R. J., 1974, Model for predicting plant growth as influenced by evapotranspiration and soil water, Agron. J., 66:660.

Hanks, R. J., Keller, J., Rasmussen, V. P., and Wilson, G. D., 1976, Line source sprinkler system for continuous variable irrigation-crop production studies, Soil Sci. Soc. Amer. Proc., 40:426.

Hodges, T., 1978, Photosynthesis, growth, and yield of sorghum and winter wheat as functions of light, temperature, water, and leaf area, Ph.D. thesis, Kansas State Universitv, Manhattan.

Kanemasu, E. T., 1983, Yield and water-use relationships: Some problems of relating grain yield to transpiration, in: "Limitations to Efficient Water Use in Crop Production", H. M. Taylor, W. R. Jordan and T. R. Sinclair, eds., American Society of Agronomy, Madison, Wisconsin.

Passioura, J. B., 1977, Grain yield, harvest index, and water use of wheat, J. Aust. Inst. agric. Sci., 43:117.

Rasmussen, V. P., 1979, Modelling winter wheat yields as affected by water relations and growth regulants, Ph.D. thesis, Kansas State University, Manhattan.

Rasmussen, V. P., and Hanks, R. J., 1978, Model for predicting spring wheat yields with limited climatological and soil data, Agron. J., 70:940.

MODELS OF GROWTH AND WATER USE OF WHEAT IN NEW ZEALAND

D. R. Wilson and P. D. Jamieson

Crop Research Division, DSIR
Christchurch, New Zealand

INTRODUCTION

Most agronomic research on wheat in New Zealand has consisted of empirical field trials, with conventional statistical analyses producing information specific to the time and location of the trial or series of trials. We have attempted to overcome this limitation by using simple models to analyse the growth and water use of wheat crops so that responses to agronomic treatments can be explained, and separated from site and seasonal variability. In this paper we briefly describe the models, present the results of the analyses, and consider their practical implications and limitations.

The models were used to analyse results from eleven crops grown in three experiments at Lincoln (latitude 43° 38'S, altitude 11 m) in the 1980, 1982 and 1983 seasons. Two New Zealand cultivars, Rongotea and Oroua, were sown on two or three dates in each experiment. All the experiments were conducted on a soil consisting of 0.3 m of silt loam overlying a sandy loam subsoil to a depth exceeding 1.5 m, the maximum rooting depth of the crops. The soil profiles retained 18% by volume of plant-available water at saturation.

CROP GROWTH

Yield was analysed as the product of three factors: the mean rate at which dry matter accumulates, the duration of growth, and the harvest index (Monteith, 1977). We have made the simplifying assumption that harvest index is constant, so the analyses were confined to the rate and duration of growth to produce total above-ground dry matter (DM). For each crop studied, growth was related linearly to the amount of energy available (PAR intercepted by the crop canopy during active growth; $\int Q.dt$) and the efficiency, A, with which Q is used to produce dry matter (Monteith, 1977; Gallagher and Biscoe, 1978; Kumar and Monteith, 1981):

$$DM = A \int Q\, dt \quad (1)$$

This simple model described satisfactorily the growth of all the crops, even though yield varied considerably among sowing times and seasons. The value of A was constant: 2.38 ± 0.03 g of dry matter were produced per MJ of PAR intercepted (Fig. 1), which agrees with British results where most determinations of A for arable crops fall between 2.2 and 2.8 g MJ^{-1} (Kumar and Monteith, 1981). Thus variation in crop yield depends primarily on variations in $\int Q.dt$, caused by variations in the proportion of incident PAR intercepted and the duration of active growth. The important component processes of canopy expansion, duration and senescence, are all affected by many factors, but the overall practical implication is that crop management should aim to maximise the opportunity for crops to intercept PAR.

CROP WATER USE

Water use was analysed using an evapotranspiration model which accounts for water deficit or incomplete ground cover, both of which are consequences of drought and are common for wheat crops in New Zealand. The model used was proposed by Ritchie (1972). It calculates separately the two components of evapotranspiration, E, i.e. transpiration, E_t, and soil evaporation, E_s, and sums them.

E_t is calculated as the potential evaporation rate, E_p, using the Penman equation, modified by ground cover and soil moisture deficit factors:

$$E_t = E_p \cdot G \cdot F \quad (2)$$

where G is fractional ground cover and F is an empirical function of soil moisture deficit (D) with the form:

$$F = 1 \qquad D \leqslant D_\ell, \text{ or after rain or irrigation}$$

$$F = 1 - [a \cdot (D - D_\ell)] \qquad D \geqslant D_\ell$$

where D_ℓ is a limiting deficit. Values of a and D_ℓ, which are crop- and soil-specific parameters, were determined from measurements of E in separate wheat experiments on the same soil to be 0.0064 mm^{-1} and 175 mm respectively.

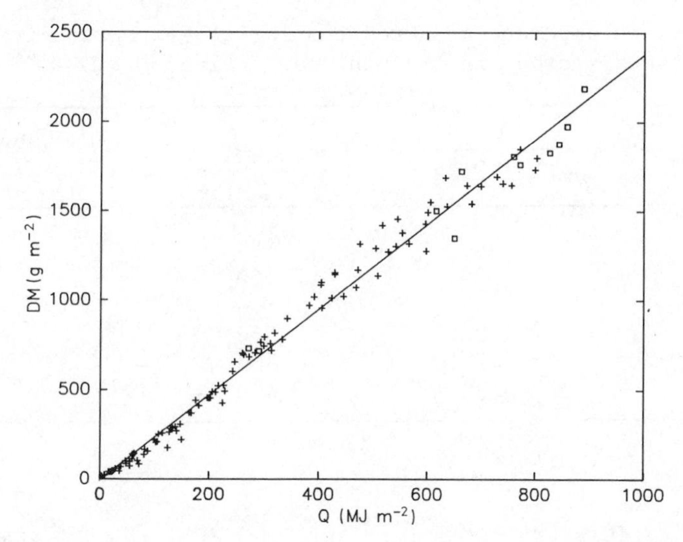

Fig. 1. Relation between cumulative total dry matter production, DM, from successive harvests, and intercepted PAR, Q. The maximum dry matter values for the 11 crops have a different symbol (□). The slope of the regression line is 2.38 ± 0.03 g MJ^{-1}.

E_s is assumed to be equal to E_p when the soil surface is wet, and to be progressively reduced after a critical available soil water content is reached. Thus E_s was taken as the smaller of two rates: either limited by the energy available at the soil surface, calculated as the product of E_p and $(1 - G)$, or limited by the rate of water vapour diffusion to the surface of the drying soil, calculated as a function of the square root of time (Ritchie, 1972; Tanner and Jury, 1976).

Totals of E were measured on six crops using the water balance method, and were generally about 10% greater than values calculated from the model (Table 1). It was not possible to establish the reasons for the discrepancies. Errors in the measured values may have been caused by water lost to drainage, because the water balance assumed drainage was zero. Since E_t and E_s were not measured separately the two parts of the model could not be tested independently. Despite the discrepancies, we consider the model is sufficiently accurate for practical purposes. Crop water use is very dependent on the amount of radiation intercepted, and so reduced E and reduced growth were both associated with decreased canopy expansion and duration, and were modified by crop management in a similar manner.

Table 1. Calculated and measured cumulative evapotranspiration, E, for six crops in Lincoln, New Zealand.

Sowing date	E (mm) Calculated	E (mm) Measured	% Underestimated
5.5.82	560	610	8.9
31.5.82	520	560	7.7
2.7.82	465	540	16.1
9.5.83	520	570	9.6
15.6.83	490	520	6.1
22.7.83	435	420	-3.6

WATER USE EFFICIENCY

A model proposed by Bierhuizen and Slatyer (1965) was used to relate crop growth and water use. Dry matter production (DM) is linearly related to the integral of the ratio of E_t and the daytime vapour pressure deficit (e* - e):

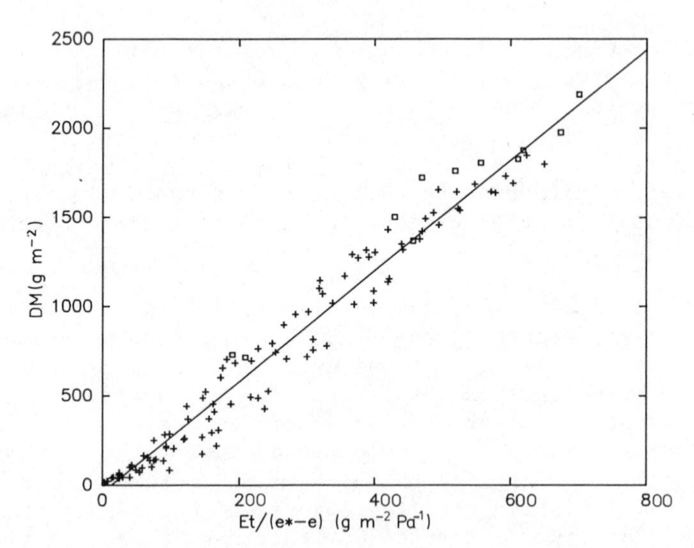

Fig. 2. Relation between cumulative total dry matter production, DM, from successive harvests, and transpiration per unit vapour pressure deficit, E_t / (e* - e). Maximum values for the 11 crops have a different symbol (□). The slope of the regression line is 3.10 ± 0.05 Pa.

$$DM = k \int [E_t / (e^* - e)] \, dt \qquad (3)$$

where k is an empirical constant which characterises the transpiration efficiency of a crop. When DM and $\int E_t.dt$ are defined as mass per unit area, k has the dimensions of pressure. Tanner and Sinclair (1983) presented values of k for several crops and though no data were presented for wheat, the results suggest that k is a stable parameter for each crop, regardless of season, location or cultural treatments.

Analyses of the results from all the experiments produced a constant value of k = 3.10 ± 0.05 Pa for wheat, regardless of season or treatment (Fig. 2). Thus there was a stable relationship between growth and water use, provided the saturation deficit during the crop's growth was taken into account, and this indicates that crop dry matter production cannot be increased without using more water by transpiration. The management required to achieve maximum yield is the same as for maximum water use. Consequently, the main prospect for improving the water use efficiency of wheat crops lies in improved management to increase E_t as a fraction of E, and the upper limit to the water use efficiency is E_t efficiency as defined by Equation 3 (Tanner and Sinclair, 1983).

DISCUSSION

The three models successfully analysed the growth, water use and transpiration efficiency of wheat crops, even though growth and water use differed appreciably among treatments and experiments. Further, they provided insights into how crop management affects growth and water use, and made it possible to separate crop responses from variable seasonal influences. However, although the models are useful analytical tools, all are limited in that they require measurements of ground cover or LAI. Variations of these characters were identified as the main causes of variations in growth and water use, and they are sensitive to crop management changes and to stresses such as water deficit. Therefore the models cannot be used to predict the effects of changed conditions; they merely describe them. It will be necessary to develop models of canopy expansion, duration and senescence, and the many factors which affect them, to make it possible to estimate growth, water use and transpiration efficiency of wheat crops from climatic information.

We have made a preliminary investigation of the potential for using the British AFRC wheat model (Weir et al., 1984) to simulate canopy development, and other aspects of wheat growth and development under New Zealand conditions. Three sections of the model, dealing with phenological development, light interception and photosynthesis, and dry matter partitioning and grain growth, simulated accurately those aspects of the growth and development of experimental crops in

New Zealand. However, results from the canopy development section did not agree with experimental observations. This model does not impose restrictions on canopy development as a result of competition for assimilate supply or nutrient or water deficits. This deficiency restricts the potential uses of the model, especially in climates where droughts are common, because predictions of crop growth and water use depend strongly on realistic simulations of canopy development.

ACKNOWLEDGEMENTS

Mr. R. Hanson is thanked for technical assistance.

REFERENCES

Bierhuizen, J. F., and Slatyer, R. O., 1965, Effect of atmospheric concentration of water vapour and CO_2 in determining transpiration-photosynthesis relationships of cotton leaves, Agric. Meterol., 2:259.

Gallagher, J. N., and Biscoe, P. V., 1978, Radiation absorption, growth and yield of cereals, J. agric. Sci., Camb., 91:47.

Kumar, M., and Monteith, J. L., 1981, Remote sensing of crop growth, in: "Plants and the Daylight Spectrum", H. Smith, ed., Academic Press, London.

Monteith, J. L., 1977, Climate and the efficiency of crop production in Britain, Phil. Trans. R. Soc., Ser. B, 281:277.

Ritchie, J. T., 1972, Model for predicting evaporation from a row crop with incomplete cover, Water Resour. Res., 8:1204.

Tanner, C. B., and Jury, W. A., 1976, Estimating evaporation and transpiration from a row crop during incomplete cover, Agron. J., 68:239.

Tanner, C. B., and Sinclair, T. R., 1983, Efficient water use in crop production: research or re-search?, in: "Limitations to Efficient Water Use in Crop Production", H. M. Taylor, W. R. Jordan and T.R. Sinclair, eds., American Society of Agronomy, Madison, Wisconsin.

Weir, A. H., Bragg, P. L., Porter, J. R., and Rayner, J. H., 1984, A winter wheat crop simulation model without water or nutrient limitations, J. agric. Sci., Camb., 102:371.

NITROGEN NUTRITION OF WINTER WHEAT

K. Vlassak and L. M. J. Verstraeten

Catholic University
Leuven, Belgium

INTRODUCTION

As wheat is one of the major agricultural crops in Europe (Table 1), its fertilizer requirements have been investigated intensively during the last decades (Scharpf and Wehrmann, 1975; Hanus, 1973; Viaux, 1980; Dilz, 1968; Ris et al., 1981; Laloux et al., 1975; Batey, 1976; Robinson, 1975). Indeed, experimental evidence has been acquired to quantify both sides of the input/output equation. The supply side consists of the inorganic nitrogen readily available in the soil profile together with an unknown amount to be mineralized during growth. On the other side is the N-demand by the crop, and this has become increasingly important within the last 15 years because of ever improving management of wheat growth (Fig. 1). This has resulted some 6 years ago in yields of 10 t ha^{-1} and more, first in the UK and the Schleswig-Holstein area of West Germany, then in other parts of Europe such as the Netherlands and Belgium, though not uniformly in all parts of Europe.

The importance of balanced nitrogen nutrition has never been neglected but the other parameters of growth have also to be optimized before full advantage of correct fertilizer treatment can be obtained. In this respect, plant breeding has been very valuable by providing short-straw cultivars with improved stem strength and harvest index and by the production of disease-resistant cultivars (Austin et al., 1980; Bingham, 1983).

Chemical control of weeds and pests and the use of growth regulators have influenced both crop yield and response to nitrogen and their impact on farming practice is no longer questioned (Sturm and Effland, 1982; Cook, 1983; Elliott, 1983). The current use of

Table 1a) Important agricultural crops in Europe (Eurostat-E.5 4/1083 (A = Area, 10^6 ha; Y = Yield, t ha^{-1}; P = Production, 10^6 t

	1982						
	Soft wheat and spelt	Rye	Barley	Oats	Grain maize	Potatoes	Sugar beet
A	10.89	0.61	9.31	1.77	2.96	1.12	1.86
Y	5.10	3.81	4.42	3.69	6.48	30.66	53.07
P	55.61	2.32	4.12	6.54	19.17	34.40	98.56

b) Situation for soft wheat and spelt in the member states of the European Economic Community. (FRG, West Germany; FRA, France; ITA, Italy; NED, Netherlands; BEL, Belgium; LUX, Luxembourg; UK, United Kingdom; EIR, Republic of Ireland; DEN, Denmark; GRE, Greece)

	EUR-10	FRG	FRA	ITA	NED	BEL	LUX	UK	EIR	DEN	GRE
A	10.84	1.58	4.73	1.63	0.13	0.18	0.006	1.66	0.057	0.18	0.74
Y	5.10	5.47	5.28	3.68	7.39	5.87	4.05	6.20	3.98	6.67	3.02
P	55.61	8.63	24.98	5.99	0.97	1.04	0.025	10.31	0.227	1.21	2.24

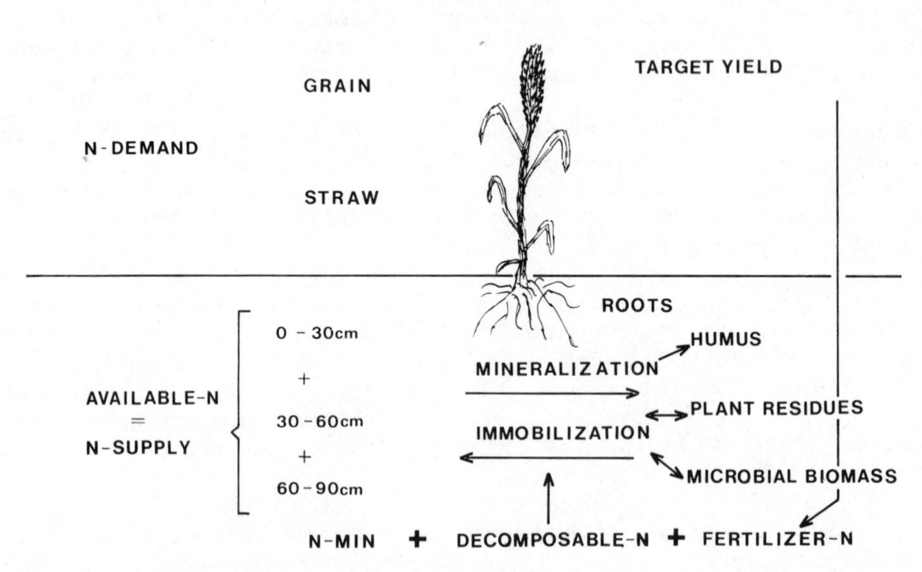

Fig. 1. A diagrammatic representation of the soil-plant system, indicating N-supply and N-demand.

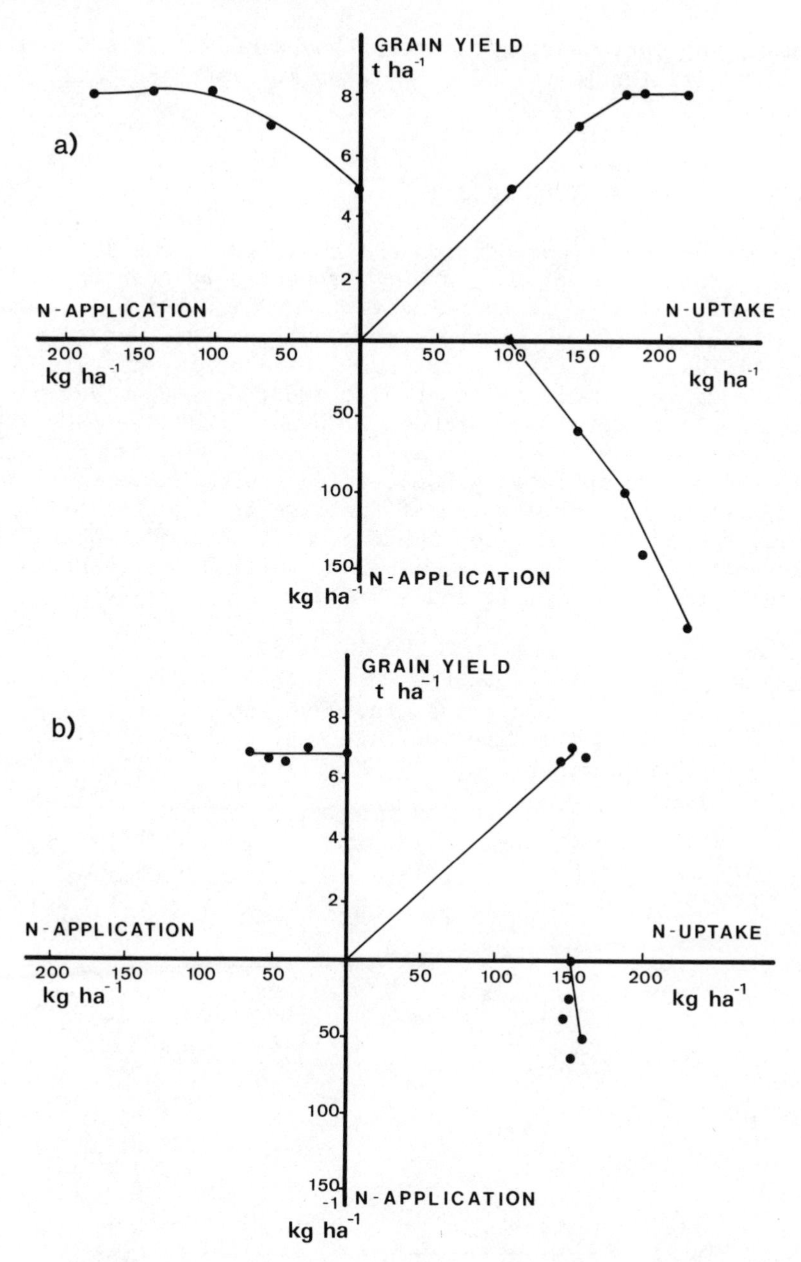

Fig. 2. Fertilizer nitrogen, N-uptake and yield interrelationships for winter wheat grown at a) Houtave and b) Heverlee.

this "technology-package" for wheat production has led to new developments in nitrogen nutrition based on crop productivity and N-requirement and using computer-based systems and modelling methods.

This has involved research into new aspects of nutrition such as fertilizer efficiency (Fig. 2), N-status in plant and soil, N stress and target yields.

FERTILIZER RECOMMENDATION

Comparing the recommendation systems used in the EEC, it is obvious that the current UK system, propagated by ADAS and based on the use of an N index, is the simplest (Needham, 1982; Sylvester-Bradley et al., 1983b). On the continent, the Dutch and Wehrmann systems include a similar concept but with profile-analysis down to 90 cm as a major feature (Dilz et al., 1982; Jungk and Wehrmann, 1978) (Table 2). Other approaches in Germany are the BASF system in which rather large amounts of N are put on in the first dressing, and the Heyland method which is a combination system, with sowing densities and plant performance at the time of anthesis being important factors for decision making (Deutsche Landwirtschafts-Gesellschaft, 1981). The "Balance-sheet" method in France is another approach that is both simple and extremely ambitious (Viaux, 1983).

Table 2. Recommendation systems in relation to crop growth stages. The numbers above the line give the dresings in kg ha^{-1}, and below the line indicate the depth of the soil profile in which mineral nitrogen is estimated.

	Seedling Growth		Tillering			Stem elongation			Ear emergenc	
Zadoks growth stage	12	13	21	25	29	30	32	39	51	61
Prof. Wehrmann		120-N_{min} / 0-90 cm					20-30		50-60	
BASF		80					20		60	
The Netherlands		140-N_{min} / 0-90 cm					0-60			
Prof. Heyland		70-N_{min} / 0-60 cm				80-N_{min} / 60-90 cm		50-60		

Simple because not all aspects of the N cycle in the soil are covered, but also quite ambitious because of the difficulties involved in including the right values for each of the parameters (Meynard et al., 1982). The current Belgian system propagated by the BDB (Soil Service) is centred around the mineral N in the profile and also takes note of the carbon content of the soil (Boon, 1983). Small adjustments are made on the basis of plant factors such as cultivar and sowing date as well as previous cropping or additions of organic materials. The "index of nitrogen" derived is then the basis for the fertilizer requirement.

In spite of all these sophisticated recommendation systems, there is still a need for a reliable method for the assessment of N-status. For example, in the N input/output equation mentioned above, several variables are unknown or difficult to evaluate. Calculated amounts of nitrogen may give a general view but give no indication of the rate of release nor of the concentration in which nitrogen will be available to the plant during the various phases of plant development. As Tinker (1979) has said: "Nitrogen is unique amongst the major nutrients in that there is no generally accepted method of soil analysis to indicate N status". Attempts have been made to improve this situation by the development of methods of leaf analysis but although a number of correlations have been established the methods are not yet commonly used (Finck, 1963; Chapman, 1967).

NITROGEN STATUS

In order to discuss the concept of N-status and its assessment it is worthwhile developing the philosophy of Greenwood on N-stress (Greenwood et al., 1965; Greenwood, 1966, 1976). Whereas N-status means the actual condition from the viewpoint of nitrogen nutrition of the plant, it is evident that deficiency of nitrogen may exist without manifesting itself by visual symptoms but still may cause a quantitative reduction in yield. In this respect, Greenwood (1976) has discussed the use of N status as "... an estimator of the intensity of current N deficiency and therefore an indicator of the size of the current or potential response to nitrogen".

The first step is to define a reference point for zero stress. This is done by relating different yields to the observed maximum yield as the standard, assuming that maximum growth rate would be attained with a non-limiting supply of nitrogen. The symbol S_N is used for the shortfall in growth rate and may be expressed as:

$$S_N = 100\frac{(\text{growth rate at maximum N response})-(\text{growth rate at deficiency})}{(\text{growth rate at maximum N response})}$$

This approach reflects the current shortfall in supply by the soil and all environmental factors affecting growth and uptake. Growth rate on a dry weight basis may be expressed in two ways:

1) Crop growth rate, $C = \dfrac{W_2 - W_1}{t_2 - t_1}$ $(g\ m^{-2}\ d^{-1})$

2) Relative growth rate, $R = \dfrac{\ln W_2 - \ln W_1}{t_2 - t_1}$ $(g\ g^{-1}\ d^{-1})$

where W_2 and W_1, are dry weights in g m^{-2} at times t_2 and t_1 (d) respectively. From these formulae, two measures of N stress can be defined.

$$S_{NC} = 100\ (C_m - C)/C_m$$

$$S_{NR} = 100\ (R_m - R)/R_m$$

where m signifies growth rate when N is not limiting.

With increasing emphasis now being placed on high yield production, and growing interest in modelling growth on a day-by-day basis, this approach to nitrogen nutrition may improve our understanding of the interrelationship between supply and demand and thus lead to improvements in our recommendation systems at work.

N-Status from Dry Matter Growth

Plant dry weight determination is a simple procedure but it has the disadvantages of being destructive, needing many replicate samples and requiring a significant time period between samplings for differences in growth to reveal responses to treatments. It is also essential to have a specific experimental design that ensures non-limiting supply on one of the reference treatments.

The study of spring barley at Aberdeen by Batey and co-workers (O'Neill, 1980; Batey, 1982, 1983) demonstrates the usefulness of this technique. The correlation coefficients between the end-of-season parameters and the estimates of N stress are given in Table 3. Similar calculations on the basis of relative growth rate also showed high correlations. However, it became evident from these figures that even high correlations were no guarantee of good predictive ability: at high N-stress a good linear response was obtained but, at lower values, variability was much greater and predictions poor.

Table 3. Correlation coefficients between S_{NC} during early growth, and grain yield and grain N. S_{NCa} is based on C_M at N = 180 kg ha^{-1} and S_{NCc} on C_M at N = 120 kg ha^{-1}: *, significant at 5% level; **, at 1% level; ***, at 0.1% level.

Week	N Stress	Grain yield	Grain N
3	S_{NCa}	-0.938 ***	-0.795 **
	S_{NCc}	-0.962 **	-0.862 *
4	S_{NCa}	-0.943 ***	-0.832 **
	S_{NCc}	-0.939 **	-0.838 *
5	S_{NCa}	-0.968 ***	-0.916 ***
	S_{NCc}	-0.967 ***	-0.871 *
6	S_{NCa}	-0.789 *	-0.819 **
	S_{NCc}	-0.532	-0.283 N.S.

One practical conclusion from this work was that early plant diagnostic sampling gave the best results (Table 3). Indeed, most tests indicate the importance of N supply around the beginning of stem extension (Zadoks 23-25) as the crop enters its very rapid growth phase. Similar results were obtained for winter wheat by Page et al. (1977) at Rothamsted. It has also been observed that the nitrogen taken up by spring wheat at the end of tillering was as much as 90% of the final amount of nitrogen in the crop at maturity. This supports Tinker's (1979) view that "It seems important in the future that more attention should be given to the very earliest stages of growth when the root system is very small and its location with respect to nitrate in the seedbed is uncertain".

Work by Greenwood and co-workers (Greenwood et al., 1977; Greenwood, 1982) at the National Vegetable Research Station, Wellesbourne (UK) has shown that the growth of many crops may be described by the simple equation:

$$W + K_1 \ln W + W_0 = K_2 t$$

where W is weight, t is time and W_0, K_1 and K_2 are constants. They found, perhaps surprisingly, that values for K_2 were remarkably similar for many vegetable crops, but W_0 varied considerably. Consequently, most of the substantial differences in the curves relating dry matter to time can be attributed to the difference in W_0, implying the importance of optimum nutrient availability during early stages of growth.

The relative simplicity and speed of the method must be set against the destructive sampling of the crop and the need for carefu experimental design, including a non-limiting N-treatment in order t make N-stress calculations. However, work by Scaife and Barnes (1977), Batey (1977) and Vlassak et al. (1982) have indicated the potential use of growth rate depression or rate of dry matter production and of N offtake for diagnostic and predictive purposes i relation to current plant nutrient status.

N-Status from Leaf Area Growth

If destructive sampling is to be avoided, the rate of increase in leaf area (LAI) may be a useful variable, particular because of its importance in a number of physiological models of wheat growth. Leaf area is an expression of photosynthetic potential and N nutrition can influence photosynthesis through its effect on leaf area and through changes in the net assimilation rate. Evidence for a close relationship between LAI and N stress has been found by Hals et al. (1969) and Bouma (1970) and followed the familiar expression:

$$S_{NA} = 100 \left\{(A_m - A)/A_m\right\}$$

Now that electronic scanners for leaf area measurement are more readily available, including versions for field use, LAI may become one of the most simple yet useful parameters of growth.

N-Status from Plant Tissue Analysis

The general concept of N stress is used without any mathematica formulation to describe any kind of nitrogen deficiency during growth, and the term is used to mean either the stress itself or the effect of that stress on plant growth. Plant analysis has been used since the 19th century (Lundegard, 1951) to quantify N stress, based on the principle that, within certain limits, an increase of N concentration in the soil results in an increase of concentration in the plant and in a higher growth rate. The problem with techniques involving analysis of N concentration in plant tissue is the cultiva of plant organs that may be tested for several forms of N and at different stages of growth (Commonwealth Bureau of Soils, 1977). Most of the research involved has concentrated on total nitrogen, nitrate nitrogen and the nitrate reductase enzyme assay.

Total nitrogen was chosen because it can be analysed reliably, though there are substantial variations in concentration, particularly related to stage of development. It is a result of several enzyme reactions and as such only a secondary parameter. Recently, Leigh et al. (1982) at Rothamsted showed that expressing the total nitrogen as the concentration in the tissue water resulted in a fairly constant value of about 5 g kg^{-1} tissue water for most of the growing period of spring barley. Our own observations on winter wheat showed a minimum of 5 g kg^{-1} tissue water in the period between mid-May and mid-June. Highest values, in spring and at ripening, were only double the lowest but the concentration curves do show a small but distinct pattern of response to splitting the N fertilizer application (Fig. 3). It is evident that this approach reflects the N-status of the plant without large variations due to plant development.

This approach does involve extra analysis compared to the dry matter production studies and therefore it is certainly not the right answer to the problem of N-status assessment for rapid field surveys.

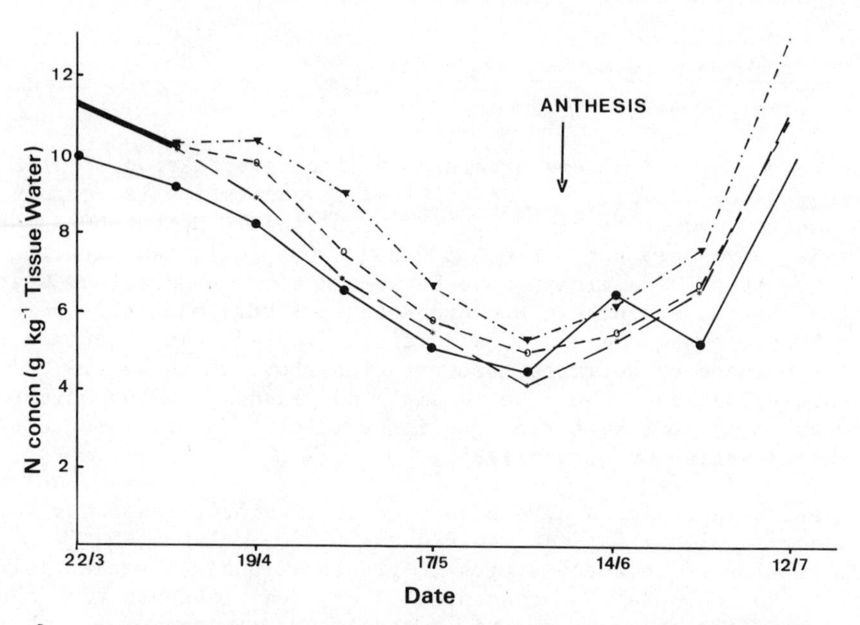

Fig. 3. Nitrogen concentration in tissue water during growth of winter wheat. N fertilizer treatments in kg ha^{-1} were divided 0:0:0, ●——●; 40:0:0, *---*; 40:40:40, ○---○; 40:100:60, ▼-•-•-▼.

On the other hand, physiologically it does reflect an optimum concentration far more fundamental than the concentration of nitrogen in dry matter, but more experiments are needed to obtain full benefit from this approach.

To be able to respond to the sudden onset of N shortage in the field requires a rapid and reproducible test such as the analysis of NO_3 concentration. This analysis does not allow quantification of N stress but does provide a measure of N status that relates to decreases in yield (Blacklow and Incoll, 1982). Earlier work has proved that nitrate analysis can be used to advantage for the assessment of N status of the crop. Concentrations of 1000 ppm NO_3-N or more gave maximum yields and consequently this amount was called the critical concentration. This concentration was independent of the crop's stage of development and may therefore be used throughout growth (Syman, 1974; Papastylianou et al., 1982; Papastylianou and Puckridge, 1981). The major drawback is the difficulty in interpreting NO_3 concentrations below this critical level. Nitrate nitrogen does not accumulate in measurable quantities, and below the critical concentration N comes increasingly from the plants organic N reserves.

The relationship between dry matter production, DM, as a fraction of maximum, DM_{max}, and NO_3 concentration, C_{NO_3}, is a rectangular hyperbola:

$$\frac{DM}{DM_{max}} = \frac{b_1 \cdot C_{NO3}}{1 + b_2 \cdot C_{NO3}}$$

Estimates of b_1 and b_2 are obtained by linear regression of 1/DM on $1/C_{NO_3}$. The accuracy of the critical concentration is largely dependent on how well this curve is established. Up to 500 ppm the curve is very steep but between 500 and 1000 ppm it has more the character of a transitional zone between a clearly deficient N status and one that is adequate. Dow and Roberts (1982) made this point clear by proposing the use of a critical nutrient range defined as: "that range of nutrient concentration above which we are reasonably confident the crop is amply supplied and below which we are reasonably confident the crop is deficient". Similar suggestions were made earlier by Ulrich (1952).

Various approaches have been used to interpret responses to low NO_3 concentrations. Gardner and Jackson (1976) related NO_3 concentration at selected stages of growth with final grain yield. However, we feel that the concept of a critical concentration should be reserved for the concentrations that maintain maximum growth at each stage of the crop's development.

N-Status from Sap Analysis

Where rapid testing is required, sap analysis has become one of the better approaches for the diagnosis and assessment of N status. Initially it was not unanimously accepted, but was adapted to complement periodic soil analysis and this is still its main function. Williams (1969) showed that the nitrate content of soils (to 90 cm) was related to the sap nitrate contents of young wheat stems, and has used this method to determine the quantity of nitrate available and hence to adjust fertilizer dressings in spring. He found a good relation that was sensitive to small (40 kg N ha^{-1}) dressings. The test was transformed into a practical tool for predicting nitrogen fertilizer requirements by Wehrmann and co-workers (Wollring and Wehrmann, 1981; Wehrmann et al., 1982). They succeeded in making the rapid sap test applicable in two different ways, firstly to determine the N fertilizer level at the booting stage and at ear emergence, and secondly, by frequent sampling, determine the optimum time for N dressings between tillering and ear emergence. Recently, Widdowson and co-workers (Tinker and Widdowson, 1982) working with the original Williams test, have shown that 200 ppm NO_3-N (in sap) is a reasonable limiting value and that the nitrate concentration in stem sap can be a practical guide of nitrogen timing.

Our work on this subject (Verstraeten et al., 1983a,b) started with the Wehrmann test (Fig. 4) but, to overcome some shortcomings, we have also developed a hand-press and included the use of the Merckoquant-strips (Scaife and Bray, 1977) (Table 4). Other sap-tests for nitrate analysis have also given difficulties; extreme sensitivity is only reached by reduction to nitrite and subsequent colour formation.

Table 4. Relation between N supply, yield increase and the occurrence of N-stress during growth.

N-treatment	% Yield increase above 0:0:0	N-Stress S_N (Greenwood)	N-Stress Period	Time when nitrate concentration first fell below 1000 ppm
0:0:0	-	68	April	April
40:0:0	43.7	40	May	early May
40:40:40	81.4	84	June	late May
40:100:60	87.6			

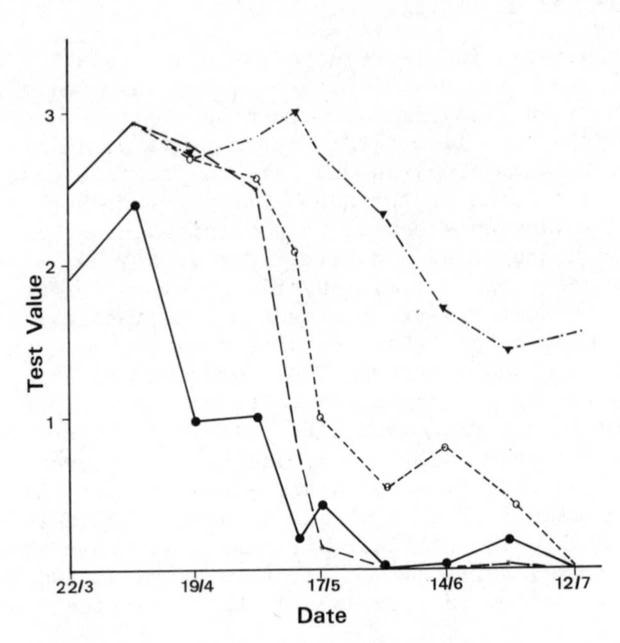

Fig. 4. Fluctuations of the NO_3-concentration (test-value) during growth in relation to N-supply. Symbols as defined in Fig. 3.

FORMS OF NITROGEN IN THE PLANT

To give some more fundamental evidence about the importance of sap concentration in relation to plant growth, we must consider aspects of the biochemistry of nitrogen assimilation (Miflin and Lea, 1977). Most soils have a large nitrifying capacity, so nitrogen uptake by the root is mainly as nitrate. Part of the nitrate influx is reduced in the root but the larger part is transported in the sap stream into the shoot. There it is transformed by the key enzyme, nitrate reductase, before assimilation into plant proteins or translocation to the grains during ripening. The nitrate reductase enzyme assay is an indirect method of measuring the amount of incoming nitrate and has been used before for assessing N nutritional status. In the assay, nitrate reductase activity (NRA) in a leaf (endogenous) is compared to the activity induced by the addition of nitrate to a comparative sample (induced):

$$NAC = \frac{NRA_i}{NRA_e}$$

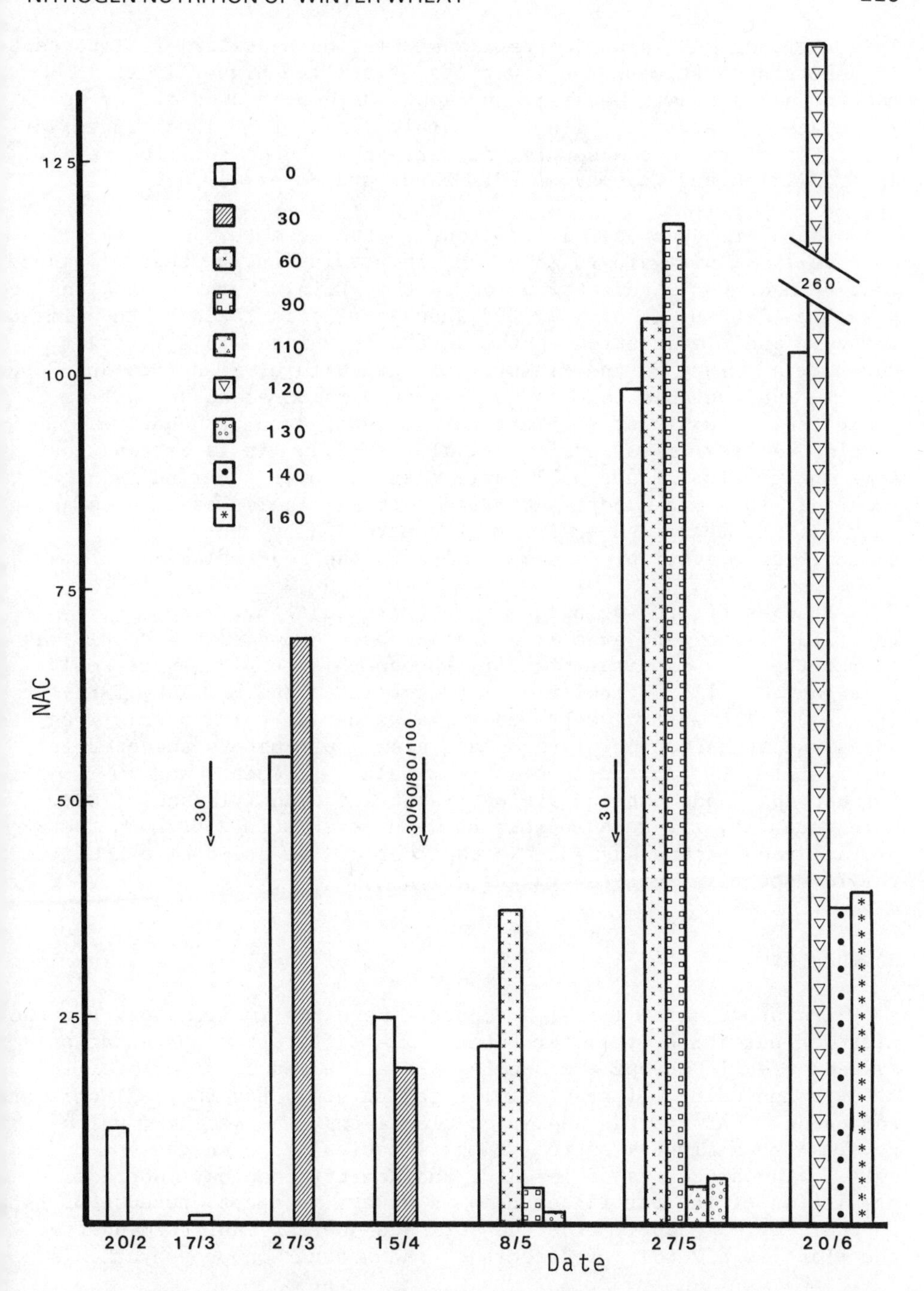

Fig. 5. Effect of N-supply (symbols = total N application to date, kg ha^{-1}) on NAC-values during growth; the lower the value, the higher is the N-status of the crop.

This ratio or NAC-value (nitrate assimilation capacity) is important in assessing N-stress; the lower its value, the higher is the N nutritional status. Several approaches have been used either to correlate its activity with final grain yield or grain nitrogen, or otherwise to reflect N supply and current N status (Dalling et al., 1975; Eilrich and Hageman, 1973; Shaner and Boyer, 1976).

There are practical limitations to the test (Goodman, 1979) especially with regard to cereal yield prediction, because allowance must be made for N redistribution to the grain. Some predictions are possible (Johnson et al., 1976), when account is taken of the time of the year and the duration of the period in which N is limiting. However, a survey by the Ministry of Agriculture, Fisheries and Food (UK) in 1980 which attempted to predict N requirement or optimum yield from an assay at stem extension, reached quite negative conclusions (Sylvester-Bradley et al., 1983a). It is evident that some integration of available soil nitrogen over a period is required. Our own experiments with this particular assay have shown that nitrate reductase activity is controlled by the nitrate concentration available and even more by the NO_3-influx.

Studies of the variation in NAC with cultivar, site, N-fertilizer level and even growth stage have shown that N supply was almost the only important factor, though this supply may be soil nitrogen as well as fertilizer N (Verstraeten and Vlassak, 1981) (Fig. 5). The NAC-value proved to be valuable, particularly when using a combination of nitrogen and growth parameters to determine the N status of plant and soil. NAC-values between 1 and 10 correspond to adequate nutrition (Verstraeten and Vlassak, 1983). More recently, the relationship between measurements of both the nitrate reductase assay and the sap-test of the stems have confirmed their respective critical ranges (Fig. 6).

CONCLUSIONS

The progress of research reported here has implications for the nitrogen nutrition of winter wheat. The different recommendation systems used in European countries are all based on site-specific effects including climate, soil, rotation etc., but they all omit one component. This is the unpredictable factor, the weather, which governs the N mineralization/ immobilization cycle in the soil as well as the processes of leaching and denitrification. Correct prediction of the fertilizer need is always in doubt, because of both unpredictable N supply from the soil and unpredictable demand from the plant as a result of effects of the weather on photosynthesis.

Recent work has shown that site-specificity is real (Hege, 1983), with the weather factor being responsible for year-to-year variations as shown by mineralization studies (Richter and Nordmeyer,

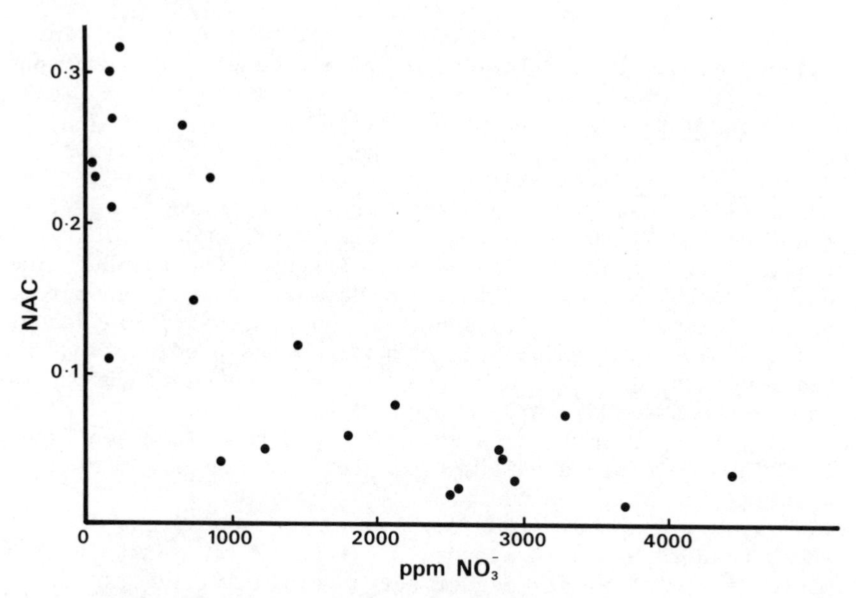

Fig. 6. Relation between the nitrate assimilation capacity (NAC) of the leaf and NO_3-concentration in stem sap.

1983; Hanschmann, 1983). It may be possible to overcome this last problem by careful modelling of all parameters involved, and to guide the development of this approach specific tests for the assessment of N-status are required. Nitrogen stress integrates the rate of nitrogen supply with all other factors essential to growth, and it is clear that under normal conditions of plant growth the availability of nitrogen throughout growth is extremely important. With the emphasis on high yield production each form of deficiency, even the smallest, should be avoided. A number of models recognize the influence of N-stress or N-status on growth rate by including some reduction factor into the simulation model (Seligman et al., 1975; Breteler et al., 1981). However as yet there are no quantitative expressions of the effect on growth processes and the effect is assumed to be of a rather general nature. On the other hand, the large variations found in a number of nutritional tests imply that there are still unknown factors which affect N nutrition and for this problem, in particular, simple and rapid testing may be useful. With current recommendation systems, the N stress concept may lead to improvements in advisory work as well as to development of useful simulation models.

REFERENCES

Austin, R. B., Bingham, J., Blackwell, R. D., Evans, L. T., Ford, M. A., Moyen, C. L., and Taylor, M., 1980, Genetic improvements in winter wheat yields since 1900 and associated physiological changes, J. agric. Sci., Camb., 94:675.

Batey, T., 1976, Some effects of nitrogen fertilizer on winter wheat, J. Sci. Fd Agric., 27:287.

Batey, T., 1977, Prediction by leaf analysis of nitrogen fertilizer required for winter wheat, J. Sci. Fd Agric., 28:275.

Batey, T., 1982, The prediction of fertilizer N requirements from plant growth and composition, in: "Assessment of the Nitrogen Status of the Soils", Research Workshop, Leuven 12-14 January.

Batey, T., 1983, Plant growth and composition as predictors of N requirements, in: "The Nitrogen Requirements of Cereals", MAFF Reference Book 385, HMSO, London.

Bingham, J., 1983, Genetic constraints on progress in wheat breeding, in: "The Yield of Cereals", Proc. Intern. Seminar, Cambridge, Royal Agric. Soc., London.

Blacklow, W. M., and Incoll, L. D., 1982, Nitrogen stress of winter wheat changed and determinants of yield and the distribution of nitrogen and total dry matter during grain filling, Aust. J. Pl. Physiol., 8:191.

Boon, R., 1983, Stikstofadvies via bepaling van de minerale stikstof in de bodem en via het opstellen van een stikstofindex voor wintergranen op leemgronden, Landbouwtijdschrift, 36:555.

Bouma, D., 1970, Effects of nitrogen nutrition on leaf expansion and photosynthesis of Trifolium subterraneum L., Ann. Bot., 34:1131.

Breteler, H., Greenwood, D. G., Petterson, I., Russell, J. S., Sauerbeck, D., van Dorp, F., van Keulen, H., and van der Meer, H. G., 1981, Soil-plant relations, in: "Simulation of Nitrogen Behaviour of Soil-Plant Systems", Pudoc, Wageningen.

Chapman, H. D., 1967, Plant analysis values suggestive of nutrient status of selected crops, in: "Soil Testing and Plant Analysis", Madison, Wis., USA.

Commonwealth Bureau of Soils, 1977, "Leaf Analysis as a Guide to Plant Nutrient Status (1960-1975)", Annotated Bibliography No. SP1839, Commonwealth Bureau, Harpenden.

Cook, R. J., 1983, Pest and disease forecasting and control: status and prospects, in: "The Yield of Cereals", Proc. Intern. Seminar, Cambridge, Royal Agric. Soc., London.

Deutsche Landwitschafts-Gesellschaft, 1981, "Weizen aktuell", DLG-Verlag, Frankfurt am Main.

Dalling, M. J., Halloran, G. M., and Wilson, J. H., 1975, The relation between nitrate reductase activity and grain nitrogen productivity in wheat, Aust. J. agric. Res., 26:1.

Dilz, K., 1968, Stikstofbemesting van granen, 16 Gedeelde stikstofgiften op winter- en zomertarwe, Stikstof, 59:471.

Dilz, K., Darwinkel, A., Boon, R., and Verstraeten, L. M. J., 1982,

Intensive wheat production as related to nitrogen fertilisation, crop protection and soil nitrogen: Experience in the Benelux, Proc. Fertil. Soc., 211:95.

Dow, A. I., and Roberts, S., 1982, Proposal, critical nutrient ranges for crop diagnosis, Agron. J., 74:401.

Eilrich, G. L., and Hageman, R. H., 1973, Nitrate reductase activity and its relationship to accumulation of vegetative and grain nitrogen in wheat, Crop Sci., 13:59.

Elliott, J. G., 1983, Herbicides bring a revolution in crop production, in: "The Yield of Cereals", Royal Agric. Soc., London.

Finck, A., 1963, Bedeutung und Anwendung der Blattanalyse in den Tropen, Landw. Forschung., 16:145.

Gardner, B. R., and Jackson, E. G., 1976, Fertilization, nutrient composition and yield relationships in irrigated spring wheat, Agron. J., 68:75.

Goodman, P. J., 1979, Genetic control of inorganic nitrogen assimilation of crop plants, in: "Nitrogen Assimilation of Plants", E. J. Hewitt and C. V. Cutting, eds., Academic Press, London.

Greenwood, D. J., 1982, Modelling of crop response to nitrogen fertilizer, Phil. Trans. R. Soc., London B, 296:351.

Greenwood, D. J., Cleaver, T. J., Loquens, S. H. M., and Niendorf, K. B., 1977, Relationship between plant weight and growing period for vegetable crops in the United Kingdom, Ann. Bot., 41:977.

Greenwood, E. A. N., 1966, Nitrogen stress in wheat. Its measurement and relation of leaf nitrogen, Pl. Soil, 24:279.

Greenwood, E. A. N., 1976, Nitrogen stress in plants, Adv. Agron., 28:1.

Greenwood, E. A. N., Goodall, D. W., and Titmanis, Z. V., 1965, The measurement of nitrogen deficiency in grass swards, Pl. Soil, 23:97.

Halse, N. J., Greenwood, E. A. N., Lapins, P., and Boundy, C. A. P., 1969, An analysis of the effects of nitrogen deficiency on the growth and yield of Western Australian wheat crop, Aust. J. agric. Res., 20:987.

Hanschmann, A., 1983, Einfluss von temperatur und feuchtigkeit auf die mineralisierung von bodenstickstoff, Arch. Acker - u. Pflanzenbau u. Bodemkd., 27:79.

Hanus, H., 1973, Witterung-Dungung-Ertrag, Mitt. DLG, 88:1410.

Hege, U., 1983, Erfahrungen mit der N_{MIN}-methode in Bayern, Landw. Forschung., 36:82.

Johnson, C. B., Whittington, W. J., and Blackwood, C. C., 1976, Nitrate reductase as a possible predictive test of crop yield, Nature, Lond., 262:133.

Jungk, A., and Wehrmann, J., 1978, Determination of nitrogen fertilizer requirements by plant and soil analysis, in: "Proc. Eight Int. Coll. on Plant Analysis and Fertilizer Problems", Government Printers, Wellington, New Zealand.

Laloux, R., Poelaert, J., and Falisse, A., 1975, Stikstofbemesting bij graangewassen, Landbouwtijdschrift, 28:1155.
Leigh, R. A., Stribley, D. P., and Johnston, A. E., 1982, How should tissue nutrient concentrations be expressed?, in: "Plant Nutrition 1982", Proc. Ninth Inter. Plant Nutrition Coll. Vol. 1, Commonwealth Agric. Bureau, Slough.
Lundegard, H., 1951, "Leaf Analysis", Hilger and Watts, London.
Meynard, J. M., Boiffin, J., and Sebillotte, M., 1982, Provision of nitrogen fertilizer for winter wheat - test of a model, in: "Plant Nutrition 1982", Proc. Ninth Int. Plant Nutrition Coll. Vol. 2, Commonwealth Agric. Bureau, Slough.
Miflin, B. J., and Lea, P. J., 1977, Amino acid metabolism, A. Rev. Pl. Physiol., 28:299.
Needham, P., 1982, The role of nitrogen in wheat production: Response, interaction and prediction of nitrogen requirements in the UK, Proc. Fertil. Soc., 211:127.
O'Neill, E. J., 1980, Assessment of nitrogen status of soils with respect to the growth of cereal crops, Ph.D. Thesis, University of Aberdeen.
Page, M. B., Lister, A., and Talibudeen, O., 1977, The effect of spring-applied nitrogen on the growth and nutrient uptake of winter wheat, Rothamsted Exp. Stn Report for 1978, Part 1, 279.
Papastylianou, I., Graham, R. D., and Puckridge, D. W., 1982, The diagnosis of nitrogen deficiency in wheat by means of a critical nitrate concentration in stem basis, Comm. Soil Sci. Pl. Anal., 13:473.
Papastylianou, I., and Puckridge, D. W., 1981, Nitrogen nutrition of cereals in a short-term rotation, Aust. J. agric. Res., 32:713.
Richter, J., and Nordmeyer, H., 1983, Stickstoffmineralisation und verfugbarkeit in loss-ackerboden, Landw. Forschung., 36:121.
Ris, J., Smilde, K. W., and Wijnen, G., 1981, Nitrogen fertilizer recommendations for arable crops as based on soil analysis, Fertil. Res., 2:21.
Robinson, J. B. D., 1975, The soil nitrogen index and its calibration with crop performance to improve fertilizer efficiency on arable soils, Spec. Publ. Commonwealth Bur. Soils, 1:53.
Scaife, M. A., and Barnes, A., 1977, The relationship between crop yield and petiole nitrate concentration at various growth stages, Pl. Soil, 46:705.
Scaife, M. A., and Bray, B. G., 1977, Quick sap test for improved control of crop nutrient status, ADAS Quart. Rev., 27:137.
Scharpf, H. C., and Wehrmann, J., 1975, Die Bedeutung des Mineralstickstoff-vorrates des Bodens zu Vegetationsbeginn fur die Bemessung der N-Dungung zu Winterweizen, Landw., Forschung., SH 32/I:100.
Seligman, N. G., van Keulen, H., and Goudriaan, J., 1975, An elementary model of nitrogen uptake and redistribution by annual plant species, Oecologia, 21:243.

Shaner, D. L., and Boyer, J. S., 1976, Nitrate reductase activity in maize (Zea mays L.) leaves, Pl. Physiol., 58:499.

Sturm, H., and Effland, H., 1982, Nitrogen fertilisation and its interaction with other cultural measures: Experience in the Federal Republic of Germany, Proc. Fertil. Soc., 211:8.

Sylvester-Bradley, R., Barnard, P. A. J., and Hart, P. F. W., 1983a, An assessment of nitrate reductase activity as a predictor of nitrogen requirement of winter cereals, in: "The Nitrogen Requirement of Cereals", MAFF Reference Book 385, HMSO, London.

Sylvester-Bradley, R., Dampney, P. M. R., and Murray, A. W. A., 1983b, The response of winter wheat to nitrogen, in: "The Nitrogen Requirement of Cereals", MAFF Reference Book 385, HMSO, London.

Syman, G., 1974, "Nitrogen Status in Growing Cereals", The Royal Agricultural College of Sweden, Uppsala.

Tinker, P. B., 1979, Uptake and consumption of soil nitrogen in relation to agronomic practice, in: "Nitrogen Assimilation of Plants", E. J. Hewitt and C. V. Cutting, eds., Academic Press, London.

Tinker, P. B., and Widdowson, F. V., 1982, Maximising wheat yields, and some causes of yield variation, Proc. Fertil. Soc., 211:151.

Ulrich, A., 1952, Physiological basis for assessing the nutritional requirements of plants, A. Rev. Pl. Physiol., 3:207.

Verstraeten, L. M. J., Duyck, M., and Vlassak, K., 1983a, Plantsaptesten en N-behoefte bij wintertarwe, Landbouwtijdschrift, 36:585.

Verstraeten, L. M. J., Stallen, M. van, Duyck, M., and Vlassak, K., 1983b, N-status of winter wheat by means of sap tests. Presentation of a sap press, in: "The Yield of Cereals", Royal Agric. Soc., London.

Verstraeten, L. M. J., and Vlassak, K., 1981, Nitrogen-stress and plant growth in relation to the nitrogen-status of plant and soil, Pedologie, 31:379.

Verstraeten, L. M. J., and Vlassak, K., 1983, Nitrate reductase activity and the assessment of nitrogen status, in: "The Nitrogen Requirement of Cereals", MAFF Reference Book 385, HMSO, London.

Viaux, P., 1980, Fumure azotee des cereales d'hiver, Perspectives Agricoles, 43:10.

Viaux, P., 1983, Nitrogenous fertilization of winter wheat in France, in: "The Nitrogen Requirement of Cereals", MAFF Reference Book 385, HMSO, London.

Vlassak, K., Verstraeten, L. M. J., and Batey, T., 1982, Dry matter production and N-offtake as parameters in the assessment of N-status for winter wheat, in: "Plant Nutrition 1982", Proc. Ninth Inter. Plant Nutrition Coll. Vol. 2, Commonwealth Agric. Bureau, Slough.

Wehrmann, J., Scharpf, H. C., Bohmer, M., and Wollring, J., 1982, Determination of nitrogen fertilizer requirements by nitrate analysis of the soil and the plant, in: "Plant Nutrition

1982", Proc. Ninth Inter. Plant Nutrition Coll. Vol. 2, Commonwealth Agric. Bureau, Slough.

Williams, R. G. B., 1969, The rapid determination of nitrate in crops, soils, drainage and rainwater by a simple field method using diphenylamine or diphenylbenzidine with glass fibre paper Chemy Ind., 1735.

Wollring, J., and Wehrmann, J., 1981, Der nitrat-schnelltest-entscheidungshilfe fur die N-spatdungung, DLG-Mitteilungen, 8:448.

THE ROLE OF CROP SIMULATION MODELS IN WHEAT AGRONOMY

R. A. Fischer

CSIRO, Division of Plant Industry
Canberra, Australia

INTRODUCTION

Agronomy is the science of crop management for greatest benefit. It proceeds from an understanding, albeit imperfect, of the crop, through prediction of better management techniques, to verification of prediction in controlled field experiments, and finally, to demonstration of benefit to farmers. Crop management decisions deal with tactical issues, i.e. questions relevant to a particular crop and season, such as a mid-crop fungicide application. There are also strategic issues applying to long-term management; for example, at the farm level, whether to long fallow or continuously crop or, nationally, how to allocate scarce water resources or to utilize new lands. The resolution of these issues is complicated by the fact that crop responses to fungicides, water or new environments interact with other agronomic factors, and with soil type, genotype and particularly weather.

Crop simulation modelling implies the representation of crop performance by mathematical relationships. It has two important and related purposes: to improve our understanding of the crop, and to predict how a crop will perform in defined but relevant agricultural environments thereby aiding tactical and strategic decision making. It is useful to consider two contrasting types of crop model, the mechanistic or explanatory model, and the descriptive or empirical one (Monteith, 1981; de Wit, 1982). The relevant example of the mechanistic model is the dynamic crop simulation model, based on interconnected physiological and physical relationships which require stochastic environmental inputs and are solved numerous times (hourly or daily) throughout the life of the crop in order to calculate crop development, growth and final yield. This model contrasts with the empirical model which in its simplest form, the regression model, contains a single function relating crop yield to environmental variables, a relationship usually derived from regression analysis o previous yield data.

In theory, dynamic simulation models appear more likely to enhance our understanding of crops than do regression models. Their advantages arise because they allow the integration of knowledge on all relevant processes and responses within an appropriate framework leading to improved understanding and information to direct research Dynamic simulation models should also, in theory, be more powerful predictive tools, permitting the interpolation and even extrapolatio of results in time and space with greater confidence, because the potential for interaction between environmental variables is part of the mechanistic relationships upon which the models are based.

These are the theoretical advantages of dynamic simulation models, but in practice things are somewhat different:

(a) The mechanistic-empirical distinction is not clear cut. Relationships used in dynamic simulation models are often based on empiricisms at the explanatory level, while those used in regression models almost always make some sense physiologically, and often have a priori origins.

(b) Comprehensive dynamic simulation models are usually large and complex, making their validation difficult. This also reduces their acceptability to people other than their creators, and especially to agronomists. Summary models derived from comprehensiv models may be a solution to this problem but their preparation does not challenge model developers (Penning de Vries, 1982).

(c) For many reasons, including (a) and (b), there have arisen independently a whole host of intermediate models, which are hybrids of the approaches represented by the mechanistic and empirical models. Such models often appear to succeed because the target environments are restricted, often to situations where single environmental factors limit and control crop performance (e.g. water temperature, nitrogen).

Therefore, while recognizing that fully-fledged dynamic simulation models of the wheat crop might provide the ultimate modelling tool for agronomists, I will not exclude empirical models from discussion of the role of modelling in agronomy. The important issues are whether modelling has led to better understanding of crop function, and of existing results and direction in basic and applied research, and secondly, whether it has led to improved crop management. I shall discuss examples concerning time of sowing, tillage, irrigation and chemical control, before attempting to draw general conclusions. Fertilizer application, a most important agronomic question, is dealt with elsewhere (see Chapter 29).

DATE OF SOWING

Many environmental variables can change with sowing date so that understanding the outcome for yield is complex. Often, however, it is helpful to consider three issues: optimum flowering date, balance between pre- and post-flowering growth, and constraints on possible sowing dates. Genotype, which controls in particular the sowing-to-anthesis duration, is inevitably another factor.

In one environment (irrigated high fertility conditions of Ciudad Obregon in north-west Mexico, lat. 27°N), a simple deterministic model was used to explain quite variable effects of sowing date on wheat yield over five years of experiments (Fig. 1).

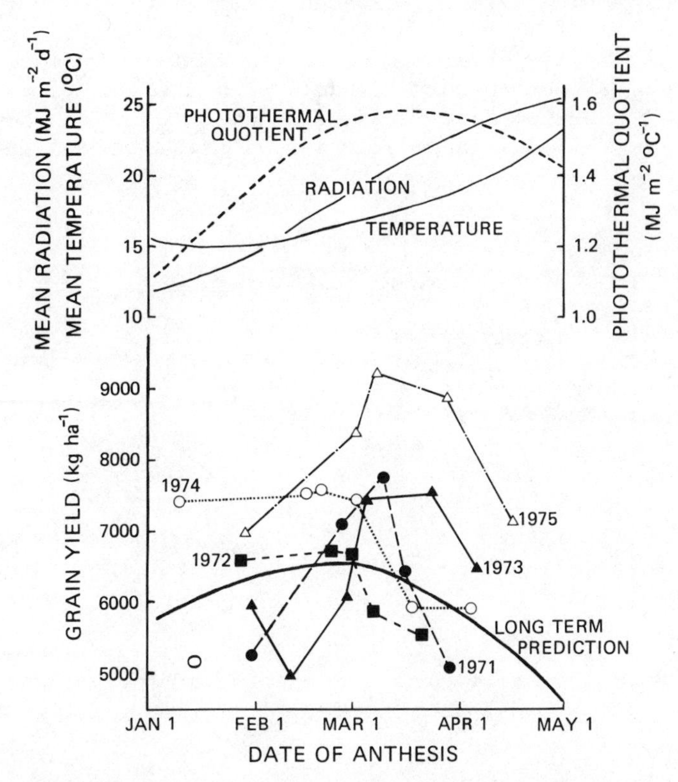

Fig. 1. Actual grain yield as a function of anthesis date for irrigated wheat (cv. Yecora 70) over 5 years at Ciudad Obregon, Mexico (lat. 27°N), and yield predicted from long-term climatic means (R. A. Fischer, unpublished data).

For crops of a single variety, variation in grain number m^{-2} was closely related to the photothermal quotient (mean solar radiation divided by mean temperature less 4.5°C) in the 20 days preceding anthesis (r = 0.83) during which all the crops had full ground cover. Variation in grain weight for the same variety was closely related to mean temperature in the 40 days following anthesis (r = -0.88). The product of these two functions was used to calculate a predicted yield which fitted actual yields reasonably closely (r = 0.73). The product function was also used to predict long-term mean yield as a function of flowering date, using long-ter radiation and temperature means for Ciudad Obregon (Fig. 1). Optimu flowering date was predicted to be early March; this is a compromise between the opposing effects of rising solar radiation and rising temperature. Optimum sowing date follows simply from optimum flowering date at Ciudad Obregon, because even the fastest developin cultivars (latest sowing date) appear to give full ground cover before the onset of the critical pre-anthesis period and drought doe not constrain sowing date there because irrigation is available.

The work of Stapper et al. (1983) is a unique example of the application of a comprehensive dynamic simulation model to obtain guidance on the sowing date in a region where there had been little prior experimentation. They used Stapper's wheat model which is

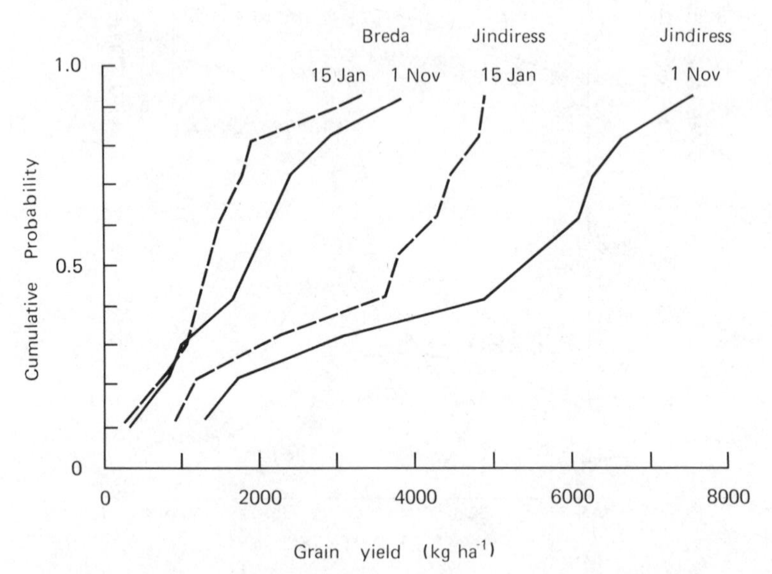

Fig. 2. Cumulative yield probability functions for Breda (280 mm rainfall) and Jindiress (470 mm) in north-west Syria (lat. 36°N) for the earliest (1 November) and latest (15 January) sowing dates simulated over 20 years by Stapper et al. (1983).

partly based on the CERES model (see Chapter 28) and which has been extensively validated under dryland and irrigated conditions, giving reasonably accurate crop development and yield predictions. The sowing date simulations were done for a medium maturity genotype at a dry (Breda, 280 mm annual rainfall) and wet (Jindiress, 470 mm) site in the mediterranean zone of north-west Syria (lat. 36°N). Twenty years of continuous cropping were simulated using historical weather records thereby giving stochastic or probabilistic results. These are expressed as cumulative frequency distributions of yield, in order to illustrate the riskiness of various strategies (Fig. 2). Intermediate sowing dates generally fall in the expected order between the dates shown. Median yield fell by 40 and 122 kg ha^{-1} for each week's delay in sowing at Breda and Jindiress, respectively. Experimental results from Wagga Wagga in eastern Australia (lat. 35°S) indicate a decline in mean wheat yield with each week's delay in sowing date (146 kg ha^{-1}) similar to that in Jindiress (Kohn and Storrier, 1970). Lack of moisture precludes earlier sowing in Syria, while the risk of spring frost damage precludes it in eastern Australia. Rising spring temperatures and greater dryness are believed to cause the steep yield decline with later sowing in Australia, but the study of Stapper et al. (1983) suggests that reduced growth before anthesis and a greater proportion of water loss by evaporation rather than transpiration also contribute to the decline, at least in Syria.

Figure 2 indicates other perhaps unexpected aspects of yield response to sowing date. It is clear that the advantage from early sowing is greater in the wetter (= higher yielding) years and at the wetter site. The former result arises partly because in drier years of lower yield, early sowings do not necessarily germinate upon sowing but must wait until rains commence, and partly because in general the potential for response to sowing date (or any other factor) is less when yields are less. Lower yield potential also probably explains the lower response to sowing date at the drier site. Nevertheless it is of some value to the agronomist to see that the simulated yield expectation from earlier sowing was always superior to that from late sowing.

Optimum sowing date therefore appears to depend on constraints affecting flowering date (late frost, high temperatures) and constraints on sowing date (soil moisture, temperature, trafficability, time, etc.). If some of the constraints are lifted then optimum sowing date may change. For example, one recent advance in wheat agronomy has been an increase in the probability of being about to sow when desired, leading to somewhat earlier sowing than in the past. This has come about through better moisture conservation techniques, better drilling machinery (e.g. deep furrow drills), faster sowing and better herbicides. These changes arose without the benefit of modelling, but the value of future changes, such as those which might arise if the resistance of wheat to late spring frosting could be improved genetically or if new reduced tillage techniques facilitate double cropping, could be anticipated through modelling.

TILLAGE

Much wheat is grown under semi-arid conditions and a major agronomic question in the wheat farming system is whether wheat should be grown after a long period of fallow that precludes the growing of wheat or alternative crops in the previous year. The alternation of wheat with long fallow from harvest to sowing (13-15 months for autumn-sown winter wheat, 19-20 months for spring-sown wheat) is a common production system in North America, USSR and Turkey. In Mediterranean regions and in Australia, long fallow tend to commence in early spring and last only nine months through to the autumn sowing.

In semi-arid regions the primary purpose of a fallow period is to conserve moisture. The comparison of wheat-fallow with continuous wheat has been an obvious target of simulation models, given the dominant role that erratic or stochastic precipitation input is likely to play. One example is the study by Baier (1971), which focusses on the drier parts of the Canadian prairie (Swift Current, mean annual precipitation 350 mm, lat. 50°N). A soil moisture budget was calculated daily for the fallow and crop (spring wheat) conditions, and run continuously for 20 years' weather data. Wheat yield was calculated as the sum of simple linear functions of available water at sowing (8.4 kg ha^{-1} mm^{-1}) and rainfall from sowing to harvest (8.0 kg ha^{-1} mm^{-1}); this was applied to crops after fallow as well as those following wheat. The difference between the two strategies is summarized by the difference in predicted average available soil moisture at sowing (107 mm cf. 67 mm) for fallow wheat and continous wheat respectively, leading to predicted average yields of 1204 and 854 kg ha^{-1}, respectively. Considering a yield of less than 336 kg ha^{-1} a failure, continuous wheat failed two years in ten while wheat after fallow only failed one year in ten (Fig. 3a). However, the most interesting results were the probable net returns per unit area of arable land, a calculation that takes into account the doubling of cropping frequency and the higher production costs per unit area of land with continuous wheat (Fig. 3b). For continous wheat at Swift Current, we appear to have a good example of an agronomic strategy which has a higher average return (+39%) but also a higher risk (greater chance of negative returns) than the alternative strategy represented by fallow wheat.

As one moves from the Canadian prairie south to the Great Plains and west to the Pacific North-West, it appears that continuous wheat yields fall to less than half those of fallow wheat and the risk of failure steadily increases (Smika, 1970; Bolton, 1981). These are

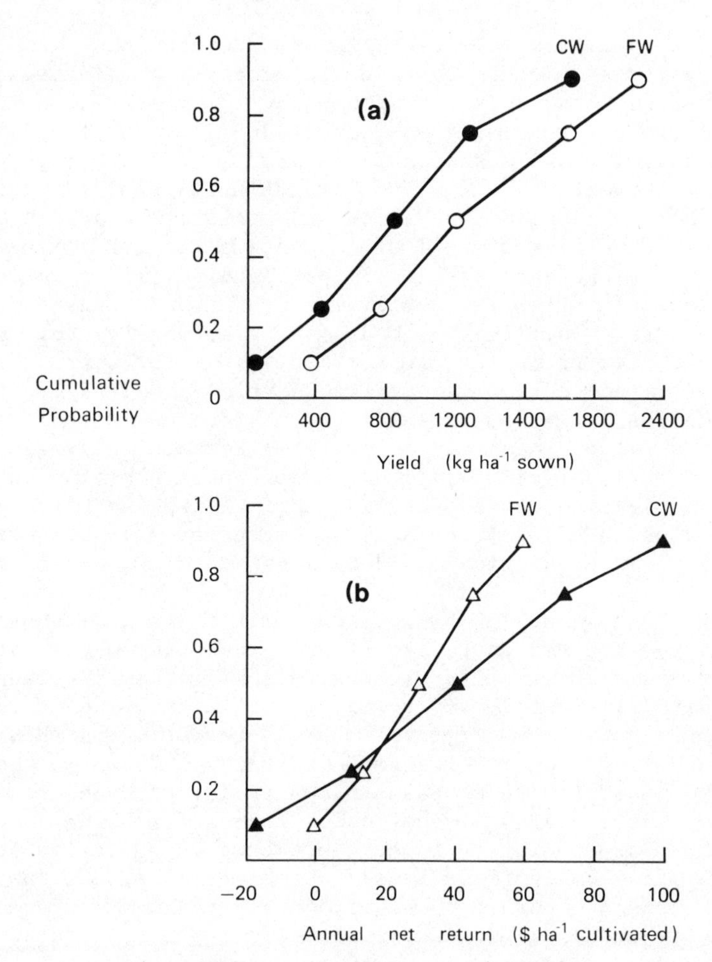

Fig. 3. Cumulative yield probability functions (a) and net returns per arable hectare (b) for fallow (FW) and continuous (CW) wheat at Swift Current, Canada (350 mm rainfall, lat. 50°N) as simulated by Baier (1971).

reliable results since they are derived from tillage experiments running for several decades at various locations. Modelling hardly seems necessary, but the difference between spring wheat in the Canadian prairie and winter wheat in the Great Plains raises an interesting question. It appears to be related to the different zero-yield intercepts in the simple linear yield *versus* water supply relationship applying in each situation; about 100 mm is needed to get a crop in Canada (Baier, 1971) but more than 200 mm is needed in the Great Plains (Greb, 1979).

One cannot leave the Great Plains without noting also that ther
has been a remarkable increase in the water storage efficiency of
fallow over the last 40 years, from only 20% of fallow precipitation
with traditional dust mulch methods to 30-35% with stubble mulching
(Greb et al., 1979). More fertilizer and semi-dwarf cultivars,
combined with better seeding and harvesting machinery, have more tha
doubled yields in the Great Plains and Pacific North-West over this
period while the risk of soil erosion has actually declined (Greb,
1979). This represents an example of positive factor interaction,
and is one of the greatest achievements of modern wheat agronomy.
Needless to say modelling has had nothing to do with this progress.
However, in view of further improvements in water storage efficiency
(up to 50%) expected from no-tillage herbicide fallow (Greb et al.,
1979), the possibiliy that yield returns from additional water
storage for fallow wheat may be smaller (Smika, 1979) and the risk o
aggravated dryland salinity, it would be an interesting modelling
exercise to see whether these changes might tip the balance in favou
of continuous wheat, particularly in wetter areas. The consequences
for world wheat production could be substantial.

In dryland wheat regions at lower latitudes like southern
Australia and the Mediterranean region, where the major rainfall is
in winter and summers are dry, long fallowing has also been the
traditional practice. However, in Australia at least, this practice
is no longer very common for a number of reasons, and this change of
practice is supported by experiment (French, 1978). In the
Queensland wheat region of eastern Australia, at even lower
latitudes, summer rainfall dominates. Short fallowing during the si
month period from harvest of one wheat crop to sowing the next is
commonly sufficient for continuous wheat cropping, and temperatures
are high enough to permit double cropping when and where moisture
supplies are better (this is also true in Argentina and southern
USA). In the absence of long-term experimental data, simulation
models based on soil water balances have proved useful in this regi
of especially erratic rainfall to compare alternative cropping
strategies (Berndt and White, 1976) or to explore the cropping
potential of virgin lands on the drier fringe of the wheat zone
(Hammer and McKeon, 1983).

Queensland is the region of the original and well known work of
Nix and Fitzpatrick (1969); their method of relating yield to water
supply index at anthesis forms the basis of yield prediction in the
comparison of cropping strategies by Berndt and White (1976). These
authors used a continuous weekly soil water budget over a 30 year
period to compare continuous wheat against an opportunistic strategy
under which wheat (winter crop) and sorghum (summer crop) were sown
whenever three soil moisture criteria were met within a defined and
feasible sowing interval for each crop. These criteria comprised
current rainfall within germination and trafficability limits, plus
for the opportunistic strategy only, total stored soil moisture in

Table 1. Results of a simulation-based comparison of two cropping systems (continuous wheat cf. opportunistic double cropping) over 30 years at two locations in southern Queensland (Berndt and White, 1976).
Key: W = Wheat, S = Sorghum

Cropping strategy over 30 years	No. of crops		Mean yield (kg ha^{-1} sown)		Mean annual run off	Mean gross margin	Years with negative gross margin
	W	S	W	S	(mm)	(\$ ha^{-1} yr^{-1})	
Cambooya, lat. 28°S; mean annual rainfall, 672 mm							
W	28	-	1245	-	197	49	2
W/S	24	23	1035	2418	133	87	2
Roma, lat. 27°S; mean annual rainfall, 519 mm							
W	26	-	886	-	89	25	9
W/S	20	15	737	1065	70	23	13

excess of a minimum value (20% of maximum or 32 mm proved approximately optimum) and a compulsory minimum two week period between harvest and sowing when double cropping. Run off, important in this region of high intensity summer storms, was also estimated using a locally-fitted function depending on soil water storage and rainfall amount. Table 1 gives the results for two locations at the wet and dry edges of the southern Queensland wheat belt. The opportunistic cropping strategy gave more crops and less run off at each location, but only greater average returns and less financial risk at the wetter location.

The advent of broad-spectrum herbicides has added a new option to tillage, namely zero-tillage or direct-drilling, where green growth is killed within one or two weeks of seeding. In semi-arid south-eastern Australia this strategy must stand up against the traditional short (2-5 months) cultivated fallow, commencing whenever substantial rain falls after harvest in the late summer or autumn. With direct drilling, pasture weeds and/or volunteer wheat, which grow with the onset of such rain, provide grazing at some cost in terms of reduced total stored moisture. The chances of adequate moisture in the seeding zone are lower and there is hence greater dependence on current rainfall at sowing time and greater possibility of delayed sowing (despite improved trafficability when, rarely, it is excessively wet). In addition it is widely observed that direct drilled crops show reduced early vigour, though whether this result in lower or higher final yields depends on the extent of water stre in the spring. These views are largely my own, derived from 5 year field experimentation in semi-arid southern New South Wales in the 400 to 600 mm rainfall zone with these alternative tillage strategies. Clearly this is not long enough to assess experimental even these relatively simple issues and we have an obvious situatio in which a simulation model could help.

Not only is it a strategic question of whether to direct drill but also there is the possibility of a tactical decision, whether t direct drill in a particular season depending on how the rains have come in the autumn, what the pasture supply is for sheep, etc. As start, we have attempted to model soil water for the direct drill (= weeds) and traditional (= no weeds) strategies in the harvest-to sowing period, making the assumption that wheat follows wheat and that all plant extractable water in the soil profile has been used at harvest (1 December). It is also assumed that, for the direct drilling option, weeds which germinate following post-harvest rain

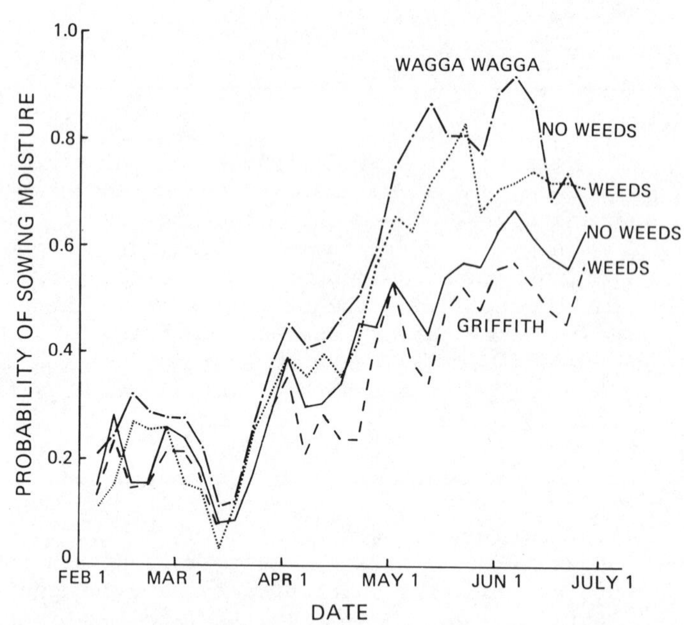

Fig. 4. Probability of topsoil moisture being adequate for sowing derived from a 17 year simulation of soil moisture in the presence (direct-drilling) and absence (traditional practice) of weeds in the period from harvest (1 December) until next sowing. Wagga Wagga (upper pair of lines) has a annual rainfall of 500 mm and Griffith (lower lines) 375 mm both locations are in southern New South Wales, Australia.

behave like the CERES wheat crop at 200 plants m^{-2}, and that a sowing day for the real wheat crop occurs whenever there is more than 10 mm extractable soil moisture in the top 40 cm (or after 15 April, more than 5 mm in the top 20 cm) for 5 consecutive days. Results of a 17 year simulation at two sites suggest less drying with weeds than was observed in the 5 years of experiments and a smaller penalty in lost sowing days with direct drilling than was expected (Fig. 4). It appears that the wheat model does not simulate summer weeds well, and more research in this area is needed.

I have been emphasizing tillage models whose effects are dominated by the consequences of tillage for water supply to the crop. Tillage and fallowing can affect the crop in many other ways, probably the most important of which are via nitrogen nutrition, in-crop weed populations, soil-borne disease incidence and soil physical properties (e.g. the early vigour problem with direct drilling). Some progress has been made with process modelling in this last mentioned area (American Society of Agronomy, 1982), but ultimately all these aspects will need to be considered if we are to have a complete understanding and make accurate predictions of wheat performance. In the meantime, agronomists must interpret the results of tillage models which consider only soil water, in the light of experimental and farmer experience with these other issues.

WATER MANAGEMENT

The major tactical question which has occupied irrigation agronomists over the years is that of irrigation timing. Theory and experiment appear to confirm that, for maximum yield, wheat must be irrigated before transpiration rate falls below potential, except possibly in early stages of development and in lodging-prone situations. Many techniques have been used in applying this understanding, but one which is common now is the soil water budget, driven by information on potential evaporation, rainfall, crop factors and the amount of extractable soil water, and this is probably the best example of a model, albeit a very simple one, having widespread acceptance in wheat management. With ready access to computers, people can now employ more sophisticated irrigation timing models, which can allow for the stage of development or amount of crop growth or for the short-term weather forecast, and which can calculate the quantity of water to apply to refill the root zone at any time. On waterlogging or leaching-prone soils, or with fully controlled sprinkler systems, such models are especially appropriate.

A common strategic question concerns the area to crop where water is limiting or expensive to apply. This becomes especially complicated where rainfall provides a significant but variable proportion of the crop water requirement. Israeli scientists have been among the first to apply simulation modelling to these questions

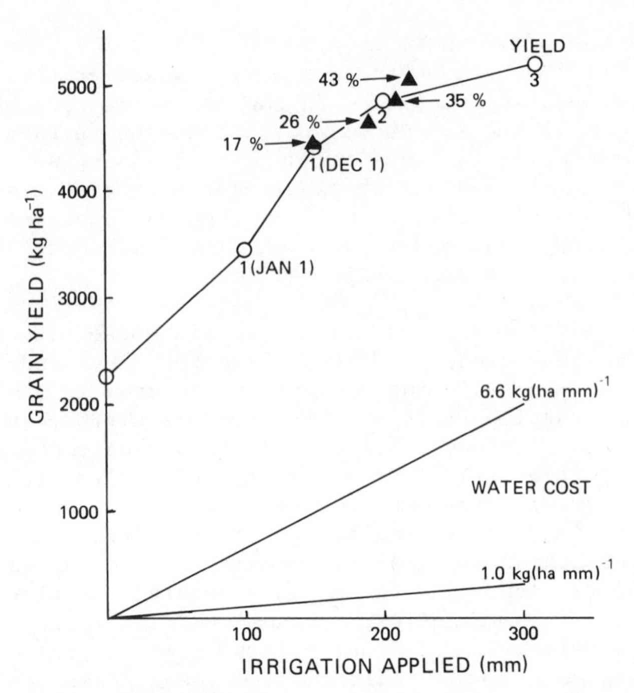

Fig. 5. Mean grain yield of wheat (cv. Florence x Aurore) versus water given as supplemental irrigation following various strategies, as simulated over 16 years for the northern Negev, Israel (200 mm rainfall, lat. 31°N) by Yaron et al. (1974).

(Yaron et al., 1974). Fixed and flexible supplemental spray irrigation strategies were compared for wheat in the northern Negev (lat. 31°N) where the annual average rainfall was 200 mm. Starting with a locally-calibrated daily soil water budgeting procedure, non-stress days (available soil water ⩾ 43%) were calculated. The grain yield of wheat (cv. Florence x Aurore) in irrigation experiments was related to a Mitscherlich function of the number of non-stress days and of germination date ($r^2 = 0.91$). This function was used with 16 years' weather data, to compare nil, one, two or three irrigations (optimally timed) or irrigation when modelled soil water fell below 17, 26, 35 or 43% available water. Figure 5 shows mean results for some strategies in terms of yield and supplemental water; net return is given by the distance between the yield curve and the water cost line, and for expensive water (6.5 kg (ha mm)$^{-1}$) was greatest (\$278 ha^{-1}) for irrigation at the 43% available water limit. This return exceeded that from the nil irrigation and best one, two or three irrigation strategies by 132, 14, 4 and 21 \$ ha^{-1}, respectively. The stochastic element in this study is revealed by a coefficient of variation of net return of 20-30% for supplemental irrigation with expensive water, compared to 82% for nil irrigation.

The relationship between yield and water supplied (Fig. 5) and its variability around the mean are also fundamental to the question of the area to sow when water supply is limited, and to broader planning isues such as the allocation of scarce water amongst different crops on a given farm or given irrigation scheme, or amongst irrigation schemes. Whether it is drawn from experience or experiment or more likely nowadays calculated from a model, this agronomic relationship probably influences investment more than any other in agriculture.

DISEASE AND PEST MANAGEMENT

In humid and irrigated areas where there is limited genetic resistance to disease or where such resistance has suddenly broken down, recourse is frequently made to chemical application. Recommendations from commercial firms and extension agencies often tend towards excessive usage thereby encouraging the evolution of resistance in the pest or pathogen and leading to unnecessary expense for the farmer and pollution of the environment. Chemical application guided by a computer model simulating disease and its effect on the crop is one solution which is gaining widespread attention.

The disease-crop loss model must be able to predict both the progress of the disease epidemic and its consequences for crop yield. The initial level of inoculum, crop stage and structure, and especially weather are vital inputs to model the epidemic. One important factor for the germination of spores of many pathogens, the duration of leaf wetness, may well be easier to model than to measure. Understanding crop loss arising from a given level of disease is not so easy. The simplest models use linear relationships of percentage loss to disease severity at a given stage of development or to severity integrated over different stages and/or weighted by stages (Teng and Gaunt, 1980). However, where a disease affects leaf area only in the grain-filling period (e.g. leaf rust), the effect on yield will obviously depend on the source-sink relations in the particular situation. Where a disease affects water use (e.g. stem rust or root rot), yield loss *versus* severity might be expected to interact with soil water supply and atmospheric demand. Comprehensive simulation models combining disease development and effects on crop growth are now being developed (Rabbinge, 1982).

Despite these complexities, at least one disease management model is used for commercial wheat crops, namely EPIPRE in The Netherlands (Zadoks, 1981). This model is a summary of a more comprehensive crop-disease model. Five diseases are currently handled: stripe and leaf rust (*Puccinia* spp.), leaf blotch and glume blotch (*Septoria* spp.) and powdery mildew (*Erysiphe graminis*). The participating farmers are taught to make their own observations on the incidence of these diseases. This data is fed into a central computer and recommendations are produced which optimize financial returns. Four per cent of Dutch wheat was covered in 1980 and was sprayed on average 0.8 times compared to the national average of 1.5 times, leading to a considerable saving according to the computer model. EPIPRE provides an important precedent in model application to wheat agronomy.

The wheat crop has a number of insect pests (aphids, hessian fly, stem sawfly, suni bug, cereal leaf beetle) and in some cases, especially for aphids, large-scale chemical control is employed. Pest-management models would seem appropriate here and EPIPRE does consider aphid control along with disease control.

Herbicides remain however a major agrochemical input in wheat growing. Despite the costs involved, tactical decision-making in herbicide use does not seem to have taken advantage of crop models. One could envisage such a model helping a farmer decide whether it was worthwhile to control a given population of a certain weed given the current growth and potential crop yield. A model, taking into account the wheat's stage of development, and the soil moisture and temperature conditions, would be an especially suitable aid to timing herbicide application.

GENERAL DISCUSSION

In the past, some researchers have seriously doubted the relevance of crop simulation models in agronomy (Passioura, 1973), while many were naively optimistic. Today a more balanced view probably prevails. There seems little doubt that modelling can help in understanding crop responses as illustrated in my two sowing date examples (cf. Figs 1 and 2). Understanding is advanced largely because the complexity of the crop system defies intuition or "back of the envelope" calculations; part of this complexity for the agronomist derives not from the crop itself but rather from the variability of the environment in which it must be assessed (especially weather but also soil). It seems we have reached the stage where the more basic agronomic researchers would benefit considerably from access to crop modellers (not to mention the benefit modellers would derive from access to experimenters). The former access is probably best achieved by having a "tame" physiologist-modeller working within major agronomic research teams: this ensures realistic and clear-cut objectives for the modelling.

But what about modelling to assist crop management decisions? There still appear to be few good examples of crop model application at the farm level: EPIPRE is one exception. In general there appears to be a major credibility gap between modelling and practical decision-making. As Anderson (1974) has pointed out, the chance of the conclusions from simulations crossing the threshold of relevance depends on the acceptability of the concepts in the model and of its accuracy, and on the presentation of the model and results. Potential users at the farm level are discouraged by the simplifying assumptions of models, the lack of real-world model testing, or the inevitable inaccuracies which arise if such testing is done. They are also put off by the naivety, opportunism and even arrogance of modellers, and the newness of their technique.

However, I believe this must change because crop modelling provides unique advantages in many situations. Firstly, it is often the only way of obtaining a quick result required, for example, when a new technology arises such as a new cultivar or agricultural chemical, or when new lands are to be developed; speed here counter-balances, to some extent, inaccuracy. Secondly, modelling seems to be the easiest way to account for weather stochasticity, an essential consideration in management decisions. The alternative is expensive long-term experimentation; the more the practice interacts with weather, the longer the term. For example, in the study of Smika (1970) at Nebraska over 27 years, the coefficient of variation of the difference between the yield of fertilized fallow wheat and that of twice fertilized continuous wheat was 160%. Even given the large mean difference (940 kg ha^{-1}), many years' experimentation (12 for $P \leq 0.05$) are required to prove that the difference is greater than zero. Similarly, many years' experimentation are needed in dryland situations to prove that one practice is less risky than another. Thirdly, modelling seems uniquely suited to the extrapolation of results from experiments, necessarily controlled in nature and restricted in location, to farms where climatic resources and especially edaphic conditions may differ.

These are still theoretical advantages however. It is useful to try to list agronomic situations most likely to be receptive to future modelling initiatives:

(i) Where there are no experimental results and it is difficult, slow, expensive and/or impossible to obtain them. A good example here is estimating the likely yield response to extra rainfall through cloud seeding, or to agronomic factors consequent upon increasing global CO_2 levels. Most strategic issues for dryland wheat such as fallowing _versus_ continuous wheat, or sowing date are very expensive and slow to resolve experimentally, yet must be continually revised as technology and prices change.

(ii) Where the biology, physics and/or soil chemistry of the situation is reasonably simple and is well understood at the explanatory level, but nevertheless sufficiently complex and interactive to preclude facile generalization. For example, a wheat crop model could help determine the best sowing and fertilizer rate of dryland wheat for crops grown largely on a variable amount of soi moisture stored at sowing (as in Queensland and the Indian subcontinent). On the other hand, it appears to be straightforward to determine sowing and fertilizer rates for irrigated wheat by generalization from specific experiments.

(iii) Where short-term predictions are required and especially when measurements on the actual growing crop to be managed provide the starting point of the model, as is the case for many tactical in crop decisions. Examples include early crop grazing, supplemental fertilizer application, herbicide, fungicide and insecticide application, and salvage grazing in droughts.

(iv) Where crop responses interact strongly with weather and soil characteristics in a manner which is well understood. Most commonly for wheat, this interaction arises through changes in the water supply to the crop. Strategic and tactical decisions on tillage, sowing date and cultivar, and fertilization are good examples, and risk analysis via modelled responses would be especially useful here. Should better monthly and seasonal weather forecasts become available, then the value of crop modelling in tactical decision making will be greatly enhanced.

(v) An additional advantage of guiding decision-making with a computer-based crop model is the fact that the computer link provide an efficient means of transferring other more mundane but no less important information to the decision maker, e.g. to update the user on new data about agricultural chemicals including their sensitivities and resistances, compatibilities and regulations. The system can also be programmed to prompt the farmer for information a key stages of crop development. This results in more skilful and observant farmers (Zadoks, 1981).

Finally, in the context of decision-making, I wish to comment o the importance of biological accuracy in modelling, and of research organization for relevant modelling. Whilst biologists and most basic researchers are concerned about achieving a high degree of predictive accuracy in models, they should not ignore the fact that in agronomy considerable doubts and errors arise from issues outside the scope of the particular crop model. Some variability will arise because of extraneous factors in farmers fields, such as weeds, disease and inadequate nutrition (where these are not themselves being modelled), but by far the greatest source of uncertainty in a prediction is the future weather. In addition, the variation in input and output prices, in the farmer's management skills and in his aversion to risk are all issues of particular importance for small farmers in developing countries. Convention and rules of thumb suggest that the yield responses that farmers can expect should be adjusted downwards relative to experiments by up to 50% due to management inadequacies, and a 20% risk premium should be added to the cost of inputs when determining the predicted yield response (Perrin et al., 1976). Such considerations are poorly quantified at present, and this would appear to reduce the need for great accuracy on the biological side of the crop model. At the same time, the need to consider the influence of decisions on the expected minimum level of returns over a number of seasons (Perrin et al., 1976), reinforces the likely benefit to be expected from a stochastic crop modelling approach.

Agronomic research is very much a regional activity, while crop modelling has tended to be centralized or even academic. Modelling does not have the sense of mission or even urgency that is found in regional agronomy. If agronomists are not to be rushing in where wise men fear to tread, as de Wit (1982) complains, or simulation modelling is not to continue to create false hopes (Monteith, 1981), then there is a need for closer association between agronomists and modellers. This can now be best achieved by having modellers and modelling activity in regional research centres. Modelling there should concentrate on agronomic problems and, by having regional targets, would then be more likely to achieve accurate and useful predictions, even if this might occur at the expense of generality and in the face of incomplete understanding of underlying processes.

ACKNOWLEDGEMENTS

I would like to thank Mr. J. Armstrong for the computer simulations relating to direct-drilling, and Ms. Y. Stockman for preparation of the figures.

REFERENCES

Anderson, J. R., 1974, Simulation: methodology and application in agricultural economics, Rev. Mktg agric. Econ. Sydney, 42:3.
American Society of Agronomy, 1982, Predicting tillage effects on soil physical properties and processes. Special Publication Number 44, ASA Madison.
Baier, W., 1971, An agroclimatic probability study of the economics of fallow-seeded and continuous spring wheat in southern Saskatchewan, Agric. Meteorol., 9:305.
Berndt, R. D., and White, B. J., 1976, A simulation-based evaluation of three cropping systems on cracking-clay soils in a summer-rainfall environment, Agric. Meteorol., 16:211.
Bolton, F. E., 1981, Optimizing the use of water and nitrogen through

soil and crop management, in: "Soil Water and Nitrogen in Mediterranean-type Environments", J. L. Monteith and C. Webb, eds., Nijhoff/Junk, The Hague.

de Wit, C. T., 1982, Simulation of living systems, in: "Simulation of Plant Growth and Crop Production", F. W. T. Penning de Vries and H. H. van Laar, eds., Pudoc, Wageningen.

French, R. J., 1978, The effect of fallowing on the yield of wheat. II. The effect on grain yield, Aust. J. agric. Res., 29:669.

Greb, B. W., 1979, Technology and wheat yields in the central Great Plains. Commercial advances, J. Soil Wat. Conserv., 34:269.

Greb, B. W., Smika, D. E., and Walsh, J. E., 1979, Technology and wheat yields in the central Great Plains, J. Soil Wat. Conserv., 34:264.

Hammer, G. L., and McKeon, G. M., 1983, Evaluating the effect of climatic variability on management of dryland agricultural systems in north-eastern Australia, in: "Need for Climatic and Hydrologic Data in Agriculture of Southeast Asia", UN Univ. Workshop, 12-15 December 1983, Canberra.

Kohn, G. D., and Storrier, R. R., 1970, Time of sowing and wheat production in southern New South Wales, Aust. J. exp. Agric. Anim. Husb., 10:604.

Monteith, J. L., 1981, Epilogue: themes and variations, in: "Soil Water and Nitrogen in Mediterranean-type Environments", J. L. Monteith and C. Webb, eds., Nijhoff/Junk, The Hague.

Nix, H. A., and Fitzpatrick, E. A., 1969, An index of crop water stress related to wheat and grain sorghum yields, Agric. Meteorol., 6:321.

Passioura, J. B., 1973, Sense and nonsense in crop simulation, J. Aust. Inst. agric. Sci., 39:181.

Penning de Vries, F. W. T., 1982, Systems analysis and models of crop growth, in: "Simulation of Plant Growth and Crop Production", F. W. T. Penning de Vries and H. H. van Laar, eds., Pudoc, Wageningen.

Perrin, R. K., Winklemann, D. L., Moscardi, E. R., and Anderson, J. R., 1976, From agronomic data to farmer recommendations. Information Bulletin 27, International Maize and Wheat Improvement Centre, Mexico.

Rabbinge, R., 1982, Pests, diseases and crop production, in: "Simulation of Plant Growth and Crop Production", F. W. T. Penning de Vries and H. H. van Laar, eds., Pudoc, Wageningen.

Smika, D. E., 1970, Summer fallow for dryland winter wheat in the semi-arid Great Plains, Agron. J., 62:15.

Stapper, M., Harris, H. C., and Smith, R. C. G., 1983, Risk analysis of wheat yields in relation to cultivar maturity type and climatic variability in semi-arid areas, using a crop growth model, in: "Need for Climatic and Hydrologic Data in Agriculture of Southeast Asia", UN Univ. Workshop, 12-15 December 1983, Canberra.

Teng, P. S., and Gaunt, R. E., 1980, Modelled systems of disease and yield loss in cereals, Agric. Syst., 6:131.

Yaron, D., Strateener, G., Shimshi, D., and Welsbrod, M., 1974, Wheat response to soil moisture and the optimal irrigation policy under conditions of unstable rainfall, Water Resour. Res., 9:1145.

Zadoks, J. C., 1981, EPIPRE: a disease and pest management system for winter wheat developed in The Netherlands, EPPO Bull., 11:365.

CROP PHYSIOLOGICAL STUDIES IN RELATION TO MATHEMATICAL MODELS

P. V. Biscoe* and V. B. A. Willington

Broom's Barn Experimental Station
Bury St. Edmunds, U.K.

INTRODUCTION

Over recent years, the yields of winter wheat in the United Kingdom have been consistently increasing and in 1983 the national average yield was estimated as 6.2 t ha^{-1} (HGCA, 1984). At the same time, specialist wheat producers in England record average farm yields of about 9 t ha^{-1} with individual crops yielding more than 10 t ha^{-1} of grain at 16% moisture content. The achievement of these heavy yields tends to be erratic, despite apparently small differences in weather between seasons and similar management regimes. The challenge for the crop physiologist is to understand how the processes governing yield are influenced by the weather and husbandry, so that advice given to farmers can be effective in their quest for consistent quantity and quality of yield.

Often crop physiologists have investigated the performance of specific processes by imposing treatments and then observing and measuring the consequent changes in each process. This approach has several limitations. First, it can only explain the response to the specific treatment levels used; extrapolation cannot be fully justified and may lead to mistaken conclusions. Second, many supplementary measurements are usually needed to explain how the measured changes in a particular physiological process influence crop productivity. Third, in field experiments the weather also influences the process being studied, complicating interpretation of the responses and often making it almost impossible to determine the precise effect of the treatment (Patterson et al., 1977). These

* Present address: ICI Agricultural Division, Billingham, Cleveland.

various limitations to the application of a detailed analytical approach are a major reason why many investigations of crop response to particular treatments are simply expressed as the average results of a number of seasons' equivalent experiments (Mycroft, 1983).

The general availability and relatively low cost of modern computers is currently providing many scientists with the opportunit to construct mathematical models of crop physiological processes. Modelling has the benefit of being cheaper and less labour intensive than field experimentation and can easily accommodate many different circumstances. The intellectual exercise of clearly defining the vital characteristics of a process can highlight areas of ignorance not obvious from the results of field experiments. Also, the modelling exercise introduces a firm discipline into the way physiological processes are defined and measured. This discipline i a valuable product of cooperation between physiologist and modeller; experimenters have tended to make measurements and express results i ways that are of practical convenience rather than of direct relevance to understanding the importance of particular processes. The formal framework for assessing the relevance of measurements should also enable the experimenter to define more clearly the direction of future research. But the modelling exercise is of only limited value until it is validated and to do this successfully also requires close cooperation between the modeller and the physiologist. Finally, it must never be forgotten that a model is intended to be used, either to predict or to explain the performance of a process or crop under specified circumstances. It must not be confined to computer memory, only providing employment and enjoyment for the scientists developing the model. Instead it should be presented so that potential users can associate with it and not feel intimidated by the complexity and jargon often accompanying the description of otherwise useful models (Lemon, 1973).

To examine the relationship that does, and the authors think should, exist between crop physiological studies and mathematical modelling, three examples have been chosen dealing with aspects of winter wheat production familiar to all crop physiologists: leaf growth, ear number and ear size. The specific examples have been selected from the research programme on winter wheat being carried out at Broom's Barn Experimental Station. The aim of this work is t identify the factors governing the yield of winter wheat crops sown on different dates and then to examine the possible role of husbandr operations, particularly the timing of nitrogen fertilizer applications, in mitigating the adverse effect of late sowing. In this context a model is seen as an aid to interpreting the physiological measurements and a convenient way to examine the likel effects of changing the times at which specific husbandry operations are applied to the crop.

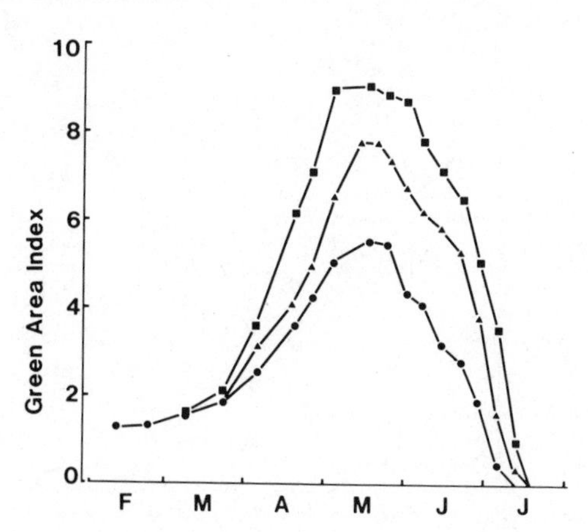

Fig. 1. Green areas of winter wheat given three different nitrogen treatments, 90 kg ha^{-1} (●), 180 kg ha^{-1} (▲) and 330 kg ha^{-1} (■).

LEAF GROWTH

In most crop physiological and agronomic experiments where leaf area growth is measured, it is usually expressed as green leaf area index, GAI (Legg et al., 1979). It is well known that a major effect of nitrogen fertilizer is to increase GAI (Thomas et al., 1978) and Fig. 1 is typical of the way that results have been presented, showing changes in GAI with time for crops with different nitrogen treatments. Judging from the frequency with which such data have been published this approach would seem to have satisfied the need for an understanding of the way nitrogen fertilizer influences leaf growth. However, when examined from the more exacting viewpoint of a modeller, such data are totally inadequate because they do not define the individual processes controlling leaf area growth or the ways these processes interact to give the actual leaf area through the season; e.g. why was GAI largest in the crop receiving most nitrogen, was it because leaves were produced faster or were the individual leaves larger? These questions can be readily answered if the physiologist and the modeller discuss the problem when the experiment is being planned, so that the most useful measurements are made.

In nearly all studies where GAI is measured, plants have to be sampled in the field, the green leaves removed from these plants and the areas measured. This was the basis of the approach to be adopted in the research programme at Broom's Barn, but after discussion with the AFRC Wheat Modelling Group at an early stage, the method for recording GAI was changed without increasing the labour requirement. Instead of pooling all the leaves removed from a plant, all the leaves from the same group, i.e. leaf 1, leaf 2 etc., were put

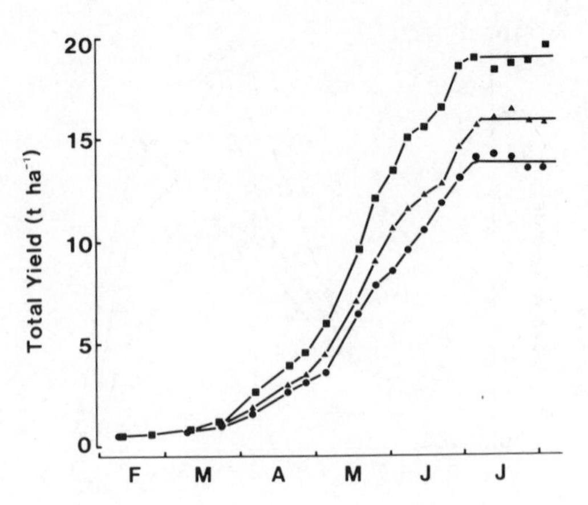

Fig. 2. Dry matter growth of the three crops shown in Fig. 1.

together. By implementing these changes the data collected were consistent with the modeller's view of the processes governing the development of the crop's leaf area, but did the physiologists directly benefit from the change?

Not only did the well fertilized crops have a larger GAI (Fig. 1) but they also had a faster rate of dry matter growth early in the season (Fig. 2) when the differences in GAI were first apparent. Because the growth rate of all crops was similar during the period of rapid growth in May and June, it was these early differences in growth which were largely responsible for producing

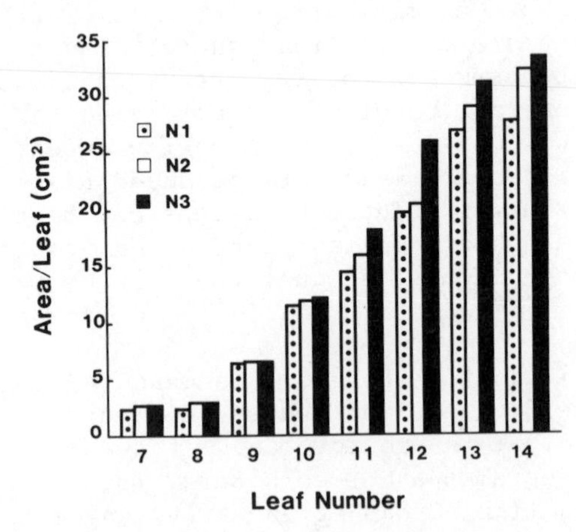

Fig. 3. Effect of fertilizer on the lamina areas of successive leaves emerging on the main stem; N1 (90 kg N ha^{-1}), N2 (180 kg N ha^{-1}) and N3 (330 kg N ha^{-1}).

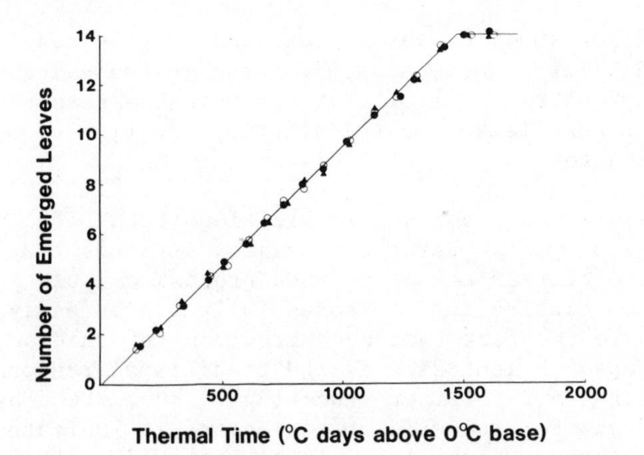

Fig. 4. Rate of emergence of leaves on the main stem related to accumulated mean air temperature above a base of 0°C.

the heavier crops. If we wish to achieve similar results with future crops, it is essential that the reasons for the early season response shown in Fig. 1 are understood. Figure 3 shows that the rate at which individual leaves appeared on the main stem, when expressed in relation to the accumulation of mean air temperature above a base of 0°C (Gallagher, 1976), remained constant and was unaffected by the nitrogen treatments. Figure 4 shows that the area of the individual leaves on the plant at and immediately after applying the treatment was the same for both crops. It was only after another 3.5 leaves

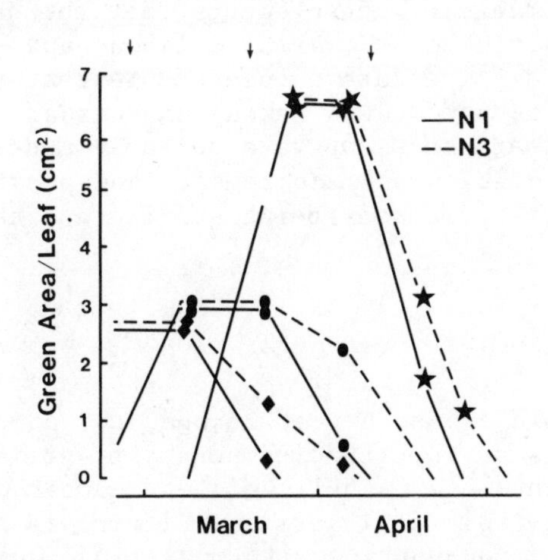

Fig. 5. Green areas of individual leaves on the main stem at and immediately after nitrogen was applied to N1 (90 kg N ha^{-1}) and N3 (330 kg N ha^{-1}).

had appeared, or about 8 weeks later, that any increase in the size of individual leaves was measurable. The increase in GAI after the application of nitrogen fertilizer early in the season was because the fully expanded leaves on the plant at the time of application remained green for longer (Fig. 5).

This result has important implications both for the modeller and the crop physiologist. For the modeller, it shows that in response to nitrogen fertilizer the senescence process can be the most important when considering increases in leaf area early in the season, despite the fact that much previous information suggested that an increase in leaf size is the traditional response to increased nitrogen fertilizer (Puckridge, 1963, cited by Bunting and Drennan, 1966). For the crop physiologist, it indicates that a rapid response to nitrogen fertilizer can be achieved early in the season even though the growth rate is very slow. In response to these initial results, we increased the harvesting frequency at this time in subsequent seasons; in the following year a 20% increase in dry matter growth rate was measured only 5 days after the application of nitrogen fertilizer at the beginning of March. This improvement in our methodology and knowledge arose as a direct result of the contact between the crop physiologist and the modeller.

This example illustrates the potential benefits of a dialogue between crop physiologist and modeller before measurements are made. The need is as great once data have been collected because complete analysis and effective interpretation are greatly assisted by the formal framework of a model. Use of such a framework may provide the only realistic approach to improve our understanding of a crop's response to environmental and cultural treatments. This is because the frequency and range of measurements that the physiologist can make are often restricted by available labour and space. Sensible use of a model to examine data should ensure that valuable experimental time is not wasted making unnecessary measurements. The following examples are cases where a suitable model would greatly assist the interpretation of measurements and provide a better understanding of the processes being studied and their effect on yield.

EAR NUMBER

For crops grown in north-west Europe, the number of ears per unit ground area is an important component of grain yield in cereals. Once plants are established, ear number depends on the production and survival of tillers. Tillering is a process ideally suited for study in conjunction with modelling; accurate measurements are very labour intensive so that it has been an unpopular subject for research programmes and even where it has been studied few ancillary measurements have been made (Masle-Meynard, 1981).

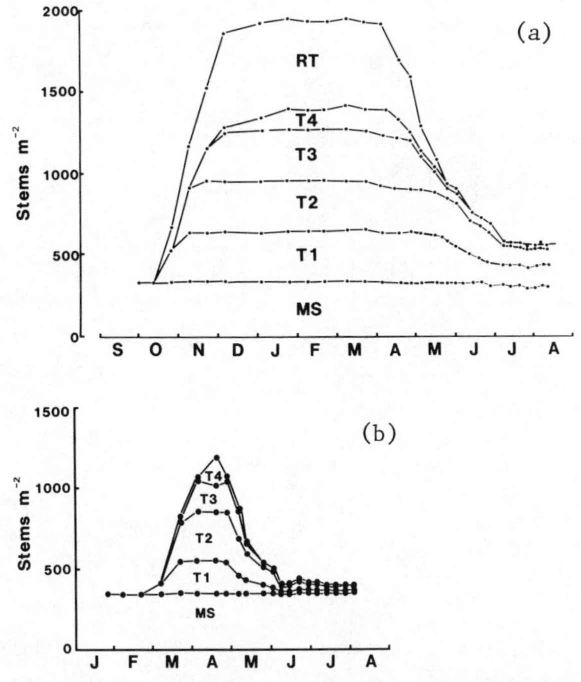

Fig. 6. Patterns of tillering for winter wheat crops sown in mid-September (a) and mid-November (b), showing the numbers of main stems (MS), tiller 1 (T1), tiller 2 (T2), tiller 3 (T3), tiller 4 (T4) and remaining tillers (RT).

Detailed measurements have been made at Broom's Barn of the number and origin of the tillers produced and of those surviving to form ears. Figure 6a shows the pattern of tiller production and death for a September-sown crop of cv. Avalon, typical of winter-sown wheat crops in north-west Europe (Darwinkel, 1978). The maximum number of tillers was reached before the end of December, and tiller death started in March and continued until anthesis in June. At harvest, grain yield was primarily from the ears on main stems and tiller 1 (the tiller arising from the axil of the first leaf). A later-sown crop (Fig. 6b) showed a similar tillering pattern, although compressed in time, and final ear number was about the same but the origin of the ear bearing tillers was very different. Tiller 1 formed on less than half the plants and few survived to bear an ear at anthesis, grain yield coming from the main stem, tiller 2 and tiller 3 ears. A more superficial study of these crops would not have identified tillering as a reason for any potential differences in the yield components between crops, as final ear numbers were similar in both. Measurements of grain weight per ear showed that main stem ears were always the heaviest, the ear weight progressively

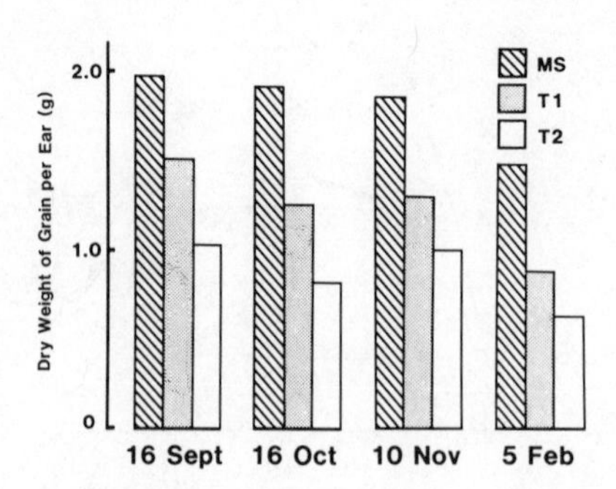

Fig. 7. Mean grain yield of wheat ears on stems of different origins for crops sown from mid-September to early February; main stem (MS), tiller 1 (T1) and tiller 2 (T2).

decreasing in ears originating from lower order tillers (Fig. 7). Consequently, the types of ears present in the later-sown crop would result in a lighter mean ear weight, independent of any other differences in crop growth etc. (Darwinkel, 1980).

In terms of practical crop management, it is therefore important that the processes governing tillering are understood. If conditions that lead to the failure of tiller 1 could be predicted, possible alternative husbandry practices could be defined to minimize the yield losses that would result from such failure. However, in most experimental programmes, little additional time is available for the accurate monitoring and recording of tiller numbers and sizes that is essential if the relationship between the tillering process and crop productivity is to be understood from experimentation alone. An alternative approach is for the physiologist and the modeller to tackle the problem together. An initial model could be developed from existing data, and then used to determine the type and priority of measurements required to validate the original assumptions within the model. This should lead to improvements in the model, eventually providing an understanding of the reasons for the observed differences in tiller production and survival (Fig. 6).

EAR SIZE

Ear size, defined as the number of grains per ear, is the product of the number of fertile spikelets produced and the average number of grains per spikelet.

The dates of the beginning and end of spikelet production on the apex and anthesis of the ear mark stages in the development of the crop and, recently, much progress has been made in quantifying the factors influencing development and their incorporation into models

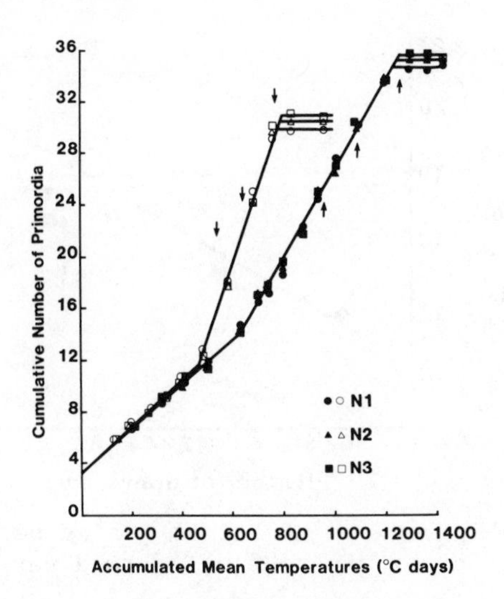

Fig. 8. Rate of leaf and spikelet primordia production on the apex of wheat sown in mid-September 1981 (solid symbols) and mid-November 1981 (open symbols) given three different nitrogen treatments; N1 (90 kg N ha^{-1}), N2 (180 kg N ha^{-1}) and N3 (330 kg N ha^{-1}).

(Weir et al., 1984). Thus development is concerned with the progress of a crop to maturity, whilst the size of the organs at maturity is primarily a function of the growth processes. The need, therefore, is to investigate the interaction between these two major processes, development and growth, in order to understand how final ear size is determined. Developmental processes influence the length of time during which grain sites are forming and the rate at which specific organs are being produced (Rawson and Bagga, 1979). For example, Fig. 8 shows the rate of leaf and spikelet primordia production for Avalon wheat sown on different dates and given different treatments. The present understanding of developmental processes enables the dates when these different phases began and ended to be predicted from a knowledge of sowing date, variety and weather. It does not extend to a prediction of the final numbers of spikelet primordia, which are increased by the application of nitrogen fertilizer (Fig. 8).

The nitrogen fertilizer also stimulated different rates of dry matter production while spikelets were forming (cf. Fig. 2) and, although final spikelet numbers were only slightly greater, these small increases can be related to differences in ear size (Fig. 9). The reason for these small, but important, differences in ear size are not clear but Fig. 10 shows that they can be related to dry matter production by the stem during the period of spikelet initiation on the apex.

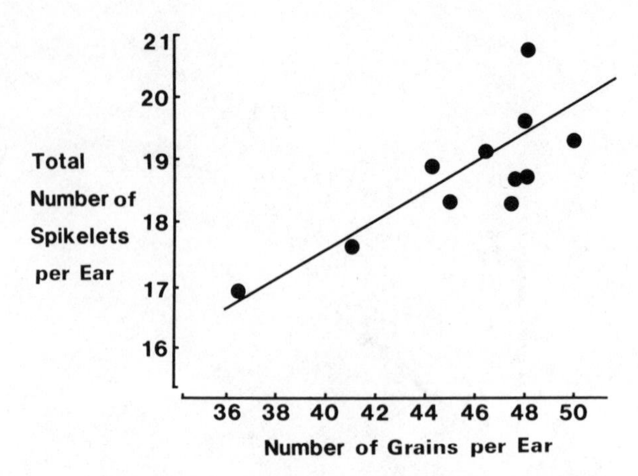

Fig. 9. Relationship between number of grains per ear at harvest and the total number of spikelets formed per ear of winter wheat.

Thus ear size depends on both development and growth (Fischer and Stockman, 1980). One method for determining the relative importance of the processes experimentally would be to create large differences in growth rate at specific developmental stages or different developmental rates when the growth rate is similar. This would suggest use of controlled environment facilities, so that any

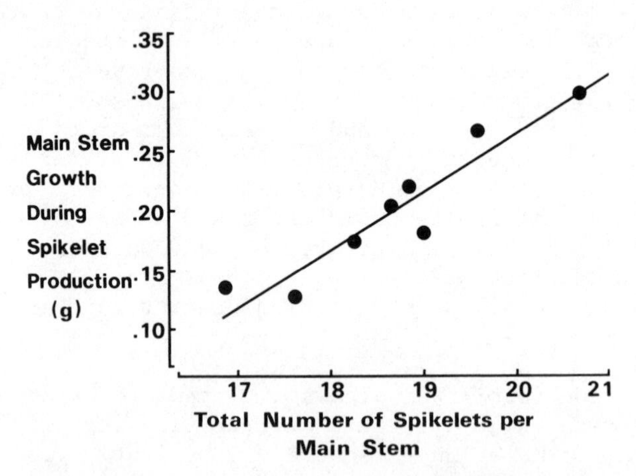

Fig. 10. Relationship between dry matter growth from the time nitrogen was first applied (when approximately half the spikelets had formed on the main-stem apex) to the end of spikelet initiation and the total number of spikelets formed per main stem apex.

one experimental variable could be altered independently of the remainder. In theory this may be attractive, but in practice there are very few controlled environment facilities which enable temperatures to be varied while radiation receipts remain similar to those normally experienced in the field. In practice, cereal plants grown in controlled environments often bear little resemblance to their counterparts in the field (Whingwiri and Kemp, 1980). These are important drawbacks when the objective is to determine the principles which govern ear size. First, we need to be able to extrapolate the results to the weather conditions which prevail in the field if required and, second, we need to determine relationships between growth and development that are valid for plants of the size and structure of those grown in the field.

An alternative approach is to combine existing models of the processes involved with a series of specific observations that seek to provide further information on areas of weakness and uncertainty. This approach has several potential advantages, if it can eliminate unnecessary measurements whilst validating an improved model that can be applied in a wide range of situations.

CONSPECTUS

The climate in north-west Europe is non-extreme and provides the potential for, by world standards, a substantial yield in nearly all years; over the last three decades, developments in cereal breeding, agrochemicals and machinery have progressively improved yield. Occasionally we record very heavy wheat yields but at present are neither able to explain nor reproduce them consistently. Under these circumstances the challenge facing the agricultural scientist is not to determine the probability, in any year, of getting a yield but to ensure consistent yield of good quality grain despite seasonal variations in weather. In the past the response to this challenge has relied almost exclusively on well designed and constructed experiments aimed at answering a specific question and minimizing experimental error. However, the errors are frequently as much as 5%, which is larger than the yield increases expected from some modern husbandry techniques. For example, the average response to plant growth regulators is about 0.2 t ha^{-1} (Woolley, 1981), an increase of only 2% assuming a yield of 8 t ha^{-1}. In the past the yield response to the major husbandry practices, e.g. sowing date or amount of nitrogen fertilizer, has been large and clearly determined by traditional experiments. Currently, cereal farmers are seeking small but consistent yield increases from better timing of husbandry inputs, relating them more to crop requirements than practical considerations. The small yield increases expected, about 0.5 t ha^{-1}, are difficult to determine as significant but financially the increase can represent a significant improvement in profitability.

Mathematical modelling in conjunction with a well organised experimental programme may provide an alternative response to this challenge. The principles governing growth and development of wheat are formally expressed in a model as a series of discrete relationships precisely defining the effect of specific variables on one of the processes controlling yield. These individual relationships can be validated more accurately than measured yield responses of a crop. Therefore, by analysing weather patterns, we could estimate first, the frequency with which a particular combination of plant and weather events is likely and second, the effect on yield of any change, however small, in one or more of the processes. From such analysis, potential yield responses could be estimated. Given a suitable experimental basis for the model and any predicted responses, such a combined approach should lead to appropriate advice for farmers to assist them in the management of the wheat crop.

REFERENCES

Bunting, A. H., and Drennan, D. S. H., 1966, Some aspects of the morphology and physiology of cereals in the vegetative stage, in: "The Growth of Cereals and Grasses", F. L. Milthorpe and J. D. Ivins, eds., Butterworth, London).

Darwinkel, A., 1978, Patterns of tillering and grain production of winter wheat at a wide range of plant densities, Neth. J. agric. Sci., 26:383.

Darwinkel, A., 1980, Ear development and formation of grain yield in winter wheat, Neth. J. agric. Sci., 28:156.

Fischer, R. A., and Stockman, Y. M., 1980, Kernel number per spike in wheat (Triticum aestivum L.). Responses to pre-anthesis shading, Aust. J. Pl. Physiol., 7:169.

Gallagher, J. N., 1976, The growth of cereals in relation to weather Ph.D. Thesis, University of Nottingham.

HGCA, 1984, "Cereals Statistics", Home Grown Cereals Authority, London.

Legg, P. J., Day, W., Lawlor, D. W., and Parkinson, K. J., 1979, The effects of drought on barley growth: models and measurements showing the relative importance of leaf area and photosynthetic rate, J. agric. Sci., Camb., 92:703.

Lemon, E., 1973, Predicting crop climate and net carbon dioxide exchange, Photosynthetica, 7:408.

Masle-Meynard, J., 1981, Relations entre croissance et développement pendant la montaison d'un peuplement de blé d'hiver. Influence des conditions de nutrition, Agronomie, 1:365.

Mycroft, N., 1983, Variability of yields in cereals times fungicide trials, J. agric. Sci., Camb., 100:535.

Patterson, H. D., Silvey, V., Tarbot, M., and Weathorp, S. T. C., 1977, Variability of yields of cereal varieties in UK trials, J. agric. Sci., Camb., 89:239.

Puckridge, D. W., 1963, The influence of competition for light on the dry matter production and ear formation of wheat plants, M. agric. Sci. thesis, University of Adelaide,

Rawson, H. M., and Bagga, A. K., 1979, Influence of temperature between floral initiation and flag-leaf emergence on grain number in wheat, Aust. J. Pl. Physiol., 6:391.

Thomas, S. M., Thorne, G. N. and Pearman, I., 1978, Effect of nitrogen on growth, yield and photorespiratory activity of spring wheat, Ann. Bot., 42:827.

Weir, A. H., Bragg, P. L., Porter, J. R., and Rayner, J. H., 1984, A winter wheat crop simulation model without water or nutrient limitation, J. agric. Sci., Camb., 102:371.

Whingwiri, E. E., and Kemp, D. R., 1980, Spikelet development and grain yield of the wheat ear in response to applied nitrogen, Aust. J. agric. Res., 31:637.

Woolley, E. W., 1981, Performance of current growth regulators in cereals, in: "Opportunities for Manipulation of Cereal Productivity", A. F. Hawkins and J. B. Jeffcoat, eds., Monograph 7, British Plant Growth Regulators Group, Wantage.

THE INTERNATIONAL BENCHMARK SITES NETWORK FOR AGROTECHNOLOGY TRANSFER (IBSNAT)

G. Uehara

University of Hawaii
Honolulu, Hawaii, U.S.A.

INTRODUCTION

How can a country that is poor in resources and has few trained agricultural scientists increase the usefulness of its research, quickly and economically? The International Benchmark Sites Network for Agrotechnology Transfer (IBSNAT), a new international project on technology transfer, is designed to increase research efficiency through networking, systems analysis, crop modelling and exploitation of the new wares of the information age.

While most countries in the tropics are not equipped to provide the full range of information and technical skills to meet their development needs, their combined potential research capability is considerable. This suggests that instead of upgrading the research capability of each country separately, it would be more desirable to create a network of national centres for sharing information and technologies. The IBSNAT strategy therefore depends on exploiting the communal strength of the hundreds of research stations scattered throughout the tropics, while at the same time ensuring that their research integrity is preserved.

AGROTECHNOLOGY TRANSFER

The key element of IBSNAT is networking for agrotechnology transfer. The transfer of agrotechnology is a two-way process in which a participating country adds to the pool of information and technology, and takes from the pool what it does not have. The role of IBSNAT is to foster this development by first establishing a prototype network of collaborating countries, and then creating a mechanism that allows network members to share their combined human and technical resources.

While IBSNAT's aim is to encourage free exchange of information and technology, most research results are site-specific. For this reason, one of IBSNAT's main activities is to transform on-going, site-specific agronomic research into research that will generate information that is useful not only for the centre doing the research but to all members of the network.

To achieve its goals of enabling network collaborators to share agrotechnology, IBSNAT collaborators have selected ten crops for agronomic study - these comprise four cereals (maize, rice, sorghum and wheat), three grain legumes (beans, groundnuts and soybeans) and three root crops (aroids, cassava and potato). The collaborators have developed guidelines for designing and installing agronomic experiments from which a minimum set of soil, crop, climate and management data will be collected. This minimum set of data was identified by the collaborators meeting at ICRISAT, at Hyderabad, India in 1983 (see below).

The minimum data set from all IBSNAT experiments will be stored and analyzed in a central data-base management system at the University of Hawaii. With the assistance of crop modellers from the Agricultural Research Service of the U.S. Department of Agriculture and the Commonwealth Scientific and Industrial Research Organization in Australia, the collaborating countries will test and validate simulation models of the ten IBSNAT crops. The models should enable the collaborating countries to identify the most suitable crop for each type of land, environmental and economic situation, and provide optimum prescriptions for rectifying mismatches between crop requirements and land characteristics. IBSNAT offers hope that crop models capable of predicting crop performance under a wide range of conditions will enable planners in developing countries to make better use of their land resources.

To transfer agrotechnology successfully, the models must be technology- and crop-specific. Models must also be able to predict the performance of the technology or crop in a wide range of tropical conditions and they must be validated in these different tropical environments. The combined minimum data set is used to develop models that predict crop growth and yields in the full range of environments in which they are tested.

MINIMUM DATA COLLECTION

Enthusiasm amongst collaborators for IBSNAT's approach to agricultural research remains high. IBSNAT sustains this enthusiasm by providing a forum for the collaborators to design new agronomic experiments for crop modelling and technology transfer and to specify the minimum data set to collect from each experiment.

To ensure that all collaborators collect the common minimum data set from agronomic experiments, IBSNAT has distributed a procedure for "Experimental Design and Data Collection"; this requires minimum data from each IBSNAT experiment on:

(1) general information
(2) daily weather
(3) soil characteristics
(4) soil nutrient analysis prior to planting
(5) plot management information
(6) phenological observation
(7) crop dry matter production
(8) final yield and yield components, and
(9) post-harvest soil analysis.

This minimum data set is designed to validate user-orientated models of the kind developed by Ritchie (the CERES-wheat (see Chapters 27 and 28), CERES-maize, and CERES-sorghum models) and the nitrogen dynamics model described by Godwin (see Chapter 29). A soil phosphorus and soil acidity model currently being developed by the Agricultural Research Service of the U.S. Department of Agriculture will eventually become part of IBSNAT's soil-crop-climate model.

EXTENT OF COLLABORATION

A number of countries have restructured their resources to benefit from IBSNAT. For example, Fiji has developed a Soil and Crop Evaluation Project and it has been agreed that eight Fijian experimental stations will become part of the IBSNAT network. In addition, Fiji is proposing to create an Oceania Benchmark Sites Network for Agrotechnology Transfer (OBSNAT) so that the island nations of the South Pacific can become part of IBSNAT. Pakistan has a project called SCAN (Soil Capability Assessment Network) and has agreed to add the SCAN research sites to the IBSNAT network. Venezuela has added six of its national agricultural research stations to the network, and, in addition, has agreed to co-sponsor with IBSNAT a training workshop on modelling the soil-crop-climate continuum for agrotechnology transfer. The substantial investment in IBSNAT's activities by collaborators reflects the success of this project.

In all, 16 national, three regional and three international agricultural research centres in South and Central America, Africa, the Middle East, the Caribbean, the Indian Sub-continent, South-East Asia and the Pacific participate in IBSNAT.

IBSNAT'S ROLE AND AIMS

The IBSNAT collaborators are primarily interested in using crop models to predict crop performance in new locations. For instance they wish to estimate the optimum date of sowing with the best cultivars in the proper location, and thereby make better use of their resources. IBSNAT conducts regional training workshops on crop modelling to enable the collaborators to analyze their minimum data set for crop modelling.

IBSNAT is funded for five years and is in its second year of operation. Continued support for IBSNAT-type research will depend on the degree to which crop models help us to understand agricultural systems, predict crop performance, and control and optimize crop production.

INTERFACING THE ANALYSIS AND SYNTHESIS OF CROP GROWTH PERFORMANCE

D. A. Charles-Edwards* and R. L. Vanderlip‡

*CSIRO Division of Tropical Crops and Pastures
St. Lucia, Queensland
Australia

‡Kansas State University
Manhattan, Kansas
U.S.A.

INTRODUCTION

If it is considered appropriate to construct a crop simulation model to study particular problems of crop production, the model needs to meet several disparate criteria in order to find ready acceptance with potential users. We can list some of these criteria:

(i) it must quantitatively predict both the rate and extent of crop dry matter production and describe the phenology and the morphology of the crop;

(ii) it needs not only to help us understand the performance of the crop but also to detect limitations to crop production and suggest the means by which they can be reduced;

(iii) it needs to be testable;

(iv) it needs to be simple both to use and to understand.

Our common perception of the complexity of crops, both their organization and development, is often a major impediment to the development of crop simulation models that meet all of these criteria.

A crop simulation model is a formal mathematical statement of assumptions that have been made about the cropping system being studied. Those assumptions may be entirely empirical, based on observations of the field performances of crops, or they may be based on an understanding, or knowledge, of the physiological mechanisms underlying crop growth and production. The model may not predict the correct behaviour of crops growing in the field because the assumptions are wrong, or put together in the wrong way. Equally, it may predict the correct behaviour fortuitously even when some of the assumptions are wrong because it is insensitive to them under those particular circumstances or because the in-built errors cancel out. It follows that any simulation model needs to be tested not only at the level of its final predictions, grain yield or forage yield, but also at appropriate intermediate levels. If the model represents a detailed description of the physiological processes of plant growth, the procedures of testing and validating the assumptions about the components (sub-models) become a major scientific exercise.

An holistic approach to the description and analysis of crop growth data has identified five main physiological determinants of the rate and extent of crop growth (Charles-Edwards, 1982). These determinants also provide a convenient intermediate level for testing and validating crop simulation models. They are amenable to direct measurement in the field and empirical descriptions of their responses to changes in the cropping environment can be readily developed. They can also be further analysed in terms of more basic physiological processes of plant growth, and at the present time three of them can be described quantitatively using more basic plant physiological information. The utility of this approach to both the analysis and synthesis of information on crop behaviour will be explored in this paper.

THE RATE AND EXTENT OF CROP GROWTH

The net daily rate of crop dry matter production, $\Delta W/\Delta t$, can be written as:

$$\Delta W/\Delta t = \varepsilon J - v \tag{1}$$

where J is the daily amount of photosynthetically active radiation intercepted by the crop, ε is the efficiency with which it uses the intercepted light energy in the production of new dry matter and v is the daily rate of loss of dry matter. Equation (1) is axiomatic; it represents a self-evident definition of the net rate of crop dry matter production. The net amount of dry matter, W, produced by the crop over a period of t days, can be obtained by summing the net daily increments of dry matter produced by the crop over the time period that is of interest, and can be written as:

$$W = \sum_{i=0}^{t} (\Delta W)_i = \sum_{i=0}^{t} (\varepsilon J - v)_i. \tag{2}$$

For any particular part of the plant (leaf, stem, root or grain etc.), whose dry matter we denote by W_H, we can write that:

$$W_H = \sum_{i=0}^{t} (\eta_H \ \varepsilon J - v_H)_i, \tag{3}$$

where η_H is the proportion of new dry matter partitioned to the component H. It should be noted that remobilization of assimilate from other plant parts to the part of interest is implicit in the numerical value of the partition coefficient. Similarly, if the part H is a storage organ there may be a component implicit in its rate of dry matter loss, v, that is attributable to remobilization of the assimilate stored in it.

An hypothesis on the development of plants has recently been advanced which enables the analysis to be extended to examine some aspects of plant morphology, e.g. branching, tillering, grain number (Charles-Edwards, 1984a,b). The conventional source/sink hypothesis for assimilate translocation envisages that assimilate moves along a concentration gradient from the source organ, for example the leaves or an appropriate storage organ, to the sink, perhaps a fruit or a panicle. The hypothesis derives from the original observations of Mason and Maskell (1928) that sugars appeared to move along concentration gradients in the stems of cotton plants, and a great deal of attention has been focussed upon its experimental validation. The new hypothesis argues that it is the flux (rate of supply) of assimilate to the sink organ that is essential to its continued growth and development and that each type of growing point on a plant requires a particular flux of assimilate to continue active growth and development. If the assimilate flux required by a single vegetative growing point, say a tiller, is denoted by a_M, the number of these vegetative meristems that can be subtended per unit ground area by a crop, N_M, can be written as:

$$N_M = (\eta_T \varepsilon J) \ / \ a_M, \tag{4}$$

where the term $(\eta_T \varepsilon J)$ is the gross daily rate of production of new dry matter that is partitioned to the above ground parts of the crop.

PHYSIOLOGICAL DETERMINANTS OF CROP GROWTH

A similar analysis has been described by Monteith (1977) and others (Biscoe and Gallagher, 1977; Natarajan and Willey, 1980a,b) and used for the routine analysis of crop data. The approach was outlined by Warren Wilson some years earlier (Warren Wilson, 1971). The main criticism that can be levelled at their analysis is that it does not explicitly take the losses of crop dry matter, v, into account leaving some ambiguity in the interpretation of their numerical estimates of ε, the efficiency of light use by the crop.

The present analysis defines five physiological determinants of the rate and extent of crop growth, J, ε, η, v, defined above, and t, the duration of production of a particular part of the crop. We will examine each in turn.

Daily amount of light energy intercepted, J

The daily amount of light energy intercepted by a crop depends both on the light energy incident upon it and on the proportion of the incident light that is intercepted. If we denote these components by S and Q respectively we can write:

$$J = SQ.$$

For a crop with a "closed leaf canopy" (one within which the downward light flux density incident upon a horizontal plane in any horizon within the canopy is spatially homogeneous), Q can be written as a simple function of the leaf area index of the crop, L, and light extinction coefficient, k:

$$Q = 1 - \exp(-kL). \qquad (5)$$

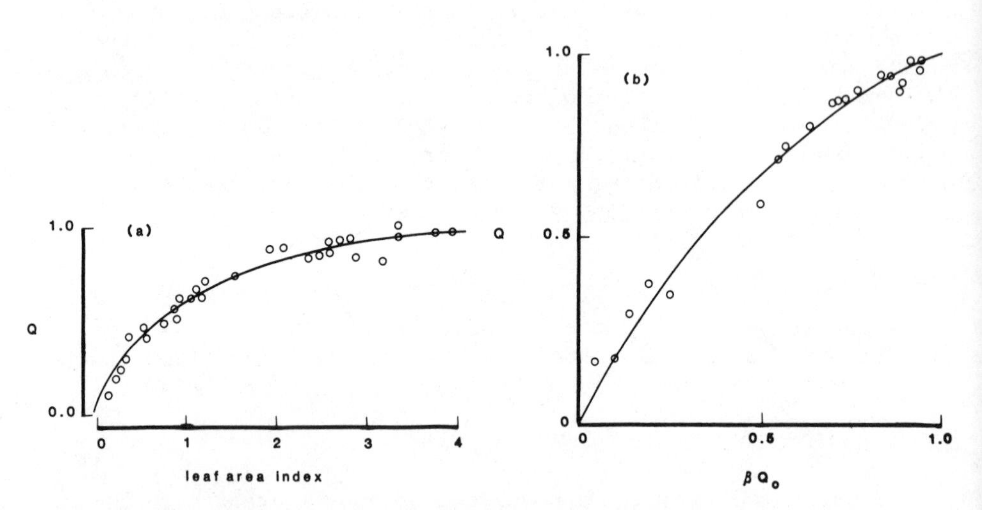

Fig. 1. (a) The relationship between the proportion of incident light intercepted (Q) and the leaf area index (L) for closed stands of Cyamopsis tetragonoloba, guar. (b) The relationship between the proportion of incident light intercepted (Q) and the product βQ_0 (see text) for a number of grain legumes grown as row crops.

The application of Equation (5) to the description of some field data on the interception of light by crops of the legume *Cyamopsis tetragonaloba* is illustrated in Fig. 1a.

However, many crops are grown in discrete rows for the whole of their lives, and almost all crops grow as discrete rows during the early phase of vegetative growth. For a row crop the proportion of the incident light intercepted, Q, depends upon several factors including the geometry of the leaf canopy and the orientation of the row. As a useful first approximation we can write:

$$Q = 2\beta Q_o \;/\; (1 + \beta Q_o), \qquad (6)$$

where β is the proportion of the ground area covered by the vertical projection of the leaf canopy downward onto the ground and Q_0 is the proportion of the incident light energy intercepted at the row centre around solar noon (Charles-Edwards and Lawn, 1984). The application of Equation (6) to a variety of field data is illustrated in Fig. 1b.

The light-use efficiency, ε

The efficiency with which a crop uses intercepted light energy in the production of new dry matter can be measured as the slope of the regression of the gross amount of dry matter produced by the crop upon the cumulated amount of light energy intercepted by it. The efficiency of light use depends upon a number of factors. We will consider here the special case of a crop with a "closed leaf canopy" for which the light use efficiency can be usefully and simply approximated by:

$$\varepsilon = qY\alpha F_M \;/\; (\alpha k I + F_M) + qb \qquad (7)$$

where q converts g (CO_2) assimilated to g (plant dry matter), Y is a yield factor describing that proportion of new assimilate that is not respired by the plant, α is the leaf photochemical efficiency, F_M is the rate of light-saturated photosynthesis by an upper unshaded leaf in the crop canopy, k is the canopy light extinction coefficient, I is the average light flux density incident upon the uppermost surface of the canopy and b is a maintenance respiration coefficient (see Section 4.4 in Charles-Edwards, 1982).

This relationship illustrates a very pertinent point. The efficiency with which a crop canopy is using the light energy intercepted by it to produce new plant dry matter depends upon more than just the photosynthetic capacities (the value of F) of its constituent leaves. For example, whereas the rates of light-saturated photosynthesis of C_4 plant types are often greater than those of C_3 plant types grown under comparable conditions, it does not follow that crops of C_4 plants will be more efficient in their use of intercepted light energy than crops of C_3 plants. The architecture of the canopy, the distribution of light energy through it, and the chemical nature of the new plant dry matter produced may have as much effect upon the efficiency of light use as the particular photosynthetic pathway. We might expect plants which produce large quantities of highly unsaturated gums or resins to have lower light use efficiencies than those which produce simple carbohydrates such as sucrose. For example, whereas a hexose molecule contains 40% elemental carbon by weight, the galactomannin gum recovered from guar contains approximately 44% carbon by weight. All other factors being equal the guar would only produce 0.62 g of galactomannin gum per gram of carbon dioxide assimilated whereas the plant producing carbohydrate would produce 0.68 g of new carbohydrate. The light use efficiency of the guar might then be about 10% less than that of the other crop simply because of the chemical nature of its main product.

The light use efficiencies of several contrasting crops, estimated simply by dividing the gross increment in crop dry weight over a given period by the amount of light energy intercepted by the crop during that period, are given in Table 1.

The partition coefficient, η

The environment under which a crop grows is known to affect the partitioning of dry matter between the different plant parts. The rate and extent of growth of roots and shoots appears to depend upon their specific activities, that is the rate at which they are able to assimilate the particular essential nutrient expressed per unit dry weight of the organ concerned. Davidson (1969) formalised this interaction by proposing the relationship, root mass x specific root activity = shoot mass x specific shoot activity. On infertile soils, when specific root activities are low, a greater proportion of the plant's new dry matter is partitioned to the roots. Davidson's relationship provides a basis for several models or predictors of crop fertilizer requirements (Greenwood et al., 1974; Thornley, 1978; Greenwood, 1981).

When a crop enters its reproductive growth phase there may appear to be discontinuous changes in the partitioning pattern. For example, leaf and root growth might cease and all new dry matter be partitioned to the reproductive part of the plant. The experimenter needs to ascertain the total weight of the harvestable part of the plant, and if it is a reproductive part (e.g. grain, seed pods or fruit), what proportion of the current assimilate has been partitioned to it and how much of its weight is the result of remobilization of stored assimilate reserves. It may also be important to know how many harvestable, reproductive units there are.

Table 1. The light use efficiencies of five contrasting crop species grown at the Cooper Research Laboratory in South-east Queensland (Fisher and Foale, personal communication).

Species	ε $\mu g\ J^{-1}$	Number of observations
Hordeum vulgare (barley)	2.4 ± 0.2	9
Helianthus annuus (sunflower)	2.5 ± 0.3	9
Glycine max (soybean)	1.4 ± 0.1	9
Sorghum bicolor (sorghum)	2.6 ± 0.1	9
Amaranthus edulus (amaranthus)	2.5 ± 0.4	10

Because we can only measure directly the net changes in the dry weight of plants or their parts, we may not be able to resolve the first of these problems in an unambiguous way from measurement of dry weights alone.

Dry matter losses, v

The physiological causes of dry matter losses by plants are amongst the least well understood areas of plant research. The most labile component of a crop or a plant is its leaf tissues, and our discussions of the physiological effectors of dry matter loss will centre on them.

Environmental factors, such as severe soil water deficits, may increase the rate of leaf abscission (cf. Fisher and Charles-Edwards, 1982). Even under optimal growing conditions most crops appear to attain a maximum standing leaf weight, or leaf area index. When they attain this maximum value the rate of loss of old leaves is in equilibrium with the rate of production of new leaves. Leaves are generally lost from the lower parts of the canopy, whilst new leaves are added in the uppermost horizons of the canopy. This is not always the case, for with many monocotyledonous plants new leaves may arise from the basal parts of the plant and grow up past the existing leaves toward the top of the canopy. There is direct experimental evidence in tomato plants that the photosynthetic capacity of leaves declines as they are increasingly shaded by newly produced leaves (Acock et al., 1978), and studies of the effect of pruning on the photosynthetic activity of leaves from different horizons within the leaf canopies of mulberry bushes suggest that this decline is reversible if the younger leaves responsible for the shading are removed (Satoh, 1982). The reduced photosynthetic activity of the lower, shaded leaves probably leads directly to their loss. They become unable to provide themselves with sufficient assimilate to maintain their metabolic integrity and deteriorate to a point where they are no longer viable entities. Water deficits or nutrient deficits are likely to exacerbate this situation.

Dry matter losses from other plant organs, such as stems, may result from the remobilization and translocation of assimilate stored in them. An extreme example is the loss of dry matter from the mother bulb of the tulip plant. Starch, stored in the mother bulb as a result of the tulip's photosynthetic activity in the summer, is remobilized during the cold winter months to provide the substrate for growth of the new shoot in the following spring. The rate of growth of the new shoot can be directly related to the amount of labile sugars (sucrose) in the mother bulb at the time of shoot growth (Charles-Edwards and Rees, 1975).

In the case of a grain crop, such as sorghum, the individual grains may be very susceptible to loss or damage, particularly through insect attack, and knowledge of the causes of loss of these tissues may be particularly important for these crops.

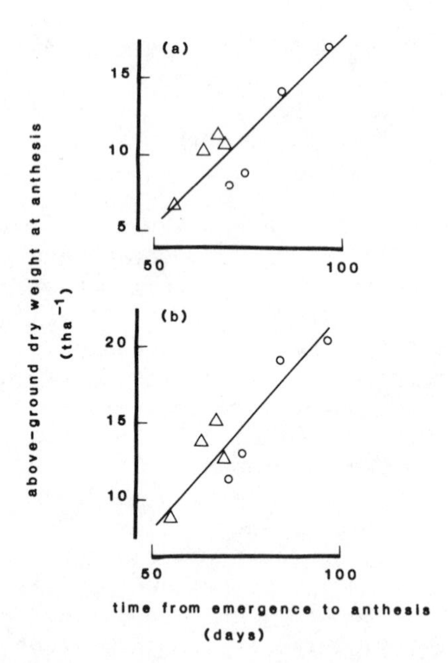

Fig. 2. The relationships between above-ground dry matter at anthesis and the length of time between emergence and anthesis for sweet (Δ) and forage (○) sorghum crops grown at four times of the year and at (a) low and (b) high plant densities.

The duration of growth, t

The length of time during which a crop grows is obviously an important determinant of its productivity. For example, the total net above-ground dry weights, at anthesis, of four crops each of sweet and forage sorghum are shown as functions of the period of time between seedling emergence and anthesis in Fig. 2 (R. Ferraris, personal communication). Similarly, the forage yield of brassica crops grown in Scotland and stylosanthes crops grown in northern Australia show similar dependencies upon the duration of their vegetative growth phases (see Fig. 2.3 in Charles-Edwards, 1982). The seed yields of mungbean, soybean and kenaf crops have also been shown to be direct functions of the duration of the reproductive growth phases of the crops (see Figs. 2.4 and 2.5 in Charles-Edwards, 1982). Knowledge and understanding of the environmental determinants of crop phenology are therefore central to our understanding and prediction of the yield of a crop.

Considerable attention has been focussed upon the development of field crops, and models based upon temperature and photoperiod have been developed (e.g. Angus et al., 1981a,b). Most of these models have been empirical, based on day degree (thermal time) relationships, although some mechanistic models for the photoperiodic responses of plants have been proposed (e.g. Schwabe and Wimble, 1976). With few exceptions, the change from vegetative to reproductive growth appears to be abrupt and discontinuous. Also there appear to be distinct qualitative differences in the flowering responses of different types of plants to changes in daylength. An hypothesis which may allow a single mechanism to describe the qualitatively different flowering responses of long-day and short-day plants has recently been advanced (Charles-Edwards, 1983).

The basic linear responses to temperature, the 'day degree' or 'thermal time' relationships can be derived with little difficulty from basic chemical considerations. Temperature is arguably one of the most difficult of the environmental variables to deal with. It affects the rates of all plant processes to a greater or lesser extent, and its prime effect on plant growth is often difficult to define in any one particular set of circumstances. Chemical rate constants vary with the thermal state of a chemical system (the temperature of the system), and Arrhenius showed that the variation could be described by a relationship of the sort:

$$\ln(k) = a_0 - a_1/T, \tag{8}$$

where k is a rate constant, a_0 and a_1 are constants and T is the temperature of the system in K (cf. pp 621-3 in Maron and Prutton, 1944). The value of the rate constant at any particular temperature T can be related to its value k_R at some reference temperature R by:

$$k = k_R \exp[a_1(T-R)/TR]. \tag{9}$$

If both T and R are within the usual range of temperature that is of physiological interest (say 273°K to 303°K), the numerical change in the product TR for a small charge in T will be small whilst the difference (T-R) will change quite markedly. We can write that:

$$k = k_R \exp(b\Delta T), \tag{10}$$

where $b=a_1/TR$ and $\Delta T=(T-R)$. Further, provided the product $b\Delta T$ is less than 0.14, this equation is approximated within 1% by:

$$k = k_R\ (1 + b\Delta T) \tag{11}$$

Now let us suppose that we are investigating the transition of plants from a state A (say vegetative growth) to another state B (say reproductive growth), and that we can write the rate of transition from state A to state B, which we will denote as $\Delta AB/\Delta t$, as some constant k, so that:

$$\Delta AB/\Delta t = k. \tag{12}$$

If k is temperature-dependent we can use Equation (11) to expand this relationship to:

$$\Delta AB/\Delta t = k_R\ (1 + b\Delta T). \tag{13}$$

The time taken for the transition of the plant from state A to state B, t_{AB}, can then be calculated by summing Equation (13) over time, and be written as a function of T:

$$t_{AB} = (B - A)/k_R - b \sum_0^t (\Delta T). \tag{14}$$

Equation (14) is a simple linear day degree relationship. The time for the plant to change from some initial state A to a final state B, t_{AB}, is linearly related to the sum of the daily temperature increments relative to the reference temperature R. The reference temperature R is equivalent to the 'base' temperature used in the normal calculations of 'thermal time', having a value between 273°K and 303°K.

PLANT MORPHOLOGY

The determinants of growth allow us to analyse or predict the rate and extent of dry matter production by the plant. They do not enable us to describe or predict its morphology. An hypothesis on the development of plants was briefly described earlier. During the early part of plant/crop growth when losses of plant dry matter are sufficiently small that they can be ignored, the number of vegetative meristems per unit of ground area, N_M, can be written as:

$$N_M = (\eta_T \varepsilon J)/a_M = (\Delta W_T/\Delta T)/a_M, \quad (15)$$

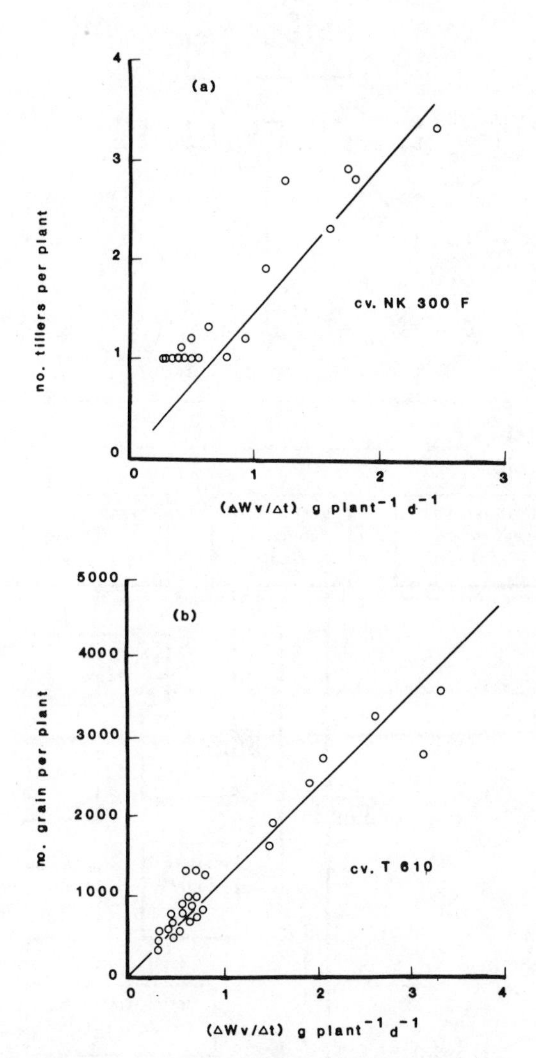

Fig. 3. (a) The numbers of tillers per plant as a function of the plant growth rate between panicle initiation and anthesis for crops of grain sorghum (genotype NK 300F) grown at a number of different sites. (b) The grain number per plant as a function of the above-ground growth rate of the plant at anthesis for crops of grain sorghum (genotype T 610) grown at a number of different sites.

where $\Delta W_T/\Delta t$ is the growth rate of the above ground parts of the crop. For example it is observed that the numbers of sorghum tillers per plant at maturity are related to the above-ground growth rate of

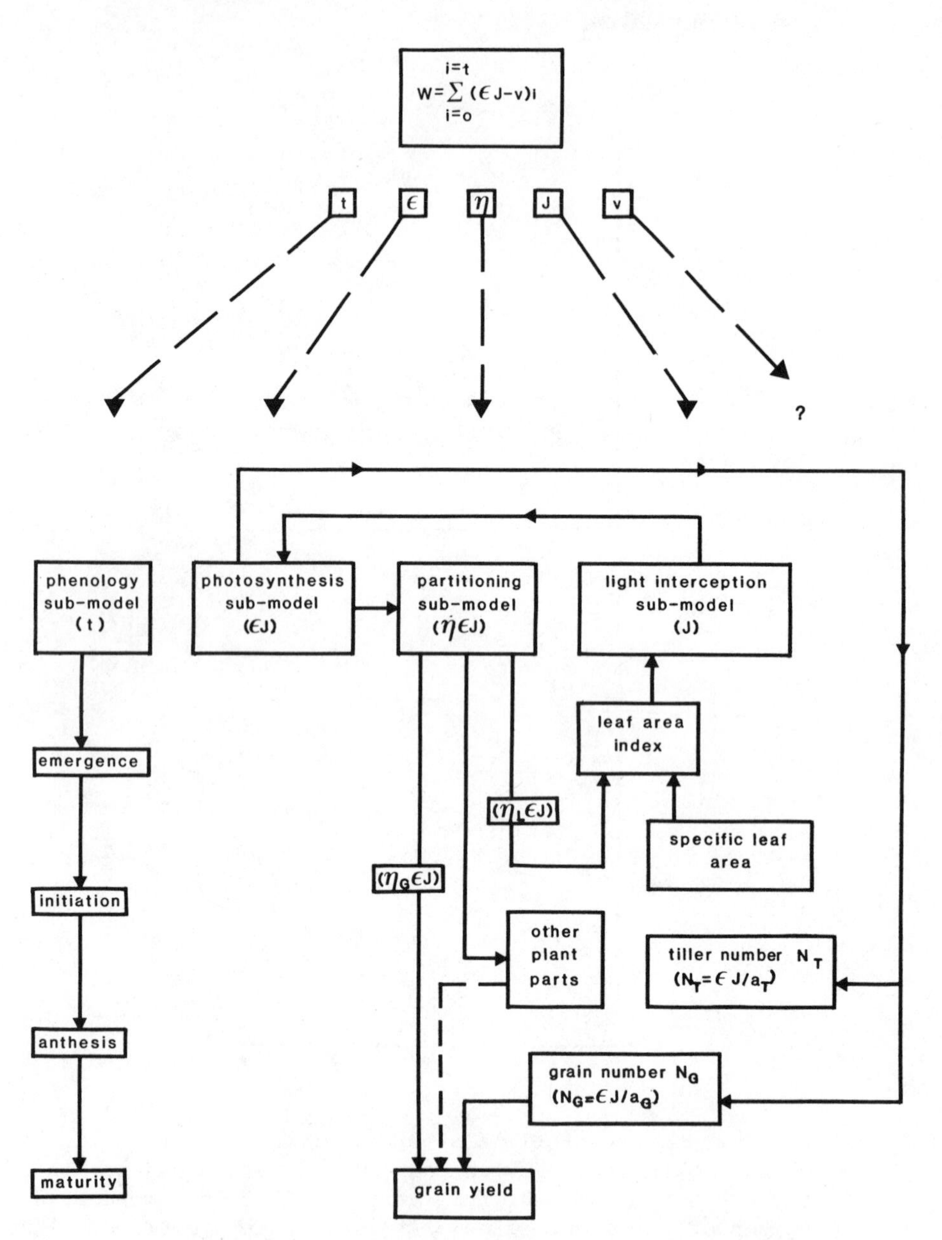

Fig. 4. A comparison of the main components (sub-models) of the sorghum growth simulation model SORGF and the growth determinants defined in this paper.

the plant between emergence and panicle initiation in this way. The data shown in Fig. 3a are for the genotype NK 300F grown under a range of conditions at a number of different sites. Similarly, the number of seeds per plant for the sorghum genotype T 610 also appears to be linearly related to the above-ground growth rate of the plant at anthesis. Data from a variety of sources illustrating this relationship are shown in Fig. 3b.

SYNTHESIS AND PREDICTION

Just as Equations (1)-(4) can be used to analyse the field growth of crops, they can also be used as the basis for plant growth simulation. An examination of the sorghum growth model SORGF (Arkin et al., 1976; Vanderlip and Arkin, 1977; Arkin et al., 1978) quickly enabled the physiological determinants described above to be identified with the main components of the model. A comparison of the determinants and the main sub-models in the model SORGF is shown in Fig. 4. The formal comparison of the analysis and the model enabled some simplification to the structure and operation of the model to be made.

The light use efficiency, ε, and the amount of light energy intercepted by the crop were already adequately modelled in SORGF (Arkin et al., 1978; Sivakumar, personal communication). However, the crop leaf area index had previously been computed on the basis of the individual development of each leaf subtended by the crop (Arkin et al., 1976). It became apparent that a simpler and more robust prediction of leaf area index could be made from a knowledge of the daily increment in new leaf dry matter produced by the crop and the specific leaf area. The specific leaf area could be computed from daily temperature and radiation data using an approach similar to that described by Charles-Edwards (1982). To enable this change in approach to be implemented, the leaf dry matter coefficients were recomputed from data collected at Manhattan (Kansas, USA) and the Kimberley Research Station (northern Australia). Tiller number and seed number were also recomputed using Equation (4).

The modified version of SORGF predicted the total dry matter production and yield of sorghum crops as well as the original version of the model. However, the inclusion of Equation (4) to predict plant morphology and development provided a simpler and more robust means of simulating both tiller number and seed number. The comparison of modelled and actual values of the numbers of tillers per plant and the numbers of seeds per plant are illustrated in Fig. 5, for observations on experiments conducted on a number of different sorghum cultivars grown at sites ranging from Narrabri in New South Wales (Australia) to the Kimberley Research Station in northern Western Australia. More rigorous testing of this modified version of SORGF is now in hand.

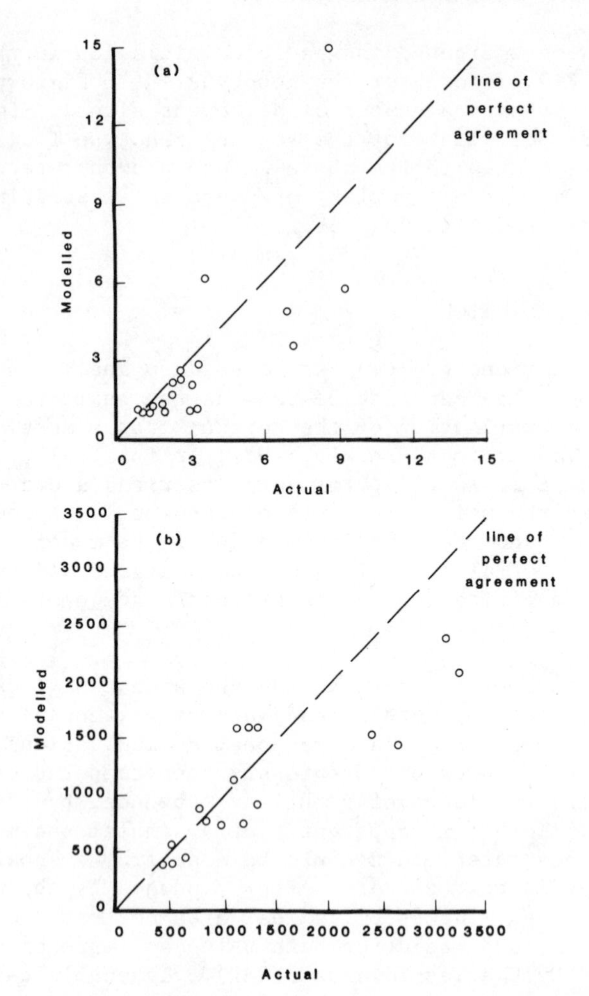

Fig. 5. A comparison of the numbers of (a) tillers per plant and (b) grain per plant predicted by the modified version of SORGF and the observed numbers for a number of grain sorghum cultivars grown at different experimental sites.

DISCUSSION

The main components (or sub-models) of most crop models could be formally identified with the growth determinants described in the first part of this paper. However, the level of mechanistic detail contained in SORGF was sufficiently simple to make the exercise almost trivial. What became immediately apparent once we had conducted the exercise was the potential value of the formal comparison in helping us identify the most appropriate methods for testing the crop model, and it is on this aspect of the work that our discussion will centre.

One of the most important criteria that must be met by any simulation model is that it is testable. The highest level, and perhaps the ultimate level, of test is of its final predictions. Even if it does predict grain yield correctly, one still needs to test whether the model is predicting the right answer for the right reasons. If the prediction is wrong, we need to establish the reasons for the error. In a detailed mechanistic simulation model, the model will predict fairly basic aspects of plant behaviour. These predictions will be about processes such as canopy photosynthetic rates, respiration rates, plant development etc. To test whether or not these predictions are correct usually involves a great deal of detailed and sophisticated experimental work. Then, if these intermediate predictions are found to be correct, yet the final prediction of yield is still in error, the methods of integrating information and understanding from the more basic physiological level to the field performance of the crop needs to be re-examined. This often requires the application of demanding and time-consuming experimental methods.

The physiological analysis of plant growth that has been described here distinguishes five determinants of crop production that are defined at a level intermediate between the basic physiology of plant processes and the field study of crop performance. The determinants are amenable to description in terms of the basic plant physiological processes yet are also amenable to direct and simple measurement in the field. They provide us with a means of integrating our knowledge of physiological processes directly into an understanding of the field performance of the crop.

However mechanistically detailed a crop model may be, it should be possible to use it to predict the total amount of dry matter produced by the crop during a particular period of time and the amount of light energy intercepted by the crop during that same time interval. By dividing one by the other the model will predict the efficiency with which the crop was using light energy in the production of new plant dry matter. This predicted value of the crop's light use efficiency can then be compared directly with measurements obtained on the dry weight changes of a crop growing in the field. Moreover, the effects of the plant's environment, such as soil water deficits and soil fertility, on any one of the growth determinants can be incorporated into the model in an empirical way until sufficient information becomes available to allow their effects on that determinant to be dealt with in a detailed mechanistic way.

REFERENCES

Acock, B., Charles-Edwards, D. A., Fitter, D. J., Hand, D. W., Ludwig, L. J., Warren Wilson, J., and Withers, A. C., 1978, The contribution of leaves from different levels within a tomato crop to canopy photosynthesis: An experimental examination of two canopy models, J. exp. Bot., 29:815.

Angus, J. F., Cunningham, R. B., Moncur, M. W., and Mackenzie, D. H., 1981a, Phasic development in field crops. I. Thermal response in the seedling phase, Field Crops Res., 3:365.

Angus, J. F., Mackenzie, D. H., Morton, R. G., and Schafer, C. A., 1981b, Phasic development in field crops. II. Thermal and photoperiodic responses of spring wheat, Field Crops Res., 4:269.

Arkin, G. F., Vanderlip, R. L., and Ritchie, J. T., 1976, A dynamic grain sorghum model, Trans. ASAE, 19:622.

Arkin, G. F., Ritchie, J., and Maas, S. J., 1978, A model for calculating light interception by a grain sorghum canopy, Trans. ASAE, 21:303.

Biscoe, P. V., and Gallagher, J. N., 1977, Weather, dry-matter production and yield, in: "Environmental Effects on Crop Physiology", J. J. Landsberg and C. V. Cutting, eds., Academic Press, London.

Charles-Edwards, D. A., and Rees, A. R., 1975, Tulip forcing: Model and reality, Acta Hort., 47:365.

Charles-Edwards, D. A., 1982, "Physiological Determinants of Crop Growth", Academic Press, Sydney.

Charles-Edwards, D. A., 1983, An hypothesis about the control of flowering, Ann. Bot., 52:105.

Charles-Edwards, D. A., 1984a, On the ordered development of plants. I. An hypothesis, Ann. Bot., 53:699.

Charles-Edwards, D. A., 1984b, On the ordered development of plants. II. Self-thinning in plant communities, Ann. Bot., 53:709.

Charles-Edwards, D. A., and Lawn, R. J., 1984, Light interception by four grain legumes grown as row crops, Pl. Cell Environ., 7:247.

Davidson, R. L., 1969, Effect of root/leaf temperature on root/shoot ratios in some pasture grasses and clover, Ann. Bot., 33:561.

Fisher, M. J., and Charles-Edwards, D. A., 1982, A physiological approach to the analysis of plant growth data. III. The effects of repeated short term water deficits on the growth of spaced plants of the legume Macroptilium atropupureum cv. Siratro, Ann. Bot., 49:341.

Greenwood, D. J., Cleaver, T. J., and Turner, M. K., 1974, Fertiliser requirements of vegetable crops, Proc. Fert. Soc., 145:5.

Greenwood, D. J., 1981, Crop response to agricultural practice, in: "Mathematics and Plant Physiology", D. A. Rose and D. A. Charles-Edwards, eds., Academic Press, London.

Maron, S. H., and Prutton, C. F., 1944, "Principles of Physical Chemistry", Macmillan, New York.

Mason, T. G., and Maskell, E. J., 1928, Studies in the transport of carbohydrate in the cotton plant. II. The factors determining the rate and direction of movement of sugars, Ann. Bot., 42:571.

Monteith, J. L., 1977, Climate and efficiency of crop production in Britain, Phil. Trans. R. Soc., Ser. B, 281:277.

Natarajan, M., and Willey, R. W., 1980a, Sorghum-pigeon pea intercropping and the effects of plant population density. I. Growth and yield, J. agric. Sci., Camb., 96:51.

Natarajan, M., and Willey, R. W., 1980b, Sorghum-pigeon pea intercropping and the effects of plant population density. II. Resource use, J. agric. Sci., Camb., 95:59.

Satoh, M., 1982, Effect of leaves retained at the time of harvest on regrowth, and changes in their physiological activity in Mulberry tree, Jap. agric. Res. Quart., 15:266.

Schwabe, W. W., and Wimble, R. H., 1976, Control of flower initiation in long- and short-day plants - a common model approach, in: "Perspectives in Environmental Biology" Vol. 2, Botany, N. Sunderland, ed., Pergamon Press, Oxford.

Thornley, J. H. M., 1978, Crop response to fertiliser, Ann. Bot., 42:817.

Vanderlip, R. L., and Arkin, G. F., 1977, Simulating accumulation and distribution of dry matter in grain sorghum, Agron. J., 68:917.

Warren Wilson, J., 1971, Maximum yield potential, in: "Transition from Extensive to Intensive Agriculture with Fertilisers", Proceedings of the 7th Colloquium of the International Potash Institute, Worbleufen-Bern, Switzerland.

A USER-ORIENTATED MODEL OF THE SOIL WATER BALANCE IN WHEAT

J. T. Ritchie

USDA-ARS, Grassland
Soil and Water Research Laboratory
Temple, Texas, U.S.A.

INTRODUCTION

In many regions of the world where agriculture is important, the season for growing crops is often characterized by large variations in water supply. These variations occur both spatially and from year to year. This uncertainty in water supply, along with other important weather variations, creates a risky environment for farmers. Models that consider the dynamics of the soil water balance as related to weather, plant and soil parameters have been proposed as a means of assisting farmers in minimizing their risks.

The evaluation of risk requires more than a model of the soil water balance. Models that can evaluate the impact of water supply on crop yield are needed to provide information for a risk analysis where the water supply is highly variable. New cultivars and other improved technology need to be risk efficient and evaluated by risk analysis. Dynamic models of crop growth that assess the importance of climatic, plant and soil properties along with farm management practices are thus needed to quantify risk.

GENERAL FEATURES OF CERES MODELS

CERES-Wheat is a wheat yield model developed by the United States Department of Agriculture, Crop Systems Evaluation Unit at Temple, Texas. It was developed to provide users with an operational model for several purposes:

- assistance with farm decision making,

- risk analysis for strategic planning,
- within-year management decisions,
- large area yield forecasting; foreign and domestic,
- policy analysis,
- definition of research needs.

In order to achieve these objectives, it was necessary to become familiar with procedures presently being used for some of these purposes and to understand some of the limitations encountered in making a model useful for a specific purpose. The main features needed for a user-orientated model appeared to be that (1) input information on weather, soils and crop genetics should be available, (2) it should be written in a familiar computer language, and (3) the computational time should be minimal. Most of the input information needed for CERES is available from routinely collected daily weather data, and soil information comes from standard soil classification data. The program is written in a familiar scientific language, FORTRAN, and runs on a main-frame computer such as the AMDHAL 470 using about one second of central processing unit time for one growing season.

In order to simplify models as much as possible, many rational empiricisms are used to incorporate information from several levels of organization into relationships needed to make the model operate in a balanced way for a community of plants growing in a field environment. CERES is a daily incrementing model and several empiricisms were developed that integrate almost minute-by-minute variations in factors like solar radiation, air temperature and rainfall into daily functions.

Because the purpose of CERES-Wheat is to provide yield estimation to users, the main features of the model deal with the factors considered to be most influential in determining final yields. These include:

- phasic development or duration of growth phases as related to plant genetics, weather, and other environmental factors,
- apical development as related to morphogenesis of vegetative and reproductive structures,
- extension growth of leaves and stems, and senescence of leaves,
- biomass accumulation and partitioning,

- the impact of soil water deficit on growth and development,

- the impact of nitrogen deficit on growth and development.

The adverse effects of weeds, pests and diseases on yield are not included because they can usually be controlled through management, and the species are so numerous that they could not be dealt with in a general way in balance with the remainder of the model. A preliminary documentation of CERES-Wheat is available (Ritchie and Otter, 1984) and a final documentation will be available in about October 1984. Some elements in the validation of the model are described in Chapter 28.

THE SOIL WATER BALANCE IN CERES-WHEAT

The soil water balance is calculated in CERES-Wheat in order to evaluate the possible yield reduction caused by soil and plant water deficits. The crop model can also be run under the assumption that the soil water balance is non-limiting for all plant processes, and in that case the water balance routine is by-passed.

The model evaluates the soil water balance of a wheat crop or fallow land using the equation:

$$S = P + I - EP - ES - R - D \tag{1}$$

where the quantity of soil water, S, is the result of the input of precipitation, P, and irrigation, I, the outputs of evaporation from plants, EP and soil, ES, and runoff, R, and drainage from the profile, D. The soil water is distributed in several layers (up to 10) with depth increments specified by the user.

Water content in any soil layer can decrease by soil evaporation, root absorption, or flow to an adjacent layer. The limits to which water can increase or decrease are input for each soil layer as the lower limit of plant water availability, the drained upper limit, and the saturated upper limit. The values used for these limits must be appropriate to the soil in the field, and accurate values are quite important in situations where the water input supply is marginal. The traditional laboratory-measured wilting point and field capacity water contents have frequently proved inaccurate for establishing field limits of water availability (Ritchie, 1981).

Our research unit has recently done a U.S.A. country-wide comparison of field-measured versus laboratory-measured limits of soil water availability and often found considerable error when laboratory methods were applied to the field (Ratliff et al., 1983). We are investigating possible reasons why laboratory pressure extraction equipment often provides biased estimates of the available water. Because of the possible bias in laboratory-measured soil water availability limits, field-measured limits are needed for best accuracy. If field limits are not available, a model developed from the U.S.A. data collected by Ratliff et al. (1983) is available for estimating the potentially extractable soil water (Cassel et al., 1983). The model uses several laboratory-measured soil physical and chemical properties as inputs.

Infiltration and Runoff

Daily precipitation, and amounts and dates of irrigation if used, are input with the weather data and with other management parameters respectively.

Infiltration of water into the soil is calculated as the difference between precipitation or irrigation and runoff. Runoff is calculated using the USDA-Soil Conservation Service (SCS) procedure termed the curve number technique (Soil Conservation Service, 1972). The procedure uses total precipitation from one or more storms occurring in a calendar day to estimate runoff. The relation excludes duration of rainfall as an explicit variable, so rainfall intensity is ignored. Runoff curves are specified by numbers which vary from 0 (no runoff) to 100 (all runoff). The SCS handbook provides a list of runoff curve numbers for various hydrological soil groups and soil-cover complexes. The SCS technique considers the wetness of the soil, calculated from previous rainfall amount, as an additional variable in determining runoff amount. The technique has been modified by J. R. Williams (personal communication) for layered soils as used in CERES. The wetness of the soil in the layers near the surface replaces the antecedent rainfall condition. This modified procedure is considered by hydrologists to be one of the most conservative models of runoff when only daily precipitation is known.

When irrigation water is applied, the runoff estimation procedure is by-passed. Thus, all irrigation is assumed to infiltrate.

Drainage

Because water can be taken up by plants while drainage is occurring, the drained upper limit soil water content is not always the appropriate upper limit of soil water availability. Many productive agriculture soils drain quite slowly, and may thus provide an appreciable quantity of water to plants before drainage practically stops. In CERES, drainage rates are calculated using an empirical relation that evaluates field drainage reasonably well.

The drainage formula assumes a fixed saturated volumetric water content, Θ_o, and a fixed drained upper limit water content, Θ_u. Thus drainage takes place when the water content, Θ_t, at any time, t, after field saturation is between Θ_o and Θ_u. The equation used is:

$$\Theta_t = (\Theta_o - \Theta_u) \exp(-K_d t) + \Theta_u \quad (2)$$

where K_d is a conductance parameter that can vary greatly between soils. Figure 1 shows a graph of the loss of water from soil with three contrasting conductance parameters using Equation (2). The value of K_d is assumed to be constant for the whole soil profile because, in many soils, the most limiting layer to water flow dominates the drainage rate from all parts of the soil profile. A problem with using Equation (2) for drainage evaluation in the field is that soils seldom reach saturation and it becomes difficult to determine an inital value for t. However, the change in water content with respect to time can be expressed in a form independent of t:

$$d\Theta_t/dt = -K_d(\Theta_t - \Theta_u) \quad (3)$$

and the drainage rate, D, from a particular layer is:

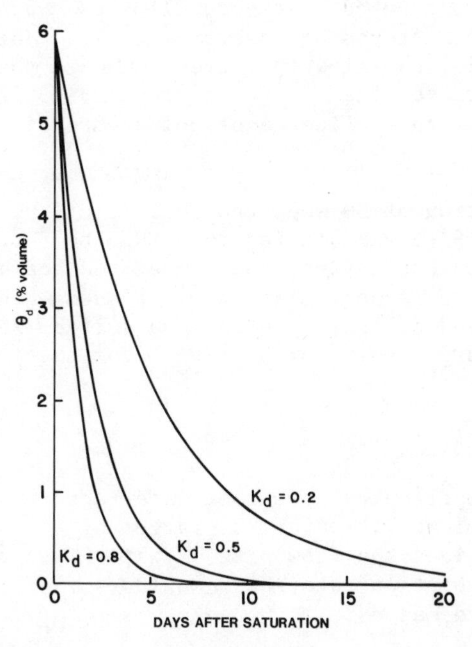

Fig. 1. A graph of the relationships used to calculate soil water drainage after soil saturation for various values of K_d (d^{-1}).

$$D = -K_d (\Theta_t - \Theta_u) z \tag{4}$$

where z is the thickness of the soil layer being considered. In the model, constant drainage throughout one day is assumed and the value of K_d represents the fraction of water between Θ_u and Θ_t that drains in one day. For Equation (4) to be used, Θ_t must be greater than Θ_u. The value of Θ_t at any depth is updated daily to account for any infiltration, water flow, or root absorption.

Evapotranspiration

Evapotranspiration, ET, is calculated using procedures described by Ritchie (1972). The procedure separates soil evaporation, ES, from transpiration, EP, for plants growing without a shortage of soil water, primarily on the basis of the energy reaching the soil, the time since the surface layer was wet, and the leaf area index, LAI. Potential ET is calculated using an equilibrium evaporation concept as modified by Priestley and Taylor (1972). A relatively simple empirical equation was developed from the complex exponential equations needed to evaluate the influence of the radiation and temperature on equilibrium evaporation. The equation calculates the approximate daytime net radiation and equilibrium evaporation, assuming that stomata are closed at night and no ET occurs then. The potential ET is calculated as the equilibrium evaporation times 1.1 to account for the effects of unsaturated air. The multiplier is increased above 1.1 to allow for advection when the maximum temperature is greater than 24°C, and reduced for temperatures below 0°C to account for the influence of cold temperatures on stomatal closure.

The calculation of ES when the soil is drying in the original model (Ritchie, 1972) was altered for CERES to further reduce ES when the soil water content in the upper layer reaches a fixed low threshold value. This modification was needed when simulating water balance for layered soils to prevent the surface soil from drying too much when roots are also removing water from near the surface.

Root Water Absorption

The CERES model calculates root water absorption using a "law of the limiting" approach in which the larger of the soil or the root resistance determines the flow rate of water into roots. The soil-limited water absorption rate, q_r, considers radial flow to single roots and is expressed as:

$$q_r = \frac{4\pi K(\Theta)\ (\Psi_r - \Psi_s)}{\ln \frac{c^2}{r^2}} \tag{5}$$

where $K(\Theta)$ is the soil hydraulic conductivity, Ψ_r is water potential at the root surface, Ψ_s is the bulk soil water potential, r is the root radius and c is the radius of the cylinder of soil through which water is moving. A useful empiricism used to calculate $K(\Theta)$ for all soils is:

$$K(\Theta) = 10^{-5} \exp (62 (\Theta - \Theta_\ell)) \text{ cm d}^{-1} \tag{6}$$

where Θ_ℓ is the lower limit water content. Comparisons from the literature suggest that, for a wide range of soil types, the values of $K(\Theta)$ at the same value of $(\Theta - \Theta_\ell)$ were within about an order of magnitude of one another for values of $(\Theta - \Theta_\ell)$ from 0 to about 7% by volume. The accuracy of measured $K(\Theta)$ values for relatively dry soil is usually in doubt by at least a factor of 5 (van Bavel et al., 1968). Thus, Equation (6) is a reasonable approximation for $K(\Theta)$ for use in water absorption calculations.

By making the usual assumption that $c = (\pi L_V)^{-\frac{1}{2}}$, where L_V is the root density (cm cm^{-3}), and further assuming that r = 0.02 cm, and $(\Psi_r - \Psi_s)$ = 21 cm water, Equation (5) becomes:

$$q_r = \frac{2.64 \times 10^{-3} \exp (62 (\Theta - \Theta_\ell))}{6.68 - \ln L_V} \tag{7}$$

The potential difference between the root and the soil is assumed to remain constant at 21 cm water for all water contents. It was determined from two experiments in which root water uptake and other factors in Equation (5) were measured or reasonably approximated (Taylor and Klepper, 1975; Gregory et al., 1978). Water absorption rates were chosen from those experiments at times when the soil was quite dry and the absorption rates were clearly being limited by soil water conductivity. This evaluation assumes that most of the water potential difference between the soil and plant leaves occurs within the plant. The resistance to water flow in plant roots has been shown to be highly dependent on flow rate (Meyer and Ritchie, 1980). Thus, such a small apparent water potential gradient of 21 cm water between the root and bulk soil is not unreasonable.

Although root resistance has been shown to have a strong relation to flow rate, there is an upper limit to the rate of flow that can occur in a root (Meyer and Ritchie, 1980). For sorghum plants growing in solution culture, the maximum absorption rate was about 0.19 cm^3 (cm root)$^{-1}$ d^{-1}. However, because flow resistance in unsaturated soil near the roots is always higher than in solution culture, and because uptake rates in CERES are the average for a day rather than instantaneous values, a flow rate of 0.03 cm^3 cm^{-1} d^{-1} was chosen as an approximate maximum plant-limited flow rate. Figure 2 is a graph of Equation (7) and shows the threshold absorption rate at which plant conductance begins to dominate flow rate. Data shown

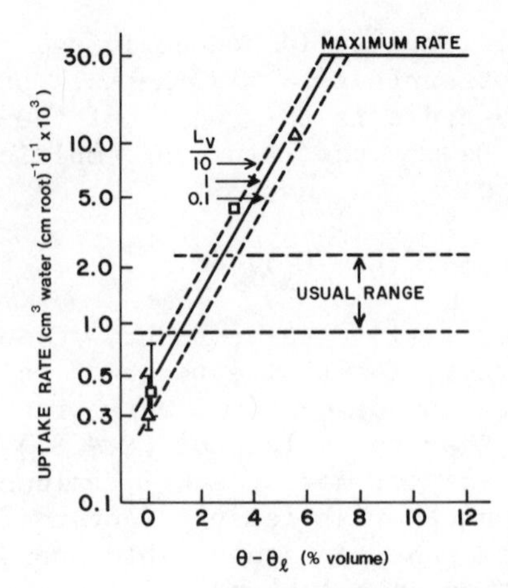

Fig. 2. A graph of the relationship used to calculate maximum root water absorption as related to $\Theta - \Theta_\ell$ (the volume water content above the lower limit) and root length density (L_V). Also shown is the assumed maximum possible rate and the usual range of absorption when all the soil profile is at an optimum water content. The observed values are from Taylor and Klepper (1975) and Gregory et al. (1978). The vertical bar lengths are twice the standard error.

in Fig. 2 are from Taylor and Klepper (1975) and Gregory et al. (1978) and represent values they measured when absorption was practically stopped as a result of dry soil ($\Theta \to \Theta_\ell$) and when a relatively few roots in deep soil were absorbing water at very high rates to compensate for low absorption by roots near the surface where the soil was dry. Figure 2 also shows the usual range of water absorption reported by Taylor and Klepper (1975) and Gregory et al. (1978) when all roots are absorbing about equally at a rate equal to the maximum transpiration demand. The graph demonstrates that for most of the range of available soil water content, root absorption for normal transpiration rates is almost 10 times less than the maximum possible rates.

In CERES, Equation (7) is used to calculate maximum soil-limited water absorption rate and this value is used unless it is greater than the plant-limited maximum rate. This soil- or plant-limited maximum absorption rate is then converted to an uptake rate for an individual soil layer using the root length density and the depth of the soil layer. Root length density and distribution in the soil are estimated in CERES, on the basis of soil properties and the amount of assimilate partitioned to roots. The sum of the maximum root absorption from each soil depth gives the maximum possible uptake from the profile. If the maximum uptake exceeds the maximum calculated transpiration rate, the maximum absorption rates calculated for each depth are reduced proportionally so that the uptake becomes equal to the transpiration rate. If the maximum uptake is less than the maximum transpiration, transpiration rate is set equal to the maximum absorption rate.

The absorption model was tested for a maize crop at Temple, Texas, in a study where root length density, transpiration and soil water were measured. The measured and calculated change in water content at several times through part of the season when the crop was growing on stored soil water are shown in Fig. 3 for two different initial soil water conditions. The results show that the model approximation of absorption provides a reasonable evaluation for the depth distribution of water content during a season.

It should be mentioned that a weak part of CERES, and of crop models in general, is the estimation of the dynamics of root growth in the soil. Some assumptions that are difficult to verify experimentally have to be used to simulate root growth patterns. The growth patterns depend on soil physical and chemical properties, the amount of assimilate transported to the roots and soil water content. More quantitative root growth information is needed before major improvement can be made in the root growth part of CERES.

USE OF CERES FOR RISK ANALYSIS

When soil water balances can be evaluated along with plant growth and yield, it is possible to simulate yields over many years

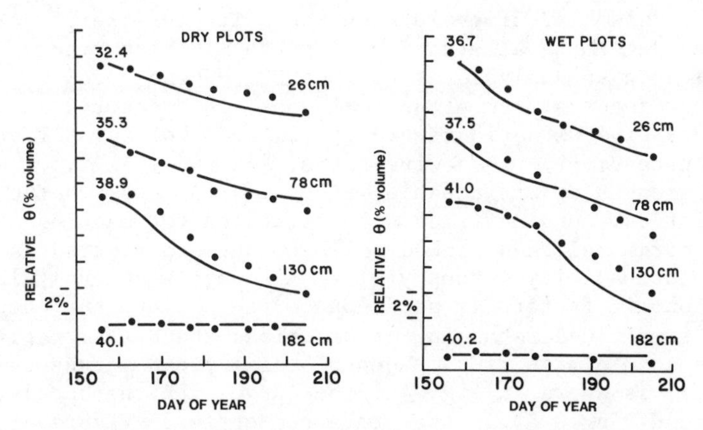

Fig. 3. A comparison of measured (dots) and model estimated (curves) soil water content at several soil depths during part of a growing season for maize. The wet and dry plots refer to the initial condition of the soil at the beginning of the evaluation. Numbers on the left of the curves are water contents at the beginning of these comparisons.

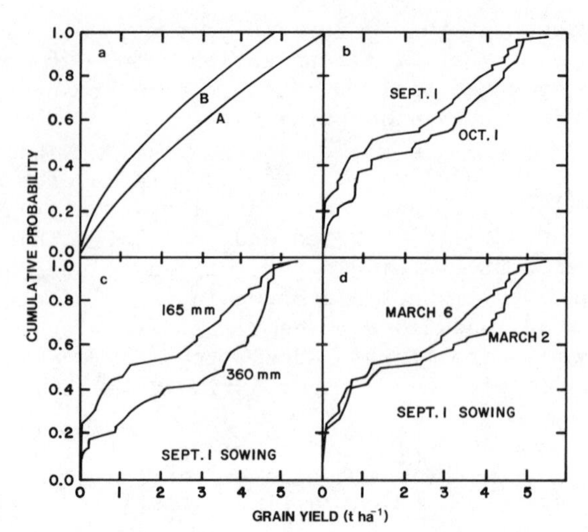

Fig. 4. Cumulative probability distribution of wheat yield at Altus, Oklahoma. When a probability curve is further to the right (higher yields at any given probability) as with curve (A) as compared with curve (B), then options chosen for curve (A) are less risky and would be preferred by farmers. The remaining figures show probability distributions of yield as influenced by sowing date (b), soil water storage capacity (c), and date of floral initiation (d).

to determine expected yield variations that result from weather variations. This kind of evaluation makes it possible to do a risk analysis to determine alternative means to minimise risks. An example of such an analysis using the CERES-Wheat model with long-term weather records from Altus, Oklahoma, is presented in Fig. 4. Three alternative situations were considered, holding all other factors constant: (1) two sowing dates, (2) two soil types with different amounts of extractable water, and (3) two genetic characteristics. Grain yields were simulated for each of 59 years of weather records and then ranked so they could be plotted as a cumulative probability versus yield. Although risk analysis should include economic factors in addition to production, the comparisons provide a simple understanding of decisions about alternative situations to minimise risk. When plotting yield versus cumulative probability, as shown in Fig. 4a, the curve with the consistently highest yield (Curve A) is the preferred option. Figure 4b,c,d show the cumulative probability distribution for the three alternative situations. Each option with the highest yields had the more favourable soil water balance during the growing season.

In the case of the sowing date (Fig. 4b), the later date is preferred because it uses less stored soil water in the autumn. In

Oklahoma, stored water is not usually replenished over winter, thus causing additional plant stress during the latter part of the season when grain filling depends on a favourable soil water supply. The extractable soil water capacity (Fig. 4c) also made a large difference in the yield cumulative probability distribution, with the larger capacity providing a better soil water balance in most years. It is common for farmers in marginal rainfall areas like Altus, Oklahoma, to use the deepest, most productive soils for crop production when there is an option. The comparison of Fig. 4d points out that a shorter-maturing genotype with an earlier floral initiation date is superior, simply because it has less chance of depleting the water supply before plant maturity, although the difference in the growing season up to floral initiation is altered by an average of only four days. Differences of that magnitude are common among maturity genotypes during cultivar selection.

The important point of such risk analysis is that it provides farmers with quantitative options, using scientific principles within the confines of highly variable weather. It is difficult to decide beween such options on the basis of 1-to-5-year research comparisons at one location, as is so often done in traditional agricultural research. In the future, I envisage the production of maps showing spatial distribution of optimum production strategies, similar to those in Fig. 4. The assembly of data bases that include climate, soil, and plant genetic characteristics will be necessary for the production of such useful information.

DISCUSSION AND CONCLUSIONS

Because weather, climate, soils, management and plant characteristics all interact to determine crop yield, an integrated approach is needed to understand crop response to soil water deficits. Evaluation of crop systems through yield models such as CERES provides a quantitative means of integrating these important production related factors.

Although all of the relationships used to calculate water balance components in CERES are quite empirical, they have all been made as general as possible to avoid using regionally-fitted parameters. The model has been tested with about 300 different measured data sets from several countries to demonstrate its generality (Ritchie and Otter, 1984; see also Chapter 28). In a few cases where soil water was measured throughout the season, the model values of soil water usually were in reasonable agreement with measured values. In cases where agreement was not good, it was generally impossible to determine the cause for the disagreement because root depth and root density were not measured. Thus in using the model where water deficit may be the dominant factor limiting yield, the measurement of root growth may be necessary to determine empirical functions needed to describe root growth dynamics for such situations.

The sub-model for root water absorption is similar to several water absorption models in that it uses the theory for flow to a single root. It differs from most other models, however, in that the water potential gradient between the root and the soil is assumed to be a constant, regardless of soil water potential or transpiration rate, for the calculation of the maximum uptake. It also differs from most models in that it calculates the maximum possible uptake and then reduces that amount if potential transpiration is less than the maximum calculated uptake. Eliminating water potential evaluations in the model without loss of accuracy simplifies the calculation, making it more practical for users.

No doubt several improvements can be made in the water balance sub-model of CERES. However, because it has provided rather reliable estimates of soil water content throughout growing seasons for wheat in several tests, its use in daily-incrementing crop yield modelling should be satisfactory until further improvements are made.

REFERENCES

Cassel, D. K., Ratliff, L. F., and Ritchie, J. T., 1983, Models for estimating *in situ* potential extractable water using soil physical and chemical properties, *Soil Sci. Soc. Am. J.*, 47:764.

Gregory, P. J., McGowan, M., and Biscoe, P. V., 1978, Water relations of winter wheat. 2. Soil water relations, *J. agric. Sci., Camb.*, 91:103.

Meyer, W. S., and Ritchie, J. T., 1980, Resistance to water flow in the sorghum plant, *Pl. Physiol.*, 65:33.

Priestley, C. H. B., and Taylor, R. J., 1972, On the assessment of surface heat and evaporation using large-scale parameters, *Monthly Weather Review*, 100:81.

Ratliff, L. F., Ritchie, J. T., and Cassel, D. K., 1983, Field-measured limits of soil water availability as related to laboratory-measured properties, *Soil Sci. Soc. Am. J.*, 47:770.

Ritchie, J. T., 1972, Model for predicting evaporation from a row crop with incomplete cover, *Water Resour. Res.*, 8:1204.

Ritchie, J. T., 1981, Soil water availability, *Pl. Soil*, 58:327.

Ritchie, J. T., and Otter, S., 1984, CERES-Wheat: A user-oriented wheat yield model. Preliminary documentation, AgRISTARS Publication No. YM-U3-04442-JSC-18892.

Soil Conservation Service, 1972, National Engineering Handbook Section 4: Hydrology, Soil Conservation Service, USDA, Washington.

Taylor, H. M., and Klepper, B., 1975, Water uptake by cotton root systems: An examination of assumptions in the single root model, Soil Sci., 120:57.

van Bavel, E. H. M., Brust, K. J., and Stirk, G. B., 1968, Hydraulic properties of a clay loam soil and the field measurement of water uptake by roots. II. The water balance of the root zone, Soil Sci. Soc. Am. Proc., 32:317.

VALIDATION OF THE CERES-WHEAT MODEL IN DIVERSE ENVIRONMENTS

S. Otter and J. T. Ritchie

USDA-ARS, Grassland, Soil
and Water Research Laboratory
Temple, Texas, U.S.A.

INTRODUCTION

Following considerable calibration and sensitivity analysis, the CERES-wheat model has been validated using independent data sets to verify model accuracy under field conditions. The main data for model validation were obtained from published experimental results, though some addional detail had almost always to be obtained through personal communication, especially concerning input data required by the model. From some experiments detailed data were also available and were used to test particular parts of the model.

For the tests, a data base was assembled to represent a diversity of wheat-growing environments, including short growing seasons in northern latitudes for spring wheat crops, environments with limited water availability, sub-tropical wheat growing areas in which little vernalisation is possible and areas where winter kill is a problem. In all, a data set of about 300 crop years was obtained for testing. It included 25 sites in the world with a range of latitudes from 36°S in Australia to 52°N in England. The value of testing depends greatly on the quality of data available. For a data set to be acceptable, there had to be soils, weather, and management information, and data on at least yield and some important phenological stages.

Goals of the validation where to establish model accuracy and limits of applicability. After evaluation of the performance of the model, general properties, such as the tendency to underestimate high yields, could be defined. The scattergram of simulated versus observed yields (Fig. 1) also shows the capability of the model to simulate yields over a wide range from 17 to 9422 kg ha^{-1}. The

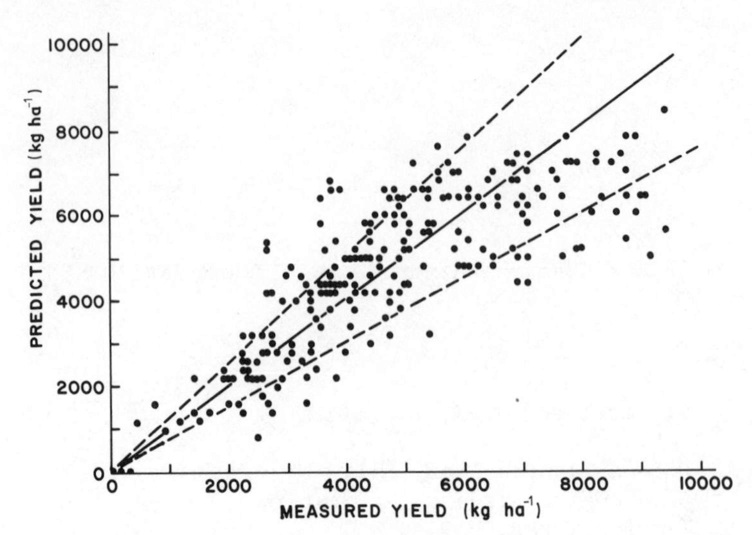

Fig. 1. Comparison of simulated yields with measured yields for 245 tests.

correlation coefficient of 0.81 for these data indicates that the model is sensitive to the factors causing yield variations. Unfortunately, the lack of detail in data did not allow an in-depth evaluation of the reasons for some large under- or over-estimates. An important aspect is the testing of the phasic development submodel because the weather conditions during certain growth phases affect growth and yield in quite specific ways. Cases where water

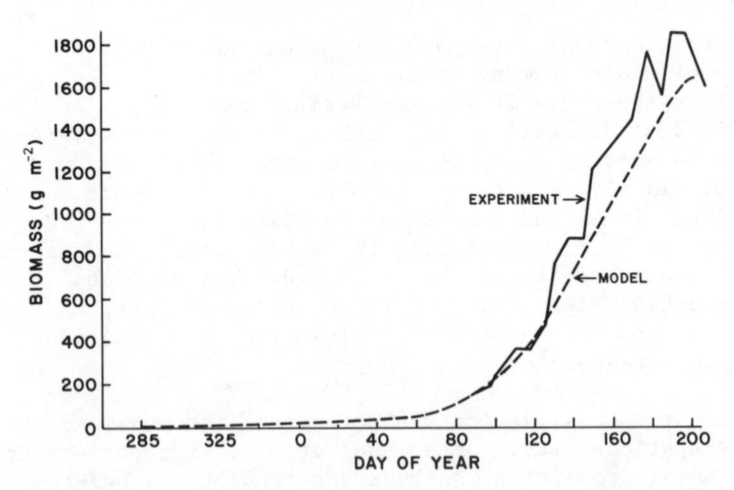

Fig. 2. Comparison of simulated with measured biomass for Maris Mardler in 1983 at Weihenstephan, Federal Republic of Germany.

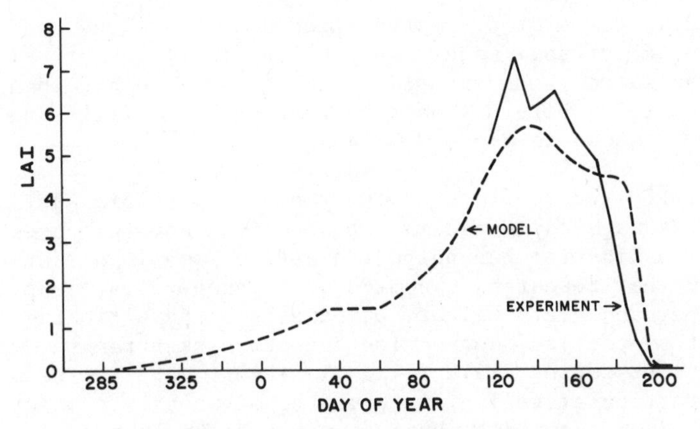

Fig. 3. Comparison of simulated with measured leaf area index (1983, Weihenstephan, Federal Republic of Germany).

deficit does not cause apparent stresses suggest that the partitioning part of the growth model could be improved. In high-yielding, dense canopies the model seems to underestimate partitioning to the shoots or to fail to produce enough assimilate per unit light interception. The data sets available were not adequate to assess this problem quantitatively.

Among the model's features are responses to different management practices such as sowing date, sowing density, cultivar and irrigation. Evaluation of the changes in simulated crop yield with variation of such input parameters has furnished valuable information about the sensitivity of the model to management. Crop response can

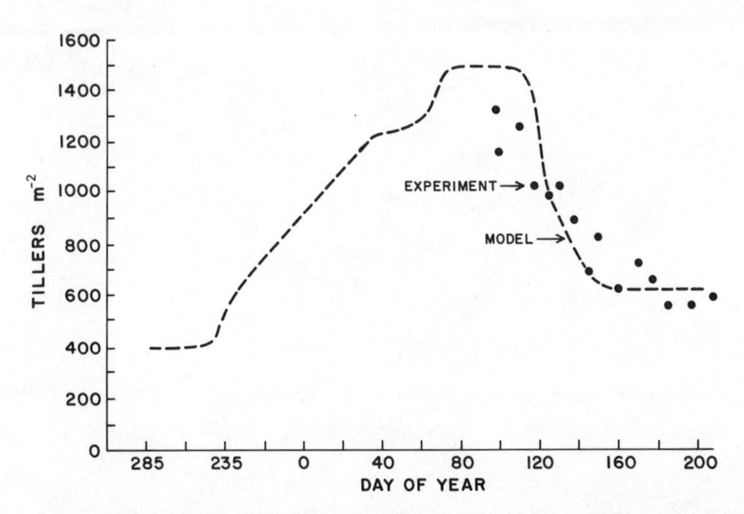

Fig. 4. Comparison of simulated with measured tiller number (1983, Weihenstephan, Federal Republic of Germany).

be checked in terms of dry matter produced, leaf area and number of tillers per m^2 or soil water availability in different soil layers through the season. An example of a test at Weihenstephan, West Germany in 1983, where dry matter, leaf area and tiller number were recorded, is shown in Figs. 2, 3 and 4.

The CERES model offers a wide range of possible applications, and the tests against experimental data indicate its potential accuracy. Validation has helped to define some important features needing further research: dynamics of organogenesis, coupling of grain number and grain weight, principles of partitioning, tillering shoot-root properties, connection between sink strength and canopy photosynthesis during grain fill. To investigate the feasibility for use in a particular region, it would be advisable to undertake a further study using a data base appropriate to that region.

SIMULATION OF NITROGEN DYNAMICS IN WHEAT CROPPING SYSTEMS

D. C. Godwin and P. L. G. Vlek

International Fertilizer Development Center
Muscle Shoals, Alabama, U.S.A.

INTRODUCTION

Annual world wheat production is approximately 450 million tons. Recent statistics compiled by Martinez and Diamond (1982) indicate that some 4.3 million tons of fertilizer N is applied to wheat annually. This survey did not include data from the U.S.S.R. and the People's Republic of China, two of the world's largest producers. Harris and Harre's (1979) statistics reveal that 9.0 million tons of fertilizer N is consumed annually in the U.S.S.R. By making some assumptions on the proportion of this that is used on wheat and making allowance for use in countries not covered by the Martinez and Diamond survey, we reach a final estimate of 8 million tons for the fertilizer N used on wheat annually.

With the approximate energy conversion factor of 12 U.S. barrels of oil per ton of fertilizer N (Mudahar and Hignett, 1982), this implies that the equivalent of 100 million barrels of oil are used each year to produce N fertilizer for the world's wheat crop. Nitrogen fertilizer consumption is projected to increase at 5% per annum (Harris and Harre, 1979). Much recent expansion in fertilizer use has taken place in the developing countries, and the need for fertilizer N will continue to increase with the growing food needs of a rapidly expanding world population striving for higher standards of living (Hauck, 1981).

The efficiency with which crops use nitrogen varies widely but is in general poor. Allison (1965) in his review suggested that the average recovery of fertilizer N in the above-ground parts of crops is commonly less than 50%. Power (1981) has indicated that usually between 20% and 90% of the fertilizer N applied is directly absorbed by the crop. The nitrogen which is not recovered by the crop may be lost from the soil:plant system through runoff, leaching, denitrification or ammonia volatilization. It may be made unavailable to the plant through immobilization in the soil, or it may become inaccessible to the plant through lack of water. The fraction that is lost from the cropping system is the source of much of the environmental pollution associated with fertilization. The magnitude of each of the various transformations affecting the use o nitrogen is influenced by many climatic, edaphic and agronomic factors. Nitrogen is thus one of the most complex of plant nutrients, and quantifying the factors affecting the crop's response to it is indeed difficult.

Craswell and Godwin (1984) attempted to delineate some of the climatic factors affecting efficiency of nitrogen use in various agroclimatic zones. They noted the extensive variability in the published data on fertilizer recovery both within and between agroclimatic zones. Defining methods of improving fertilizer efficiency and yields, given such variability, is an exacting task. Climatically-driven computer simulation models can contribute to furthering our understanding of cropping systems by quantifying most of this variability. Cooke (1979) recognized the role that models may play in future research when defining priorities for British soi science. Penning de Vries (1981) and Tejeda et al. (1981) have also highlighted the contribution simulation modelling can make in fertilizer research.

Optimizing fertilization strategies, given the uncertainties of climate, is generally difficult, but in some less developed regions of the world where fertilizer data are sparse, the problem is compounded. Where adequate climatic, soil and crop data exist, simulation madels will allow some extrapolation into these less developed areas and thus provide some insights into fertilizer behaviour in different environments.

Many different simulation models exist which describe some or all of the N cycle processes occurring in cropping systems. Some of these models have been the subject of a recent book edited b Frissel and van Veen (1981) and of a review by Tanji (1982). Some o the models cited in these reviews are concerned with specific aspect of the N cycle such as ammonia volatilization in the model of Parton et al. (1981), leaching in the models of Addiscott (1981) and Burns (1980), and denitrification in the model of Smith (1981). Most of the models are primarily concerned with the major soil processes in the N cycle and few consider the crop. Conversely, several plant-oriented models not referred to in these reviews have examined uptak processes and distribution of nitrogen within the plant, and they do not consider soil processes affecting supply of N to the plant. The N model for lettuce of Scaife (1974) and the competition model of Baldwin (1976) are of this type. Other models have been developed with the specific purpose of examining aspects of pollution from organic wastes and from excessive fertilization. The models of Rao et al. (1981), Selim and Iskandar (1978, 1981) and Donigian and Crawford (1976) are designed to address these problems. Some models, such as those of Watts and Hanks (1978), Tillotson et al. (1980), and Tanji et al. (1981), simulate most of the major soil transformations of nitrogen as well as uptake by the crop. PAPRAN (Seligman and van Keulen, 1981) is a simulation model examining annual pasture production limited by water and nitrogen.

To simulate N dynamics adequately in a range of diverse wheat cropping environments, a model capable of describing the major soil transformations, as well as the plant component, is required. The CERES-WHEAT-N model simulates growth, phenology, water and nitrogen balance, and yield, and it has widespread applicability in diverse environments.

DESCRIPTION OF CERES-WHEAT-N

The CERES model has two forms. One version is used to simulate crop growth and development under conditions where nitrogen is not limiting. This version describes evapotranspiration, soil water balance, crop development as influenced by temperature, vernalization and photoperiod, vegetative growth, root growth and grain growth. The model has been described briefly elsewhere (Ritchie and Godwin, 1983; Jones et al., 1984), and extensive documentation of it is currently in preparation (Ritchie and Otter, 1984). Aspects of the soil water balance and of the general application of this model are described in Chapters 27 and 28 respectively in this volume.

The second version of the model adds to this the description of various processes related to the dynamics of nitrogen in the soil and in the plant. This version will be described in more detail in future documentation. To provide understanding of the level at which processes are modelled, a general description follows.

The model incorporates a soil water balance component which includes calculations of surface runoff, evaporation, through-drainage and plant water extraction. The soil water balance model operates on a layer-by-layer basis with the layer depths and storage characteristics as input parameters. It includes two field-determined limits of plant-available water, a lower limit and a drained upper limit (Ritchie, 1981). Water in the profile above the drained upper limit moisture content drains to lower layers and, in so doing, leaches nitrate.

The leaching process is modelled as a function of the volume of this drainage water that moves through a layer. The nitrate and water moving out of one layer are added to the layer below, and the cascading system continues until a sufficiently dry layer or the bottom of the profile is reached. The reverse process of upwards movement of nitrate with loss of water by evaporation, and the movement associated with unsaturated flow, are also modelled.

Plant water extraction is calculated as a function of root length density, water availability and potential transpiration. A root growth submodel is included which distributes a daily increment of new root growth between layers on the basis of water and N availability. The size of the increment is a function of the carbon supply coming from the tops. Thus, root distribution is modelled as a water- and N-sensitive process, and therefore processes which are dependent on root distribution, such as water extraction and nitroge uptake, will reflect this sensitivity. This provides a valuable too for examining the effects of timing of fertilizer applications and placement of fertilizer on subsequent uptake rates and how these factors will influence other plant processes.

Light interception and photosynthesis are modelled as a functio of leaf area index and incoming solar radiation. Nitrogen deficienc in the plant reduces crop photosynthesis in two ways. Firstly, N deficiency reduces the rate of leaf expansion and this will ultimately reduce the size of the intercepting canopy; secondly, and less importantly, the photosynthesis per unit of leaf area will also be reduced. Tiller survival, stem growth and leaf senescence are also modelled as factors dependent on plant N status. The simulatio of the distribution of plant N between roots, vegetative shoot and grain encompasses factors describing the carbon balance and N status of the plant.

The model includes components that simulate plant development and thus provide estimates of the dates of important events and the duration of the distinct phases of growth in the life cycle of wheat. Plant development is defined in terms of thermal time (accumulated temperature) with modifications to cater for the plant's sensitivity to photoperiod and vernalization. Differences between cultivars in the length of the growing season are accounted for by including some genetic-specific constants in the model. The timing of the onset of N stress will thus have differing ramifications for different cultivars in the same environment. The capacity of the model to describe these important differences enables modelling studies to identify genetic attributes required to suit a specific climatic and N environment. The model calculates growth rates of leaves, stems, ears, roots and grain during phases defined by the plant's development.

Crop residue decomposition and the turnover of the soil humic fraction with subsequent ramifications for the mineralization and immobilization of nitrogen in the soil are included in the simulation. Soil temperature, soil moisture, residue composition, and N availability are the key components used in prediction of residue decay rates. The balance between mineralization and immobilization of nitrogen associated with this decay is a function of C:N ratio. The method used is based on a modification of that employed by Seligman and van Keulen (1981).

Crop N demand is determined on the basis of the nitrogen deficiency of the existing plant biomass plus a component related to the N required for new growth. Potential N uptake is a function of root length density, N availability, soil water availability and a scaled maximum uptake per unit of root length. The actual uptake rate is the lesser of N supply and N demand. The model has the capacity to differentiate between nitrate and ammonium uptake. Since organic matter decomposition, crop N uptake and growth, and soil and fertilizer N transformations all occur simultaneously, each process will affect soil N availability and will thus have feedback implications on the rates of the other processes.

Denitrification of nitrate is modelled as a function of water availability above the drained upper limit, soil temperature and availability of carbon. The method is based on that used by Rolston et al. (1980). Nitrification of ammonium is modelled as a function of substrate concentration, water availability, soil temperature, and a lag phase factor that is dependent upon recent soil history. Simplistic first-order rate equations are employed in the simulation of nitrification and denitrification. The development of routines to describe urea dissolution and hydrolysis and volatilization of ammonia is at an early stage. Rate equations dependent upon environmental variables will be included in these routines (Vlek and Carter, 1983).

The various interrelationships and feedbacks among the processes modelled are illustrated in Fig. 1.

Model Inputs

To facilitate model usage across a wide range of locations and model applications, the input data required have been kept to a minimum. The requirement for daily climatic data is in the form of precipitation, maximum temperature, minimum temperature, and solar radiation. When irrigation is applied, the dates and quantities of irrigation water are needed. The data required to define soil water characteristics for each layer include layer depth and the lower limit, drained upper limit, saturation and initial moisture contents. Similarly for the N component of the model extractable nitrate and ammonium concentrations, organic carbon content, bulk density, and pH are required for each layer. The model also requires an estimate of the amount of crop residue present and its C:N ratio, as well as fertilizer information in the form of date, rate and depth

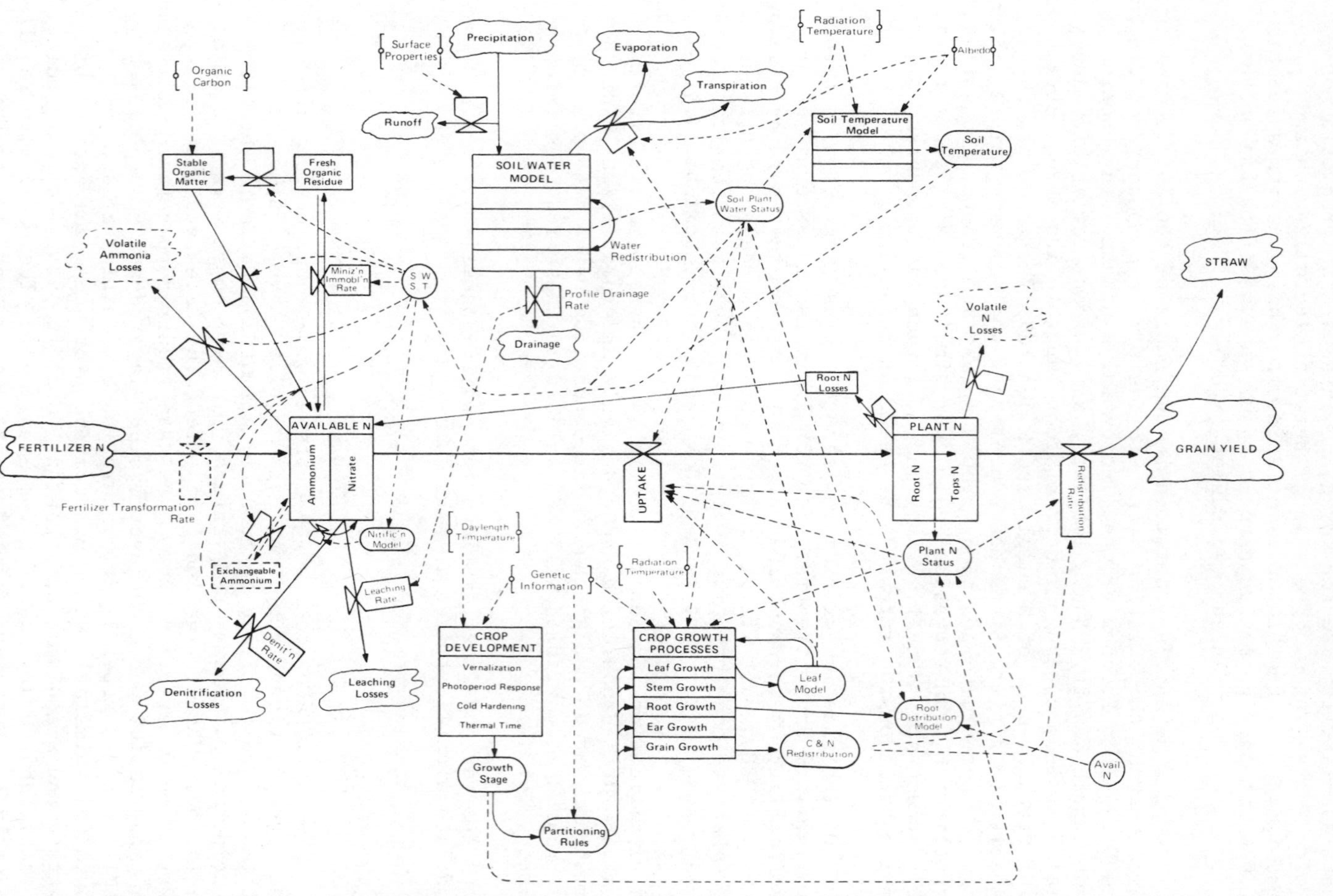

Fig. 1. Systems diagram of CERES-WHEAT-N: solid lines, material flows; dotted lines, information

of all applications made to the crop, and the type of fertilizer used.

EVALUATION OF THE CERES-WHEAT-N MODEL

The capability of the CERES model has been extensively tested (Otter et al., 1983). Since the N components of the model have undergone less development work to date, testing is less advanced. Most of the validation has been in relation to plant components of the model.

For thorough model evaluation, test data sets (published and unpublished) have been obtained representing 22 site-year combinations. These data sets span the range of winter-planted spring wheats in Syria and Australia, spring-planted spring wheats in the United States and Canada, and winter wheats in North America and Europe. Each data set had a range of N rates and often a range of other treatments, among them irrigation timing, fertilizer application timing and cultivar. In each data set, pests, diseases and nutrients other than nitrogen were considered not to be limiting growth. The observed grain yield in these data sets ranged from 0.33 t ha^{-1} at Lancelin, Western Australia, in the experiment of Mason et al. (1972), to 8.58 t ha^{-1} in the Flevopolder in the Netherlands, in the study of Spiertz and Ellen (1978).

Preliminary testing of the model against the range of data sets available indicates that the model generally gives reasonable predictions of grain yield, biomass, N uptake by the above-ground plant and the partitioning of this nitrogen into grain (Fig. 2). Some comparisons between predicted and observed response to applied N in some representative locations are illustrated in Fig. 3. The model was able to simulate the response to fertilizer in highly N-responsive conditions (Wageningen) and in non-responsive conditions (Madras, Oregon).

In the experiment of Campbell et al. (1977), plants were grown in small field-placed lysimeters at various rates of N application. Half of the lysimeters were irrigated throughout the experiment. In the dryland treatment, the response to nitrogen was limited because of the moisture stress and this effect was matched by the model. In the irrigated treatments there was a much greater response to N, but the model simulated this poorly.

The experiments of Mason et al. (1972) provide a good test of the model's capability to predict response to timing of a fertilizer application. In this experiment on a coarse sand in Western Australia, urea was applied at planting or at 2, 4 or 8 weeks after planting. The later the application date the less nitrogen was lost via leaching and the greater was both the recovery of nitrogen by the

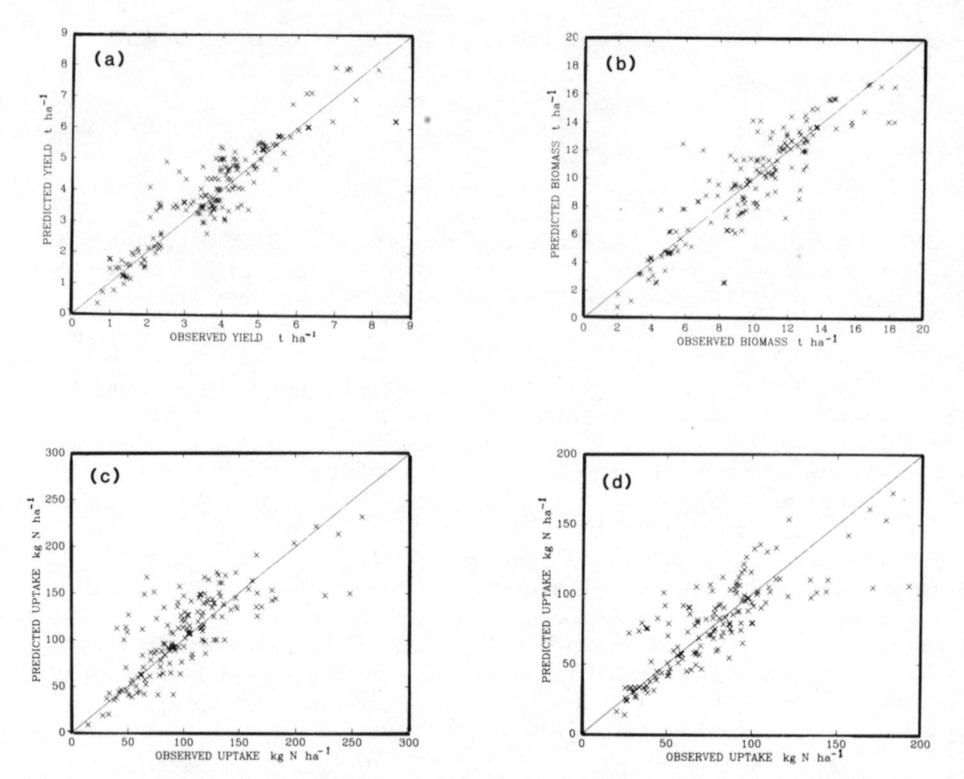

Fig. 2. Comparison of predictions of the CERES-WHEAT-N model with observed data from field experiments, for (a) grain yield, (b) total above ground biomass, (c) N uptake by above-ground plant, and (d) N uptake by grain. The 1:1 line (prediction = observation) is included in each case.

crop and the grain yield. The model was able to simulate these effects (Fig. 3).

The model can reliably account for observed yield, dry-matter production and N uptake and its partitioning under most conditions, as indicated by the correlation coefficients tabulated in Table 1. To test the ability of the model to predict the response to N, an index of relative response (RR) was calculated. Grain yields with no N fertilizer applied were given a value of unity and yields from treatments receiving N fertilizer were expressed relative to this:

$$RR = \frac{\text{Grain yield (fertilized)}}{\text{Grain yield (unfertilized)}}$$

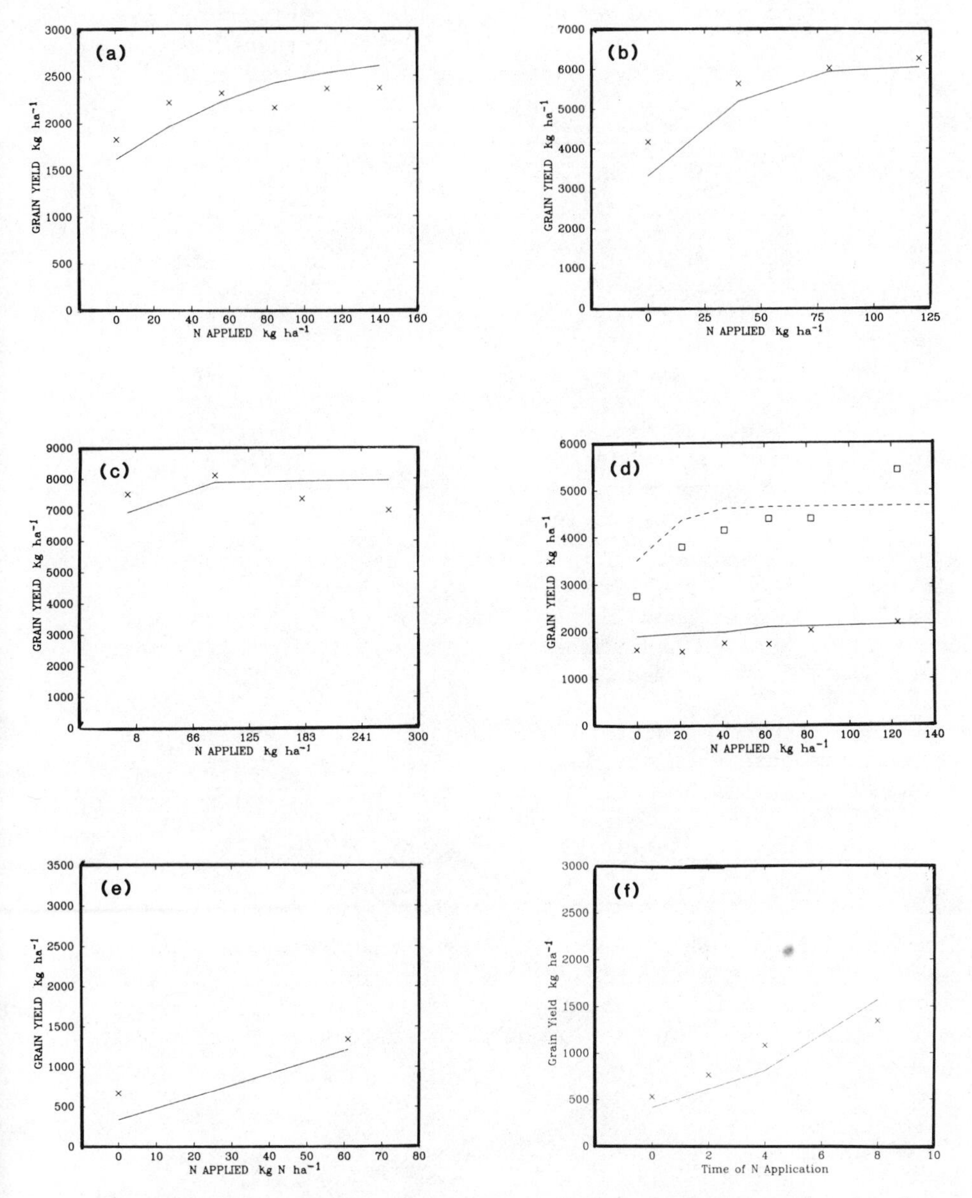

Fig. 3. Observed response (symbols) to N application rate (a–e) or timing (f) and predicted response (lines) (a) Hutchinson, Kansas (Kissel; personal communication), (b) Wageningen, The Netherlands (Ellen and Spiertz, 1980), (c) Madras, Oregon (Ambler, 1976), (d) Swift Current, Canada (Campbell et al., 1977): Irrigated (upper line, squares); Dryland (lower line, crosses), (e) Wongan Hills, Australia (Mason and Rowley, 1969), (f) Lancelin, Australia (Mason et al., 1972).

Table 1. Values of the simple correlation coefficient (r) for comparisons between predicted and observed data for several key variables

Variables	r
Grain yield	0.91***
Biomass	0.88***
Total N uptake	0.78***
Grain N uptake	0.81***
Straw N concentration	0.58***
N uptake at anthesis	0.84***
Relative response	0.89***
Apparent recovery	0.49**

Values of RR less than unity indicate a negative response to N, which can occur on dry sites (Storrier, 1965). In our data set there were negative responses, and also results from highly responsive conditions where RR had values greater than 2, and this provides a rigorous test of the model.

The uptake components of the model, as well as its ability to predict response to nitrogen, can be tested by comparing the apparent N recovery (AR) predicted with that calculated from observed data:

$$AR = \frac{\text{N uptake (fertilized) - N uptake (unfertilized)}}{\text{N rate applied}}$$

Since AR depends upon the N uptake from two different treatments, small errors in the prediction of either can lead to quite spurious predicted values for apparent recovery. These compounding errors are partly responsible for the low, though still significant, value of r in Table 1.

USING THE MODEL TO EXAMINE N DIFFERENCES IN DIVERSE LOCATIONS

Stochastic Weather Generation and Risk Analysis

Grain yield, fertilizer recovery, and the processes affecting them vary greatly from year to year in any location. Thus, to develop optimal fertilizer strategies in any location, it would be desirable to have fertilizer experiments conducted over many years. Long-term data of this type are seldom obtained, and it is in this role that modelling excels. Where long-term weather records exist, the model can be run to provide a more complete picture of fertilizer response variations over time. An alternative to the use of observed weather data is to use stochastic generation of weather data. Richardson (1981), Nicks (1974) and Stern et al. (1982) describe practical methods for this stochastic generation process. Coupling such generators to simulation models can produce a very flexible and powerful tool for rapidly examining fertilizer strategies.

Anderson (1973, 1974) has proposed methods for determining optimal strategies when operating under conditions of uncertainty and risk. These risk analysis procedures are seldom used in agronomic research but are often employed in economic research. Dowling and Smith (1976) utilized a stochastic water balance model and risk analysis procedures to determine the optimum time for pasture establishment in highly variable soil moisture regimes in the New England area of Australia. Subsequently, Smith and Harris (1981) and Stapper (1984) have utilized these techniques with wheat models to define optimal sowing times and maturity types for wheat grown under the variable rainfall conditions of the Middle East.

Initial Conditions

The CERES-WHEAT-N model, coupled with a version of the Richardson (1981) weather generator, was used to simulate N dynamics in wheat cropping systems in three diverse locations. The three sites examined were: Warooka in South Australia with a Mediterranean climate; Rothamsted, England, with a humid temperate climate; Topeka, Kansas, with a drier temperate climate. At Warooka, winter-planted spring wheats are grown with a growing season of approximately seven months and, at Rothamsted and Topeka, winter wheats are grown with growing season of 11 months and nine months, respectively. The model utilized 50 years of climatic data generated daily and representative of each of the three sites, and it commenced with the same initial conditions for each of the 50 simulations. For each of the 50 years, five fertilizer rates varying from 0 to 120 kg N ha^{-1} in increments of 30 kg N ha^{-1} were examined. The fertilizer placement depth in the simulations was 5 cm, and the application time coincided with planting. An additional two treatments, with 60 kg N ha^{-1} as either split or delayed applications, were also tested.

Water storage characteristics and initial N inputs typical of a Rhodustalf (red-brown earth) soil were used for Warooka and those typical of an Argiustoll were used for Topeka. To highlight differences between the various locations, the soil water and N characteristics on a structureless sand were used for Rothamsted: it should be noted that this is not representative of the soils actually found at Rothamsted. Genotypic data representative of wheat cultivars typically grown in each of these locations were used to match the differing growing seasons.

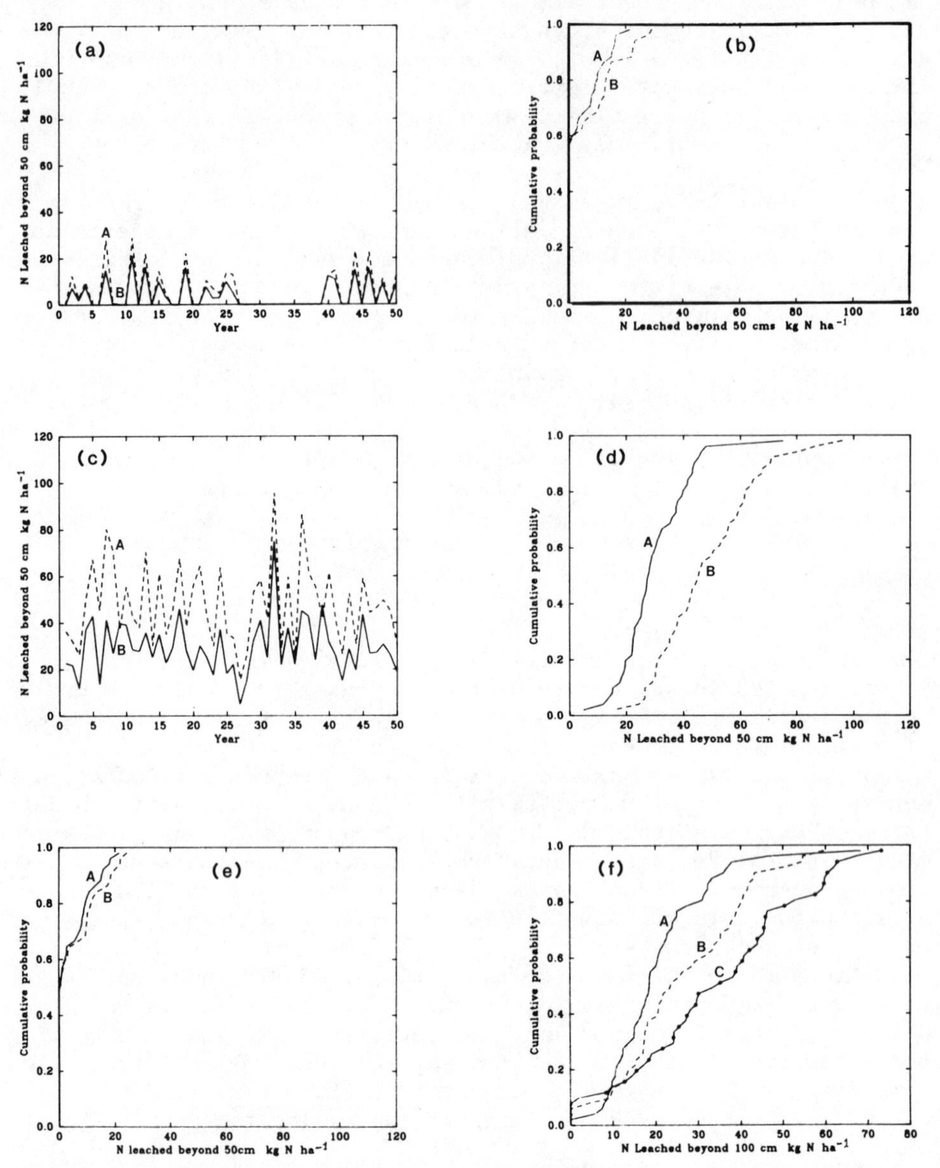

Fig. 4. (a) Simulated leaching of nitrate from a layer 50 cm deep for two fertilizer rates (30 and 90 kg N ha^{-1}; A, B respectively) at Warooka. (b) Cumulative probability distribution for data in (a). (c) Simulated leaching of nitrate from a layer 50 cm deep for two fertilizer rates (3 and 90 kg N ha^{-1}; A, B respectively) using sandy soil characteristics at Rothamsted. (d) Cumulative probability distribution for data in (c).

Fig. 4 (cont.) (e) Cumulative probability distribution for nitrate leaching from a layer 50 cm deep for two fertilizer rates (30 and 90 kg N ha^{-1}; A, B respectively) at Topeka. (f) Cumulative probability distributions for nitrate leaching from a layer 1 m deep for three fertilizer rates (30, 60 and 90 kg N ha^{-1}; A, B, C respectively) using sandy soil characteristics at Rothamsted.

←

Mineralization. The model predicted an N mineralization rate in an average growing season of 30 kg N ha^{-1} at both Warooka and Rothamsted with a range of from 15 to 45 kg N ha^{-1} over the 50-year span. The model predicted a much larger average mineralization rate of 70 kg N ha^{-1} at Topeka, resulting from the higher soil organic matter content and the higher initial mineral N content.

Nitrate deposition. Leaching of nitrate may be a significant loss mechanism in some wheat cropping systems. Some redistribution of nitrate was predicted to occur in the upper layers of the profile at each of the three sites. As shown in Fig. 4a, the amount of nitrate moving down from a layer 50 cm deep in the profile at Warooka varied greatly from year to year. In many years there may be no movement at all, but in some years up to 30 kg N ha^{-1} may be leached below this depth. When these data are expressed in terms of the cumulative probability of the event occurring (Fig. 4b), a clearer picture of the frequency and magnitude of nitrate movement is obtained. Clearly, in about 60% of the years, no leaching occurred and in 90% of the years less than 20 kg N ha^{-1} was leached beyond this depth. Increasing the amount of fertilizer N applied slightly increased the amount of nitrogen redistributed beyond 50 cm at Warooka. Leaching beyond 1 m depth occurred in less than 10% of the years simulated. These predictions are in accordance with the observations of Prescott and Piper (1930) on the disposition of nitrate in similar soils at a nearby location.

Nitrate moving down from the layer 50 cm deep may not be lost to the crop, but its movement to deeper layers may affect subsequent uptake patterns and the distribution of root growth. However, water moving beyond 1 m at Warooka can be considered as lost since water extraction from below this depth occurs infrequently.

Denitrification. The soil profile at Rothamsted was more frequently saturated than that at the other two sites and this led to a greater rate of denitrification (Fig. 5). Annual losses seldom exceeded 8 kg N ha^{-1} at Warooka, and, in the 50 years simulated, losses were always less than 17 kg N ha^{-1}. Predicted losses at Topeka were larger than at Warooka, and the amount lost was more sensitive to the rate of fertilizer applied. At Rothamsted, other processes (presumably predominantly uptake) compete for nitrate, and the losses do not necessarily increase in proportion to the rates of

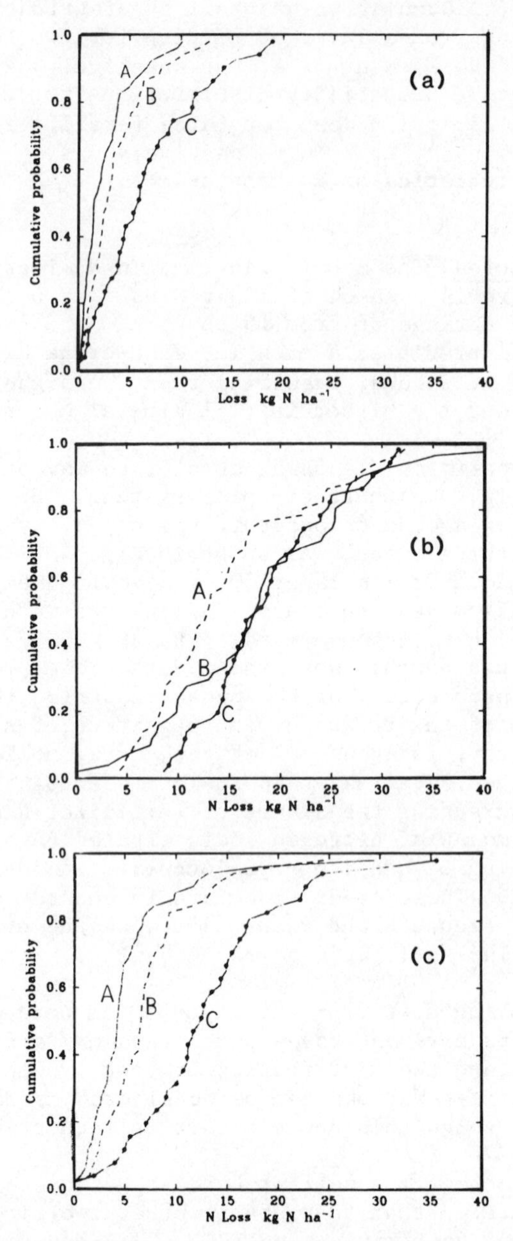

Fig. 5. Cumulative probability distributions of simulated denitrification losses from soils with three rates of fertilizer 0, 30 or 90 kg N ha^{-1} (A, B, C respectively) at (a) Warooka, (b) Rothamsted and (c) Topeka.

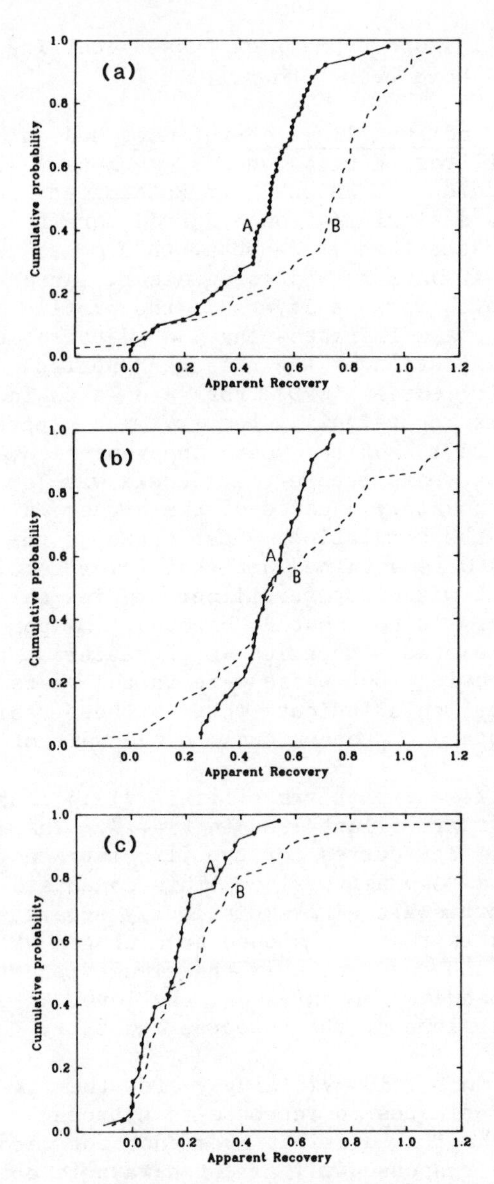

Fig. 6. Cumulative probability distributions of simulated apparent recovery of fertilizer N applied at rates of 30 (B) and 90 kg N ha^{-1} (A) at (a) Warooka, (b) Rothamsted and (c) Topeka.

fertilization. Losses at Rothamsted varied from zero to 40 kg N ha^{-1}; if the simulation had been made using soil characteristics of a more clayey soil, more typical of those occurring at Rothamsted, then losses might have been higher.

Grain yield and fertilizer efficiency. Median apparent recovery of fertilizer (Fig. 6) when 90 kg N ha^{-1} was applied was not markedly different at Warooka and Rothamsted. Given that losses from leaching and denitrification occurred more frequently and were greater at Rothamsted than at Warooka, this poses the question as to why the recovery at Warooka was not greater. An analysis of the frequency with which various layers in the profile are sufficiently dry to impede N uptake indicates that, at Warooka, periodic droughts often impede exploitation of the nitrogen applied. The cumulative probability density curves (CPDF) for Warooka do indicate, however, that in some years recovery will be very much greater and, conversely, that in 20% of the years apparent recovery will be less than 40%. In most years recovery at Topeka was lower than at the other two sites, possibly because of the higher initial soil N content used for the simulation. Most often at the three sites, the higher recoveries were attained with the lower fertilizer rates. This is consistent with most fertilizer studies surveyed by Craswell and Godwin (1984). At Rothamsted, however, in approximately 40% of the crops the simulated recoveries were greater at higher rates of fertilizer application, but these were in the years in which recovery was poorest. This would indicate that in these years some N loss was occurring which was not proportional to the rate of N application.

The very different patterns of grain yield response to nitrogen at the three sites are illustrated in Fig. 7. The closeness of the cumulative probability curves for the five rates at Topeka indicates that this is not a responsive site. This contrasts with Rothamsted where the five curves are always distinctly separate. This separation indicates that a response to nitrogen will always be obtained on the sandy soil used in the simulations for Rothamsted and the distinction between the curves at any level of probability is indicative of the slope of the response function.

At Warooka, in the low-yielding years, the five curves are not distinct, which indicates no response to nitrogen. In most years the CPDF for 30 kg N ha^{-1} is distinct from that for zero N application, suggesting that a response will almost always be obtained. Response to nitrogen at rates higher than 60 kg N ha^{-1} seldom occurs, as indicated by the overlapping CPDFs.

Given the magnitude of the N losses at Rothamsted through leaching and denitrification, strategies which reduce these losses may substantially improve recovery and yield. To examine this hypothesis, some simulations were run using an ammoniacal fertilizer source. The nitrification process was blocked for a period of 30 days to mimic the effect of nitrification inhibitor. As the model in its present form ignores leaching of ammonium N, the losses cited above are solely of nitrate. Even in this sandy soil, maintenance of the fertilizer in the ammoniacal form should then reduce losses. In most years a small reduction in denitrification and leaching was

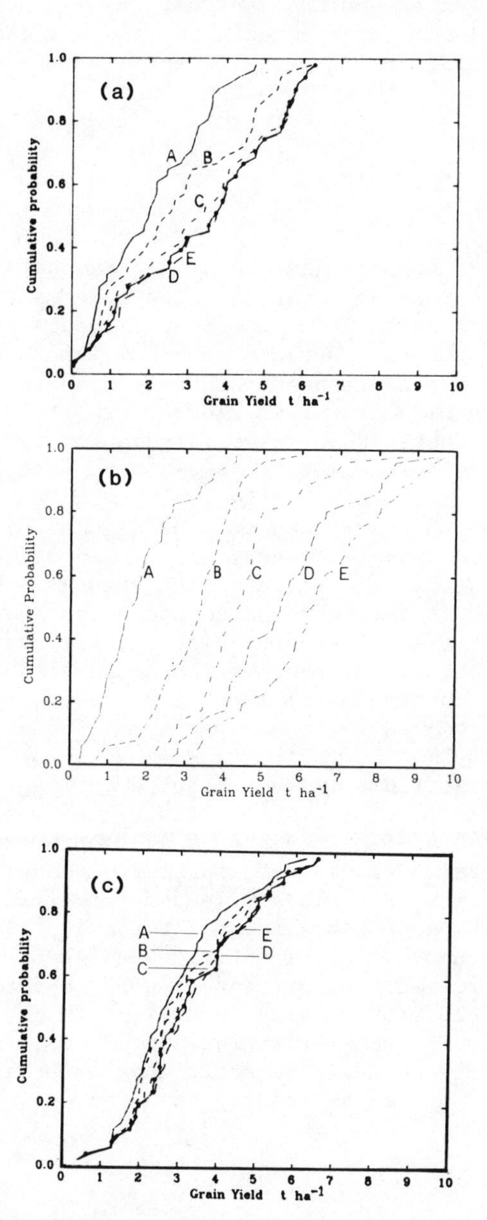

Fig. 7. Cumulative probability distributions of simulated grain yields for 5 fertilizer rates (0, 30, 60, 90 and 120 kg N ha^{-1}; A, B, C, D, E respectively) at (a) Warooka, (b) Rothamsted and (c) Topeka.

noted. This led to a small improvement in fertilizer recovery, but made no difference in yield. In many cases nitrification could have proceeded rapidly after the 30-day period, so the losses were merely delayed rather than prevented. Further examination of N management strategies showed that improvements in recovery and grain yield could be attained by split application of fertilizer N.

CONCLUSIONS

It should be stressed that the model can only account for variations in the factors defined in the model's description and this assumes that other potentially important factors such as other nutrients, pests and diseases are non-limiting. The model is unable to account for volatile losses of ammonia from the soil surface, leaching of ammonium N, or direct losses of nitrogen from plants to the atmosphere. Where these losses are substantial the model will be inaccurate. Since the soil water and N components of the model operate in a one-dimensional manner, placement patterns such as side banding and point placement can only be simulated by assuming uniform incorporation into a layer. Under certain conditions this may lead to erroneous predictions. These various problems will be the subject of future studies to further enhance the model's versatility. From the validation studies cited above, it appears that the model is able to explain most of the observed variation in yield and other key parameters where these conditions of loss and placement do not apply.

The model, particularly when coupled to long-term weather data or generated climatic data, is a valuable tool for providing insights into the behaviour of many aspects of a cropping system. Running the model with long-term weather data enables a quantification of the temporal variability in yield and response to fertilizer. The frequency and nature of N losses from the system leading to poor fertilizer efficiency can be identified and their relative significance evaluated. Experiments can be performed on the model to determine the effects of varying fertilizer rates, timing, placement depths, fertilizer sources, planting times, etc., and in this way a fertilizer strategy to maximize efficiency under the uncertainties of climatic variability can be readily obtained.

ACKNOWLEDGEMENTS

The authors wish to gratefully acknowledge the researchers who generously provided unpublished data for model development and testing: Dr. Stapper of ICARDA, Syria, and Dr. Hooker, Dr. Wagger and Dr. Kissel of Kansas State University.

REFERENCES

Addiscott, T. M., 1981, Leaching of nitrate in structured soils, in: "Simulation of Nitrogen Behaviour of Soil-Plant Systems", M. J. Frissel and J. A. van Veen, eds., Pudoc, Wageningen.

Allison, F. E., 1965, Evaluation of incoming and outgoing processes that affect soil nitrogen, in: "Soil Nitrogen", W. V. Bartholomew and F. E. Clark, eds., Agronomy Monograph No. 10, American Society of Agronomy, Madison, Wisconsin.

Ambler, J. R., 1976, Varietal differences in response of winter wheat varieties to nitrogen fertilizer and environments, Ph.D. dissertation, Oregon State University.

Anderson, J. R., 1973, Sparse data, climatic variability and yield uncertainty in response analysis, Amer. J. agric. Econ., 55:77.

Anderson, J. R., 1974, Risk efficiency in the interpretation of agricultural production research, Rev. Marketing agric. Econ., 42:131.

Baldwin, J. P., 1976, Competition for plant nutrients in soils: A theoretical approach, J. agric. Sci., Camb., 87:341.

Burns, I. G., 1980, A simple model for predicting the effects of leaching of fertilizer nitrate during the growing season on the nitrogen fertilizer needs of crops, J. Soil Sci., 31:175.

Campbell, C. A., Davidson, H. R., and Warder, F. G., 1977, Effects of fertilizer N and soil moisture on yield, yield components, protein content and N accumulation on the above-ground parts of spring wheat, Can. J. Soil Sci., 57:311.

Cooke, G. W., 1979, Some priorities for British soil science, J. Soil Sci., 30:187.

Craswell, E. T., and Godwin, D. C., 1984, The efficiency of nitrogen fertilizer applied to cereals grown in different climates, in: "Advances in Plant Nutrition", P. B. Tinker and A. Lauchli, eds., Praeger Scientific, New York.

Donigian, A. S., Jr., and Crawford, N. H., 1976, "Modelling Pesticides and Nutrients on Agricultural Lands", EPA-600/2-76-043, U.S. Environmental Protection Agency, Ada, Oklahoma.

Dowling, P. M., and Smith, R. C. G., 1976, Use of a soil moisture model and risk analysis to predict the optimum time for the aerial sowing of pastures on the Northern Tablelands of New South Wales, Aust. J. Exp. agric. Anim. Husb., 16:871.

Ellen, J., and Spiertz, J. H. J., 1980, Effects of rate and timing of nitrogen dressings on grain yield formation of winter wheat (T. aestivum L.), Fertil. Res., 1:177.

Frissel, M. J., and van Veen, J. A., eds., 1981, "Simulation of Nitrogen Behaviour of Soil-Plant Systems", Pudoc, Wageningen.

Harris, G. T., and Harre, E. A., 1979, "World Fertilizer Situation and Outlook - 1978-85", Technical Bulletin T-13, International Fertilizer Development Center, Muscle Shoals, Alabama.

Hauck, R. D., 1981, Nitrogen fertilizer effects on nitrogen cycle processes, in: "Terrestrial Nitrogen Cycles. Processes,

Ecosystem Strategies and Management Impacts", F. E. Clark and T. Rosswall, eds., Ecological Bulletin No. 33, Stockholm.

Jones, C. A., Ritchie, J. T., Kiniry, J. R., Godwin, D. C., and Otter, S., 1984, The CERES wheat and maize models, in: "Minimum Data Sets for Agrotechnology Transfer", IBSNAT/ICRISAT Symposium, Hyberabad, India.

Martinez, A., and Diamond, R. B., 1982, "Fertilizer Use Statistics in Crop Production", Technical Bulletin T-24, International Fertilizer Development Center, Muscle Shoals, Alabama.

Mason, M. G., and Rowley, A. M., 1969, The fate of anhydrous ammonia and urea applied to a wheat crop on a loamy sand in the wheat belt of Western Australia, Aust. J. Exp. agric. Anim. Husb., 9:630.

Mason, M. G., Rowley, A. M., and Qualye, D. J., 1972, The fate of urea applied at various intervals after the sowing of a wheat crop on a sandy soil in Western Australia, Aust. J. Exp. agric. Anim. Husb., 12:171.

Mudahar, M., and Hignett, T., 1982, "Energy and Fertilizer: Policy Implications and Options for Developing Countries", Technical Bulletin T-20, International Fertilizer Development Center, Muscle Shoals, Alabama.

Nicks, A. D., 1974, Stochastic generation of the occurrence, pattern and location of maximum amount of daily rainfall, in: "Proceedings of Symposium on Statistical Hydrology", Misc. Pub. No. 1275, USDA, Washington DC.

Otter, S., Ritchie, J. T., and Godwin, D. C., 1983, Tests of a wheat yield model - CERES-WHEAT - in diverse environments, in: "Agronomy Abstracts of 75th Annual Meeting", American Society of Agronomy, Madison, Wisconsin.

Parton, W. J., Gould, W. P., Adamson, F. J., Torbit, S., and Woodmansee, R. G., 1981, NH_3 volitilization model, in: "Simulation of Nitrogen Behaviour of Soil-Plant Systems", M. J. Frissel and J. A. van Veen, eds., Pudoc, Wageningen.

Penning de Vries, F. W. T., 1981, Simulation models of growth of crops particularly under nutrient stress, in: "Physiological Aspects of Crop Productivity", A. von Peter and H. Künzli, eds., Proceedings of the 15th Colloquium of the International Potash Institute, Worbleufen-Bern, Switzerland.

Power, J. F., 1981, Nitrogen in the cultivated ecosystem, in: "Terrestrial Nitrogen Cycles. Processes, Ecosystem Strategies and Management Impacts", F. E. Clark and T. Rosswall, eds., Ecological Bulletin No. 33, Stockholm.

Prescott, J. A., and Piper, G. R., 1930, Nitrate fluctuations in a South Australian soil, J. agric. Sci., Camb., 20:517.

Rao, P. S. C., Davidson, J. M., and Jessup, R. E., 1981, Simulation of nitrogen behaviour in the root zone of cropped land areas receiving organic wastes, in: "Simulation of Nitrogen Behaviour of Soil-Plant Systems", M. J. Frissel and J. A. van Veen, eds., Pudoc, Wageningen.

Richardson, C. W., 1981, Stochastic simulation of daily

precipitation, temperature and solar radiation, Water Resour. Res., 17:182.

Ritchie, J. T., 1981, Soil water availability, Pl. Soil., 58:327.

Ritchie, J. T., and Godwin, D. C., 1983, CERES-Wheat: A general, user-oriented wheat yield model, in: "Agronomy Abstracts of the 75th Annual Meeting", American Society of Agronomy, Madison, Wisconsin.

Ritchie, J. T., and Otter, S., 1984, CERES-Wheat: A user-oriented wheat yield model. Preliminary documentation, AgRISTARS Publication No. YM-U3-04442-JSC-18892.

Rolston, D. E., Sharpley, A. N., Toy, D. W., Hoffman, D. L., and Broadbent, F. E., 1980, "Denitrification as affected by irrigated frequency of a field soil". EPA-600/2-80-06. U.S. Environmental Protection Agency, Ada, Oklahoma.

Scaife, M. A., 1974, Computer simulation of nitrogen uptake and growth, in: "Plant Analysis and Fertilizer Problems. Proceedings of the 7th International Colloquium on Plant Analysis and Fertilizer Problems", J. Wehrmann, ed., Hanover, Germany.

Seligman, N. G., and van Keulen, H., 1981, PAPRAN: A simulation model of annual pasture production limited by rainfall and nitrogen, in: "Simulation of Nitrogen Behaviour of Soil-Plant Systems", M. J. Frissel and J. A. van Veen, eds., Pudoc, Wageningen.

Selim, H. M., and Iskandar, I. K., 1978, Nitrogen behaviour in land treatment of wastewater: A simplified model, in: "International Symposium on the State of Knowledge in Land Treatment of Wastewater", H. L. McKim, Co-ordinator, U.S. Army Cold Regions Research and Engineering Laboratory, Hanover, New Hampshire.

Selim, H. M., and Iskander, I. K., 1981, A model for predicting nitrogen behaviour in slow and rapid infiltration systems, in: "Modelling Wastewater Renovation by Land Disposal", I. K. Iskandar, ed., Wiley, New York.

Smith, K. A., 1981, A model of denitrification in aggregated soils, in: "Simulation of Nitrogen Behaviour of Soil-Plant Systems", M. J. Frissel and J. A. van Veen, eds., Pudoc, Wageningen.

Smith, R. C. G., and Harris, H. C., 1981, Environmental resources and restraints to agricultural production in a Mediterranean-type environment, Pl. Soil, 58:31.

Spiertz, J. H. J., and Ellen, J., 1978, Effects of nitrogen on crop development and grain growth of winter wheat in relation to assimilation and utilization of assimilates and nutrients, Neth. J. agric. Sci., 26:210.

Stern, R. D., Dennet, M. D., and Dale, I. C., 1982, Methods for analysing daily rainfall measurements to give useful agronomic results. II. A modelling approach, Exp. agric., 18:237.

Stapper, M., 1984, The use of a simulation model for the prediction of wheat cultivar response to agroclimatic factors in semi-arid regions. Ph.D. Thesis, University of New England, Armidale, Australia.

Storrier, R. R., 1965, Excess soil nitrogen and the yield and uptake of nitrogen by wheat in southern New South Wales, Aust. J. exp. Agric. Anim. Husb., 5:317.

Tanji, K. K., 1982, Modelling of the nitrogen cycle, in: "Nitrogen in Agricultural Soils", F. J. Stevenson, ed., Agronomy Monograph No. 22, American Society of Agronomy, Madison, Wisconsin.

Tanji, K. K., Mehran, M., and Gupta, S. K., 1981, Water and nitrogen fluxes in the root zone of irrigated maize, in: "Simulation of Nitrogen Behaviour of Soil-Plant Systems", M. J. Frissel and H. van Veen, eds., Pudoc, Wageningen.

Tejeda, H., Godwin, D. C., and Sidhu, S. S., 1981, Conceptual models for maximizing fertilizer use efficiency, in: "Strategies for Achieving Fertilizer Consumption Targets and Improving Fertilizer Use Efficiency", Fertilizer Association of India, New Delhi.

Tillotson, W. R., Robbins, C. W., Wagenet, R. J., and Hanks, R. J., 1980, "Soil Water, Solute and Plant Growth Simulation", Utah Agric. Exp. Stn Bull. 502, Utah State University.

Vlek, P. L. G., and Carter, M. F., 1983, The effect of soil environment and fertilizer modifications on the rate of urea hydrolysis, Soil Sci., 136:56.

Watts, D. G., and Hanks, R. J., 1978, A soil-water-nitrogen model for irrigated corn on sandy soils, Soil Sci. Soc. Am. Proc., 42:492.

MODELLING THE GROWTH OF WINTER WHEAT TO IMPROVE NITROGEN FERTILIZER RECOMMENDATIONS

C. Belmans, and K. De Wijngaert

Catholic University
Leuven, Belgium

INTRODUCTION

In this paper, we present the aims, background information, methodology and the first year's results of our project. The reactions we get from this workshop will influence the further evolution of our work. The project is run in close co-operation with a "Nitrogen Fertiliser Recommendation Group" in The Netherlands. One of the objectives of this study group is to build a field-specific management system. The institutes which are member of this study group are: PAGV, Lelystad; IB, Groningen; CABO, Wageningen and the Agricultural University, Wageningen.

The aims of the project can be divided on the basis of their time scale.

Long-term aims. We plan to produce a field-specific nitrogen fertilizer recommendation system for winter wheat which takes into account the weather pattern during the growing season. The system will be based upon the application of simulation models for the growth of winter wheat and the nitrogen and water availability in the soil profile.

Short-term aims. We are building a model for the growth and development of winter wheat. This crop growth model must be able to calculate potential growth and also the actual growth under the influence of water and nitrogen shortages, and will include submodels for water and nitrogen availability in the soil.

BACKGROUND INFORMATION

Most N-fertilizer recommendation systems for winter wheat in Western Europe are based upon the measurement of the mineral nitrogen content of the soil profile in spring. Some systems take into account the field characteristics in order to estimate the mineralization to be expected and the efficiency of nitrogen fertilizer use. The separate dressings of fertilizer are applied either on fixed dates or at development stages (e.g. tillering, jointing, heading). The optimal N-dose is determined either by long-term experiments, with a range of N-levels, or by the N-uptake to be expected to reach a target yield.

However, the availability of water and nitrogen (N-supply) and the growth of the crop (determining the N-demand) are strongly dependent on the weather conditions. An ideal N-fertilizer recommendation should therefore balance the N-demand and N-supply during crop growth. This would require a regular sampling of soil and plants, which is very expensive and time consuming. An alternative is the use of models for

i) simulation of growth and N-uptake

ii) the nitrogen and water availability during crop growth.

A RECOMMENDATION SYSTEM USING SIMULATION MODELS

Although the system is not yet worked out, because the simulation models required are not yet complete, we would like to present in outline the procedure we have in mind. The first N-application, for example in early March, can be determined as follows. The models are set up with some observations of mineral N in the soil and crop characteristics, and run for an average weather pattern up to a specific development stage of the plants, e.g. the double ridge stage. The amount of growth gives us the N-uptake by the crop during this period. The model for N-processes in the soil gives us the time course of the mineral nitrogen content in the soil profile. If the crop growth doesn't reach its potential maximum during this period, the model is run again but with a fertilizer dose added to the mineral nitrogen in the soil. In this way we can determine the N-fertilizer requirement for an average weather pattern. We can then run the model regularly during the growing season, but now using the actual weather, and find out when a minimum threshold value for nitrogen in the soil is reached, based on the simulated N-amounts in the soil. This value may be reached earlier or later than expected for the average weather pattern. If it is reached earlier one can consider a second N-application, possibly after an additional test in the field. The size of this second application can be determined by repeating the procedure, as described above.

THE SIMULATION MODELS

Nitrogen fertilizer recommendation systems based on simulation models become very complex if extensive models of the different processes involved are used. To be useful as a practical tool, the models need to contain only the most important features of the processes. The amount of input data should be as restricted as possible. For this reason we have started with rather simple models, which will be improved and extended if necessary.

Soil Water and Nitrogen Model

This model is used for two purposes: i) to estimate the amount of mineral nitrogen present in the soil profile in early spring - the calculations can start after the harvest of the previous crop; and ii) to calculate the water and nitrogen availability during the growth season. For our purposes we selected part of the PAPRAN-model (Seligman and van Keulen, 1981). The model describes soil water movement, leaching, and mineralization of humus and fresh organic material. In a workshop on 'Practical Nitrogen Models' (held in Groningen, The Netherlands, in 1983) the performances of six similar models were compared (P. Willigen and J.J. Neeleson, personal communication). Each participant in the workshop was provided with a

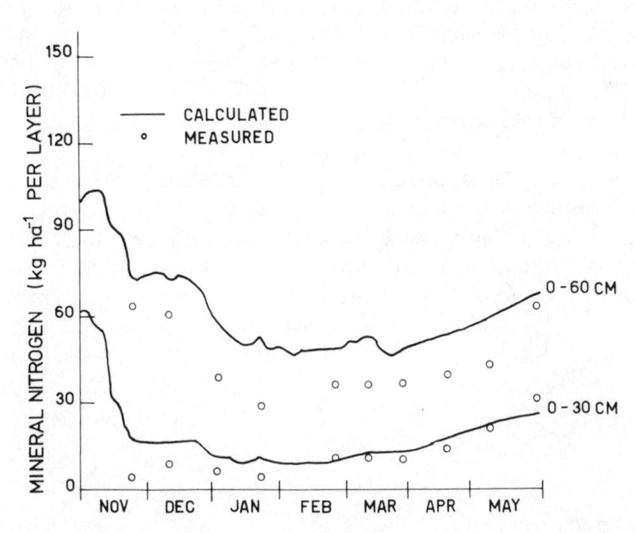

Fig. 1. Measured and simulated time course of the mineral nitrogen content of two soil layers. Experimental data from Nagele (The Netherlands) November 1977 to May 1978.

Table 1. Observed and simulated dates of development stages of winter wheat in 1983.

Development stage	Observed	Predicted
Double ridge	23 March to 7 April	4 April
Anthesis	15 June	22 June
Ripeness	19 July	24 July

set of data from an experimental field at Nagele (Groningen, The Netherlands). The initial conditions (on 1 November 1977) consisted of the mineral nitrogen and water content profiles and the amount of composition of humus and fresh organic material. The season's rainfall, potential soil water evaporation, soil temperature and the groundwater level were given. To validate the models, the mineral nitrogen distribution in the soil profile was given for a number of days during the period considered. The model we are now using predicted approximately the correct leaching of mineral nitrogen but overestimated the mineralization of nitrogen from the fresh organic material (Fig. 1). A correction for this is being developed.

More experimental data are needed to validate the model. To this end, measurements of mineral nitrogen content in the soil profile of six fields near Louvain have been made between 15 November 1983 and 1 March 1984. In one field a dose of 200 kg Cl ha^{-1} was applied, and leaching of chloride was measured to validate the leaching part of the model.

Development of Winter Wheat

The model for development of winter wheat is a simplified version of the model of Weir et al. (1984). Using data provided by R.J. Delhaye (Station de Phytotechnie, Gembloux, Belgium; personal communication), a comparision was made between calculated and measured dates of anthesis and ripeness. For crop growth during the period 1976 to 1981, the model predicted these two development stages with a mean deviation of ± 4 days. During the growing season of 1983 we measured development of winter wheat at Hélécine, identifying stages in accordance with Kirby and Appleyard (1981). The agreement between observation and prediction was satisfactory (Table 1).

Potential Growth of Winter Wheat

The model we are building - though inspired by the comprehensive spring wheat model of H. van Keulen (personal communication) - in its details more closely follows the simpler methods of Penning de

Vries and van Laar (1982). A fixed time step of 1 day is adopted, and plant growth is divided between roots, leaves, stems (including leaf sheaths), a reserve pool for carbohydrates, and grains. The main processes are:

1) Photosynthesis, according to the procedure of Goudriaan and van Laar (1978).

2) Maintenance respiration, proportional to dry matter and temperature.

3) Partitioning of assimilates. Dry matter distribution was derived from measurements and related to the calculated development stage (Fig. 2). A functional balance is not included.

4) Growth of plant organs. This is calculated as the product of

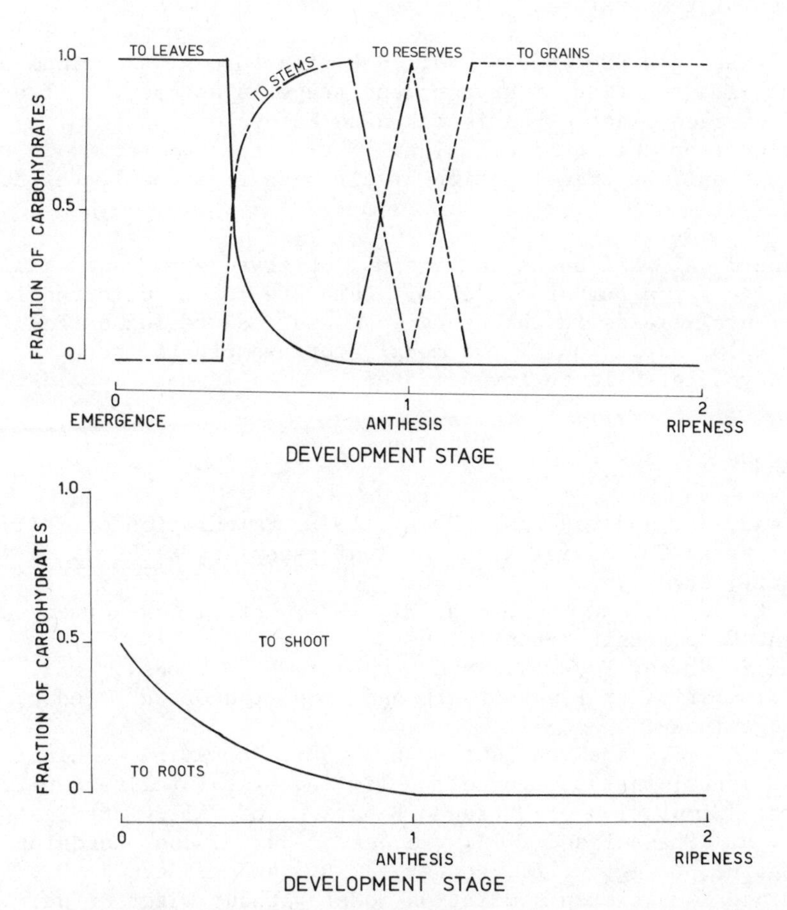

Fig. 2. Carbohydrate partitioning during the growth of winter wheat as a function of development stage.

assimilate flow to a particular organ and a conversion efficiency factor.

5) Leaf area increase, calculated as the product of a specific leaf area (m^2 leaf surface per kg dry matter in the leaves) and the increase in dry weight. A constant value of 20 is used for the specific leaf area, although experimental results in 1983 showed a decrease from 30 (in April) to about 15 (in July). In that experiment the flag leaves had a nearly constant value equal to 20.

6) Decline in leaf area, as a function of the nitrogen content of the leaves after anthesis.

7) Grain growth. The number of grains per hectare is calculated from the dry weight of the crop at anthesis. Carbohydrates from photosynthesis and translocation of reserve carbohydrates are used for grain filling. The maximum growth rate of the grains is a function of temperature.

8) Nitrogen processes. For each plant organ an optimum nitrogen content as a function of development stage is assumed. Nitrogen demand of each plant organ is taken to be proportional to the deviation from this optimum value. Nitrogen accumulation in the grains is only by translocation of nitrogen from the leaves and the stems.

The model will be tested against observed growth and grain production for a number of years. Data are being collected from experimental fields in the sandy loam and loam region of Belgium. The need for expansion of the model, for example to increase leaf and tiller dynamics will be investigated.

REFERENCES

Goudriaan, J., and van Laar, H. H., 1978, Calculation of daily totals of gross CO_2 assimilation of leaf canopies, Neth. J. agric. Sci., 26:373.

Kirby, E. J. M., and Appleyard, M., 1981, Cereal Development Guide, NAC Cereal Unit, Stoneleigh.

Penning de Vries, F. W. T., and van Laar, H. H., eds., 1982, "Simulation of Plant Growth and Crop Production", Pudoc, Wageningen.

Seligman, N. G., and van Keulen, H., 1981, PAPRAN: A simulation model of annual pasture production limited by rainfall and nitrogen, in: "Simulation of Nitrogen Behaviour of Soil-Plant Systems", J. M. Frissel and J. A. van Veen, eds., Pudoc, Wageningen.

Weir, A. H., Bragg, P. L., Porter, J. R., and Rayner, J. H., 1984, A winter wheat crop simulation model without water or nutrient limitations, J. agric. Sci., Camb., 102:371.

USING A WHOLE CROP MODEL

A. H. Weir,* W. Day,* and T. G. Sastry‡

*Rothamsted Experimental Station
Harpenden, U.K.

‡Indian Agricultural Research Institute
New Delhi, India

INTRODUCTION

A crop model is a quantitative description of crop growth. Using it involves understanding how it works, checking its performance against the growth of real crops, improving it and making simulations to explain or predict crop growth for different seasons, soils, management inputs and disease and pest damage. As confidence in the model's performance grows its scope and number of potential users also increase.

A winter wheat crop model is being developed jointly by staff of the Agricultural and Food Research Council (AFRC) at the Letcombe Laboratory, Long Ashton Research Station, Plant Breeding Institute and Rothamsted Experimental Station. The model is based on quantitative relationships between physiological processes and the crop's environment. The objective of the work is to understand the contribution of changing environmental conditions to the variation in yield of winter wheat crops in the U.K. The basic model simulates the growth of healthy winter wheat crops, free from nutrient and water stress. Such simulations are appropriate to many wheat crops in the U.K., where diseases and pests are well controlled, there are adequate inputs of N fertilizer to soils rich in P and K and the combination of large reserves of soil water and relatively small potential soil moisture deficits minimises water stress (Gales, 1983).

In this paper a simulated crop is used to explain how the model works. Simulations are then compared with actual crop data, and ways of extending or improving the model are considered. Finally, examples are given of how the model may be used to predict or explain crop performance in particular sites and seasons.

THE BASIS OF THE MODEL

The model is in the form of a FORTRAN program, 2,000 lines long. The version used in this paper is an improvement on that described for cv. Hustler by Weir et al. (1984) and Porter (1984), although the majority of the routines are unchanged. Some small changes have been made to development parameters to make the model more appropriate to cv. Avalon.

The weather data required to run the model are daily maximum and minimum temperature, total short wave radiation and air humidity (at 0900 GMT). The results from a simulation using weather data from Littlehampton (on the south coast of England) illustrate the structure of the model (Fig. 1).

Development

The development of the crop (Fig. 1a) is divided into phases by the following growth stages: sowing (S), emergence (E), floral initiation (FI), double ridges (DR), terminal spikelet (TS), anthesis (A), end of grain-filling (EG) and maturity (M). The duration of each phase is defined in thermal time (temperature above 1°C accumulated daily) such that the stages match observations on real crops. The beginning of ear growth (BE) and of grain growth (BG) are additional stages used in the model to mark the start of linear growth of the ear, before anthesis, and of grain respectively; in real crops this initial growth is exponential. The duration of each phase between emergence and anthesis depends upon photoperiod (development is faster in long days) and that from emergence to double ridges also includes a vernalization requirement (development is faster after a period of low temperatures) (Weir et al., 1984).

→

Fig. 1. The development and growth of winter wheat, cv. Avalon, simulated by the AFRC computer model, against date expressed as Julian days, for a stand of 200 plants m^{-2} sown on 16 October, 1980 with daily weather for Littlehampton, Sussex.
(a) Growth stages : Sowing S, emergence E, floral initiation FI, double ridges DR, terminal spikelet TS, beginning of ear growth BE, anthesis A, beginning of grain-filling BG, end of grain-filling EG, maturity M. The end of the dashed line is the time at which the flag leaf reached maximum size.
(b) Main stems MS, tiller classes T1, 2, 3, 4.
(c) Green area index, GAI, for main stems and tillers.

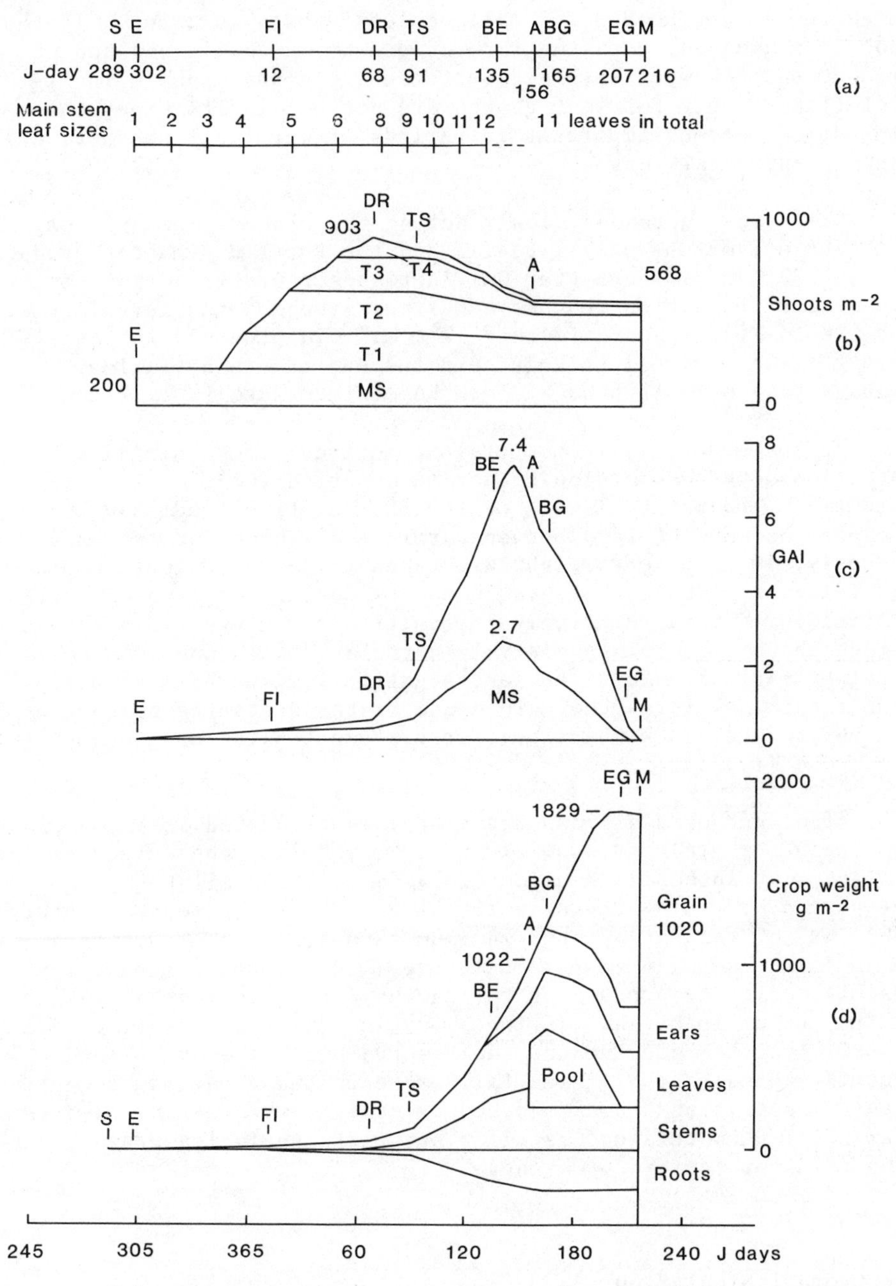

(d) Weights of plant parts, g m^{-2} D.M. The grain yield was equivalent to 12.0 t ha^{-1} at 85% D.M. Components of yield: harvest index 0.57, no. of grains 23,530 m^{-2}, weight per grain 43.7 mg, no. of ears 568 m^{-2}, grains per ear 41.4.

which the amount needed for maximum leaf growth is taken. If the pool is exhausted, both the rate of growth of new leaves and of production of new tillers are decreased. In the example in Fig. 1 assimilate supply for leaf growth did not become limiting; this limitation becomes important in periods of warm dull weather, and at large plant densities.

Leaf area increases slowly during the winter, and by terminal spikelet GAI may be only 1 to 1.5 and the total dry matter produced 100 to 200 g m^{-2}. Thereafter GAI increases rapidly and the rate of dry matter production reaches a maximum around anthesis. In the example in Fig. 1 approximately 17 t ha^{-1} is produced in the 115 days between terminal spikelet and the end of grain filling, an average rate of 0.15 t ha^{-1} d^{-1}; the maximum rate is 0.22 t ha^{-1} d^{-1}.

In the model, grain numbers are derived from assimilate partitioned to the developing ear in a period immediately prior to anthesis. The period is 45% of the thermal time (modified by photoperiod sensitivity) betwen terminal spikelet and anthesis. At anthesis, 10 mg of ear weight are taken to be equivalent to one grain set (Fischer, 1983). Ear growth occurs during the time of maximum GAI values. It is relatively insensitive to changes in peak GAI values above 6, but is positively correlated with the mean daily total of radiation received, and negatively with the mean daily temperature. This is the main cause of the differing numbers of grains produced in simulations for different sites and seasons (see below and Fig. 5).

After anthesis all new dry matter is allocated to a pool that is available for grain growth. In addition, 30% of the weight of leaves and stems at anthesis is allocated as mobile assimilate to the pool. The period of grain growth is set at 240°C days (base temperature 9°C). Each grain grows at a maximum rate of $0.045T + 0.4$ mg d^{-1} where T is the daily mean temperature (°C). Total demand for assimilate is the product of the demand per grain and the number of grains set at anthesis. If the amount of daily net assimilate is insufficient to meet demand, the reserve pool is used. If that becomes exhausted the grains fail to reach their maximum potential size. Respiration can exceed photosynthesis when GAI has declined to near zero; when this occurs all plant parts including grain lose a proportion of their dry weight.

The Overall Simulation

Figure 1 illustrates the general structure and some of the details of a crop simulation. A general understanding of the model and its relationship to physiological principles can be assessed from such simulations. A number of features of the model that may need

Leaf and Tiller Growth

Leaves appear on the main stem after fixed intervals of thermal time (Fig. 1a). Following Baker et al. (1980), the model allows this interval to vary according to the rate of change of daylength at plant emergence. Groups of tillers (cohorts) are produced at weekly intervals after the appearance of the third leaf on the main stem, the numbers produced per plant being proportional to the thermal time in the preceding week. Tillering stops at double ridges, and some tillers die between then and anthesis (Fig. 1b); the tiller survival probability is directly related to tiller age and inversely to population density. The effects of photoperiod and vernalization requirement on the early stages of plant development lead to differences in the maximum number of tillers and of leaves per shoot in different environments.

Leaf length increases linearly in thermal time during a period that is a multiple of the leaf appearance interval, and at a rate per unit thermal time that is proportional to the leaf's potential maximum size. The size of a leaf is influenced by ontogeny - leaves produced early (numbers 1 to 7) are of similar size and relatively small, whilst later leaves steadily increase in size up to leaf number 12 (Gallagher, 1979). This switch to progressively increasing leaf size is synchronized with the double ridge stage. If, as in the example in Fig. 1, only 6 main stem leaves have appeared by double ridges, then leaf size 7 is omitted.

Leaf senescence is linked to leaf production and growth so that when new leaves are appearing there are three green leaves per shoot, and after flag leaf appearance leaf area declines steadily to zero at about crop maturity (Porter, 1984).

The leaf canopy consists of the green area of the leaves and leaf sheaths of the main stems and tillers. The contribution that tillers make to the canopy varies with the numbers of plants established. The example (Fig. 1c) shows 64% of the maximum green area index (GAI) of 7.4 on tillers; this is typical for 200 plants m^{-2}.

Photosynthesis and Partitioning

Light interception by the crop canopy is calculated using an exponential attenuation coefficient of 0.44. Photosynthesis is then calculated for each layer of the canopy, and respiration subtracted to give daily dry matter increase (Weir et al., 1984). This is partitioned between plant parts in proportions that change with the phase of development (Fig. 1d).

The assimilate partitioned to leaves is held in a pool from

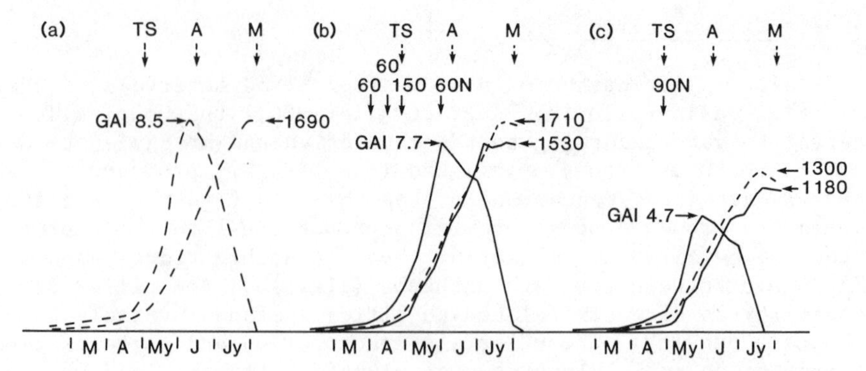

Fig. 2. Observed crop growth of winter wheat, cv. Avalon, sown 10 November 1981 at Broom's Barn and computer simulated growth for the same site and season, (a) N3 crop: simulated GAI and weight of tops, g m^{-2} D.M., (b) N3 crop: observed GAI and weight of tops ——, simulated weight of tops using observed GAI ---, (c) N1 crop: symbols as in (b). Broken arrows indicate simulated growth stages, abbreviations as in Fig. 1. Full arrows indicate timing of N applications; amounts are in kg ha^{-1}.

improvement are apparent - loss of leaf area by senescence is controlled in the model by thermal time alone, tiller loss is based on competition with other shoots not on competition for assimilate, the distribution of shoots among classes does not affect grain numbers or grain growth. These are all features that can only be considered further in relation to detailed observations on field crops.

CHECKING THE MODEL AND INTERPRETING OBSERVATIONS ON ACTUAL CROPS

The model was originally developed and tested using data for the cultivar Hustler grown in a series of multifactorial experiments at Rothamsted (Prew et al., 1983). More detailed testing required data from an experiment with more intensive and frequent crop monitoring, such as that from Broom's Barn Experimental Station (Willington and Biscoe, 1983). The analysis presented here uses two treatments from that experiment: N fertilizer at 90 kg ha^{-1} (N1) and at 330 kg ha^{-1} (N3) for a crop of cv. Avalon sown in November 1981.

The model assumes no limitation to growth due to shortage of nitrogen, and this should have been true of the N3 crop. Figure 2(a and b) shows the measured GAI and weight of the above-ground crop, and the results of a simulation that assumed a stand of 330 plants/ m^{-2} and used weather data from Broom's Barn; the agreement is generally good.

Table 1. Grain yield, top weight at anthesis (A) and maturity (M), harvest index and yield components for Avalon winter wheat sown on 10th November, 1981 at Broom's Barn and for computer simulations of that crop.

	N3 (330 kg N ha^{-1})			N1 (90 kg N ha^{-1})	
	Observed	Simulated		Observed	Simulated
		Obs.GAI	Sim.GAI		Obs.GAI
Grain yield (dry matter) g m^{-2}	845 ±85	675	689	594 ±86	612
Grains m^{-2} x 10^{-3}	19.4 ±2.8	17.3	17.7	14.4 ±2.3	15.7
Grain weight mg	43.5 ±1.8	39.0	39.0	41.4 ±0.7	39.0
Top weight (M) g m^{-2}	1543 ±150	1702	1673	1179 ±160	1255
Harvest index %	55	40	41	50	49
Top weight (A) g m^{-2}	890	930	902	620	708
Weight gain (A-M) g m^{-2}	653	772	771	560	547
Decrease in non-grain dry matter (A-M) g m^{-2}	192	0	0	34	65

In many circumstances it is the leaf area that is most affected by treatment differences (e.g. in nitrogen nutrition), and for such cases it is valuable to use partial simulations in which observed values of green area index are substituted for simulated ones. Such partial simulations indicate whether the processes of dry matter production and partitioning have been modified by the experimental treatment. Figure 2b gives the dry matter production for such a partial simulation using the observed GAI values of the N3 crop in place of the simulated ones. The two simulations gave similar dry matter growth and were in good agreement with the observations until well after anthesis, but gave significantly more dry matter at maturity (Table 1). The observed crop appears to have a slightly greater maximum rate of dry matter production than the simulated for the same GAI values. However, a sudden cut-off to dry matter production was observed which was not paralleled by the partial simulation, and, as the GAI was still 5, this indicates a sudden check to the conversion efficiency of the crop.

The simulated grain yield was too small because both grain number and grain size were small (Table 1). The resulting decreased demand for assimilate during grain-fill meant that almost 100 g m^{-2} of the assimilate produced after anthesis was not used in grain growth. In contrast, the observed check to dry matter growth noted above amounted to post-anthesis weight increment that was 120 g m^{-2} small than that simulated. The crop made good this shortfall and th demand of a larger grain number by a decrease of 190 g m^{-2} in non-grain dry matter between anthesis and maturity.

A partial simulation may be used to study the performance of th N1 crop, for which nitrogen shortage substantially restricted leaf area growth and yield, and which the full model is not yet able to simulate. Figure 2c shows that the simulated maximum rate of dry matter production was close to that observed, although in the simulation the period of rapid dry matter increase began rather earlier. As with the N3 crop, the discrepancy between simulation an observation increases in July. The difference between the grain number for the N3 and N1 crops was 5080 grains m^{-2}, which is much greater than the difference between the simulated grain numbers, 16C grains m^{-2}. Grain number is determined in the model from the pre-anthesis dry matter increase of the crop, and as this increase was well simulated, the large difference observed for the N1 and N3 crop suggests that nitrogen nutrition directly affect grain number.

Other components in these simulations can be checked against th data sets. Willington and Biscoe (1983) showed that the differences in grain size between N1 and N3 were due to a 6% difference in the rate of grain filling. The partial simulations for these crops showed a comparable difference in final grain size, but this was associated with a difference in duration of filling not in rate (Fig. 3a). The simulated rate of grain filling was very close to that of the N1 crop, but the duration of the grain filling period wa about 7% too short. Thus both rate and duration of grain filling ar slightly underestimated in the model.

A further check on grain growth in the model can be made agains observations of grain weight on field crops for which the daily maximum and minimum temperatures during crop growth are available. The development sub-model can be used to define the grain-filling period and determine the mean temperature during grain-filling. The AFRC model grows grain for 240/(T-9) days at a maximum rate of 0.045 + 0.4 mg d^{-1} where T is the daily mean temperature in °C. To a first approximation T can be replaced by $\bar{T}$, the mean temperature

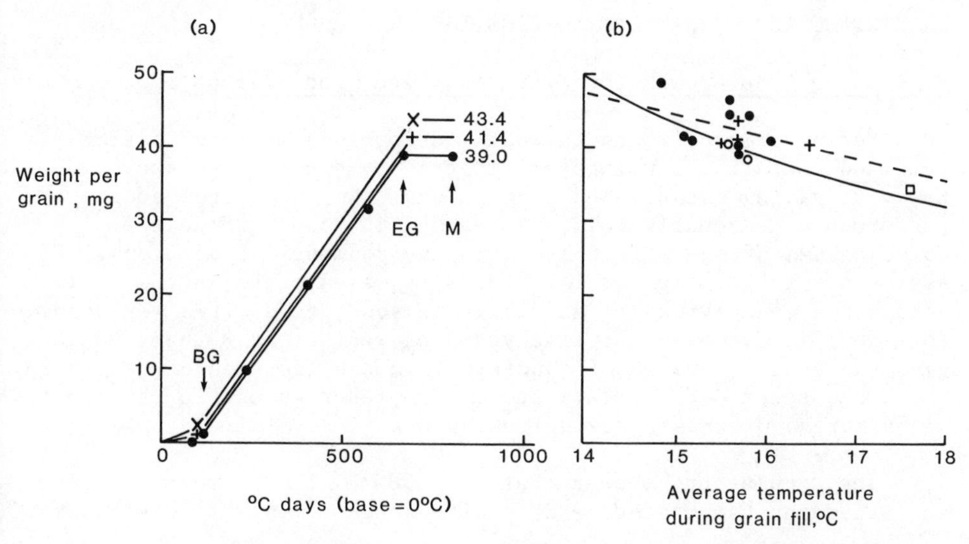

Fig. 3. (a) Grain growth for Broom's Barn crops of Fig. 2: observed N3 X——X, N1 +——+; simulated N3 and N1: ●——●. Simulated growth stages, abbreviations as in Fig. 1.
(b) Observed grain weights of winter wheat, cv. Avalon, grown at different sites plotted against average temperatures during the grain filling period, using computer simulations of the timing of that period. Experimental Husbandry Farms 1981/2 ●, Broom's Barn 1981/82 +, Rothamsted 1981/82 ○, Woburn 1982/83 □. The full line is for the equation:

$$\text{maximum grain weight} = (0.045\bar{T} + 0.4) \times 240/(\bar{T}-9) \text{ mg}$$

where $\bar{T}$ is the average temperature, °C. The dashed line (slope - 2.85 mg °C^{-1}) is the relationship found by Willington and Biscoe (1983) for the observed grain filling period at Broom's Barn.

during grain filling, to give the maximum potential grain weight as $(0.045\bar{T} + 0.4) \times 240/(\bar{T}-9)$ mg. This relationship is shown in Fig. 3b; though grain weights less than the potential maximum might be expected from actual crops, it is clear that many crops achieved larger grains.

This corroborates the detailed measurements at Broom's Barn which suggest that the model underestimates grain size. The dashed line in Fig. 3b gives the temperature sensitivity of grain size found by Willington and Biscoe (1983); this reduced temperature sensitivity (2.85 mg grain weight decrease for each one degree increase in temperature) fits the observed data at least as well as the version in the model.

PREDICTING AND EXTRAPOLATING FROM THE CROP MODEL

Simulating Crop Growth for Different Sites and Seasons

Further improvements to model performance may result from detailed comparisons with specific crops, as described above. However, having established a crop model that describes crop performance reasonably well, we can use it to determine the consequences of specific changes in environmental conditions. To assess the range of potential yields to be expected in the U.K. across the country and in different seasons, five sites were chosen for 1980/81, and also the five years of weather at Rothamsted that gave the broadest range of conditions in the last decade (Table 2). The sites spanned 6° of latitude and represented most of the north-south spread of wheat growing in the U.K.

The simulations showed that for 1980/81 the range of latitude and attendant temperature made a difference of only 14 days in the

Table 2. Details of site, sowing date, and days of growth for simulated crops

	Latitude	Sowing date	Days of growt	
			*S-M	TS-E
Mylnefield, Scotland	56.5°N	15.10.80	307	10
Caywood, Yorkshire	53.7°N	15.10.80	297	10
Sutton Bonington, Notts.	52.7°N	15.10.80	295	11
Littlehampton, Sussex.	50.8°N	15.10.80	293	11
Rothamsted, Herts.	52.0°N	15.10.80	302	11
" "	"	16.10.74	288	9
" "	"	16.10.75	278	8
" "	"	16.10.78	302	8
" "	"	16.10.81	291	9
Lincoln, New Zealand	43.0°S	5.5.82	277	11

*sowing S, maturity M, terminal spikelet TS, end of grain-fill EG.

total growing period and 8 days in the duration of the main period of dry matter production - from terminal spikelet to the end of grain-filling. Comparing different years at Rothamsted showed larger differences, 24 and 30 days for the same two periods.

The simulations had maximum GAI values varying from 5.3 to 7.2, with six of the nine between 5.7 and 6.1. For these crops, which all had 200 plants m^{-2}, the contribution of main stems to peak GAI varied little, and the main variation resulted from the different tillering patterns of the crops. The proportion of GAI from main stems can vary considerably with plant density (Fig. 4a), but the proportion of incident photosynthetically active radiation (PAR) that is intercepted follows the same relationship for different densities, sites and years. The asymptotic shape of the response is evident; between maximum GAI values of 5 and 10 the proportion of PAR intercepted increases by only 17%, from 70 to 87%.

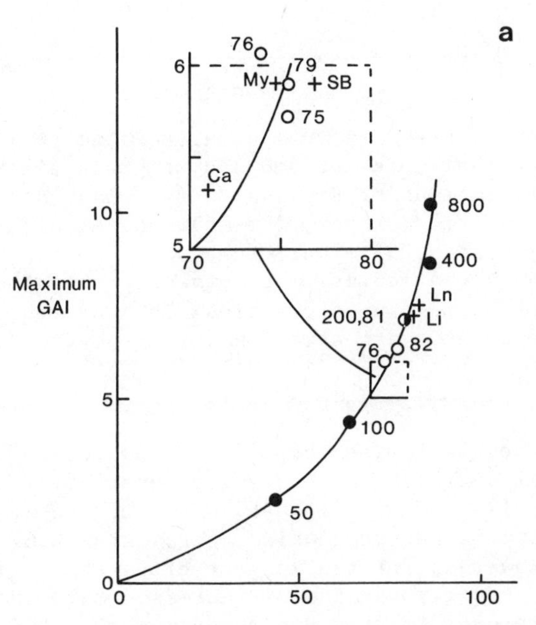

Fig. 4. (a) Maximum GAI values plotted against proportion of incident photosynthetically active radiation (PAR) intercepted between terminal spikelet (TS) and end of grain-fill (EG) for computer simulated wheat crops. The line was drawn through the points by eye. Key: plant density for Rothamsted, 50 to 800 plants m^{-2}, ●; sites, (Ca Caywood, Li Littlehampton, Ln Lincoln N.Z., My Mylnefield, SB Sutton Bonington), +; year of harvest for Rothamsted, 1975 to 1982, o.

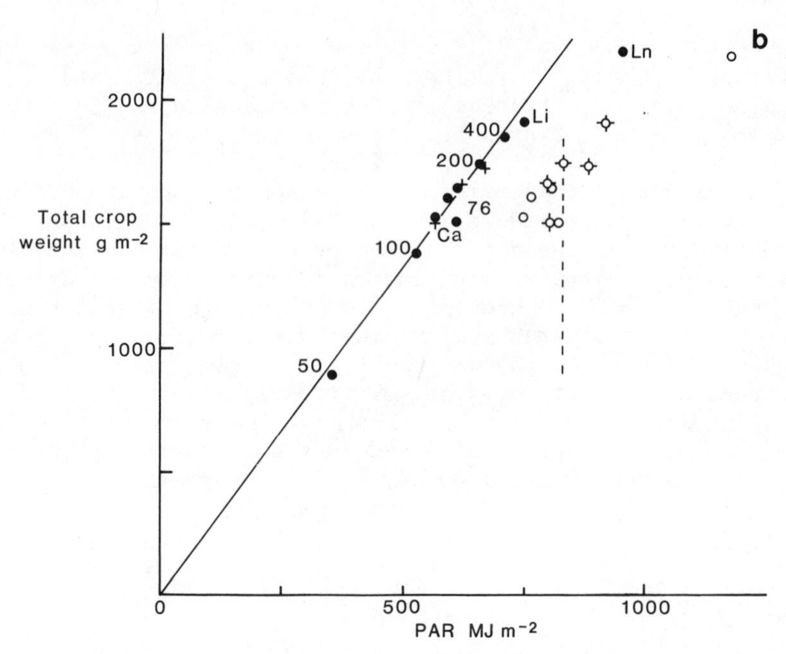

Fig. 4. (b) PAR received (○-◇-) and intercepted (● +) between terminal spikelet and end of grain filling for the simulations of Fig. 4a. Key: plant densities, 50 to 400 plants m^{-2}; sites, (Ca Caywood, Li Littlehampton, Ln Lincoln N.Z.); year of harvest, 1976. The dashed line shows the radiation received by the simulations using different plant densities. The continuous line represents a conversion efficiency of 2.67 g MJ^{-1}.

The relationship between amounts of PAR received and intercepted between terminal spikelet and the end of grain-filling and the dry matter produced is shown in Fig. 4b. Most of the crops have an efficiency of dry matter production approaching 2.67 g MJ^{-1} of PAR intercepted. However, for the largest crops the efficiency is somewhat less, probably because of increased maintenance respiration, which is proportional to crop dry weight. Dry matter production after terminal spikelet varied from 15.0 to 17.4 t ha^{-1} for the five years at Rothamsted, and for the five sites from 15.0 to 19.1 t ha^{-1}. Thus the interaction of temperature, day-length and radiation alone could result in differences across the country in total dry matter production for healthy unstressed crops in a particular year of 4 t ha^{-1} and for a single site over 10 years of 2.5 t ha^{-1}.

The potential for grain yield is made up of the dry matter produced after anthesis together with a proportion of the weight of

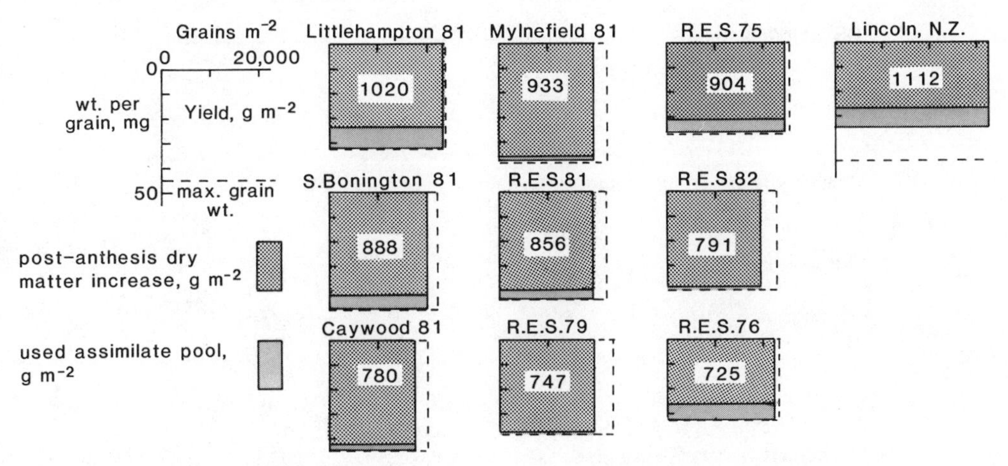

Fig. 5. Computer simulated grain yields, g m^{-2} D.M., in terms of grain number and weight. The shaded areas represent the proportion of yield equivalent to the sum of the weight of post-anthesis dry matter and used assimilate pool. Unshaded areas outlined by dashed lines show the increased grain numbers needed to utilise all the assimilate pool.

stems and leaves at anthesis. This is illustrated in Fig. 5 as the areas bounded by the dashed lines. The shaded areas are the simulated grain yields expressed as the product of the numbers of grains produced and their weight. Of the nine UK simulations, only the Littlehampton crop realised its potential yield; all the others reached their maximum grain weights, but because their grain numbers were small, part of the reserve pool of assimilate was not used. Five of the nine simulations gave grain yields exceeding 8.5 t ha^{-1} (10 t ha^{-1} at 85% dry matter). Of the remaining four, only the 1975/6 crop did not have the potential to reach this large yield; the others had a potential grain yield of 8.5 t ha^{-1} or more.

These examples show how the model may be used to investigate site and weather effects on wheat growth in the U.K. The comparisons with observed crops indicate that both dry matter production and grain growth rates may be rather underestimated by the model, but with this proviso the yields and their ranges may be taken as a guide to what might be expected from real crops growing at their potential maximum rates.

Simulating Growth in Other Environments

The model has been developed in close conjunction with observations on U.K. winter wheat crops (Weir et al., 1984). Its application to other environments must be approached with caution,

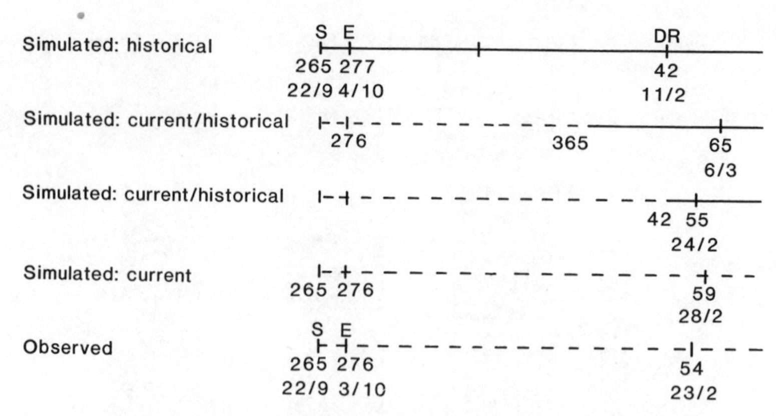

Fig. 6. Observed growth stages (Thorne and Wood, 1983) for winter wheat, cv. Avalon, at Rothamsted in 1981/82 and computer simulated dates using 10-year average weather (historical ——) and 1981/82 weather (current -----). Abbreviations for growth stages as in Fig. 1. The numbers represent dates in Julian and calendar days.

but can provide valuable insights into potential weaknesses in its formulation, as well as exemplifying the contrast between environments.

A simulation of wheat growth at Lincoln, New Zealand for 1982/83 was run to test performance in a similar climate, though at a substantially different latitude (Table 2, Figs 4 and 5). The total growing period was short, although the period from terminal spikelet to end of grain-filling was relatively long. Total dry matter production, 22.5 t ha^{-1}, was much greater than for the U.K. simulations as a result of the higher incident radiation, used with a similar conversion efficiency, over a similar period of growth. For the Lincoln crop, the dry matter increase in the period of ear growth prior to anthesis was particularly large, and resulted in 32,500 grains m^{-2}. This exceptionally large value is probably a consequence of taking the ear growth period to be a fixed proportion of the terminal spikelet to anthesis period. This latter is influenced by photoperiod, so that the shorter days in Lincoln immediately prior to anthesis extend the duration of ear growth and hence increase grain number. The large grain number may therefore be an artefact of the model: if the simulated grain numbers had been 25,000 grains m^{-2}, then the individual grains would have reached their temperature controlled maximum weight of 44.5 mg.

There are therefore interesting consequences of simulating wheat growth in an environment outside the range for which the model has been developed, but any further discussion must await the growth of a crop of Avalon at Lincoln.

Predicting Current Growth

The examples used so far deal with past weather and past crops. However, models can also be used to assist the understanding and management of crops during their growth. For our model this use has been restricted so far to phenological development, illustrated here by predicting the date of the double ridge stage for a crop grown at Rothamsted. The occurrence of this stage marks the start of a relatively safe period for spraying wheat with growth-regulator herbicides (Tottmann, 1977). The double ridge stage may occur over a range of as much as 60 days in different seasons (Porter, 1983) and can only be detected by frequent crop sampling followed by apical dissection and inspection of the apex under a microscope. Clearly, if the date could be predicted by computer model there could be considerable management benefits. Figure 6 shows how this might have been done with our crop model for winter wheat (cv. Avalon) sown on 22 October 1981. Using 10-year average weather, a date of 11 February 1982 was predicted for double ridges. The simulation was updated early in 1982 by using the recorded actual weather up to 31 December 1981. The early winter had been colder than average so the predicted date moved to 6 March 1982. When the simulation was repeated again using actual weather data up to 11 February a new date of 24 February was given, very close to the observed date of 23 February. Finally, when actual weather data for February was available, the model predicted the double ridge stage for 28 February. A check then would have shown the crop had reached double ridges and was safe to spray, as indeed it had been for several days because the final prediction was not quite accurate.

This is one example where a simulation result would be of considerable benefit to a crop manager, though it does require weather data to be readily available and quickly updated.

CONCLUSIONS

This paper has covered a number of the ways in which a model can be used to focus our understanding of how wheat crops grow. These exercises are possible even when the model deals only with crop growth not limited by water or nutrient shortage, and can be made with parts of the model as well as the whole.

If a sub-model is small, it is relatively easy to use; the development sub-model is a case in point. It is about 350 lines of Fortran in length and the program can be supplied as a listing or on a disk or tape to run on a microcomputer. It is easily understood and requires relatively simple inputs of daily maximum and minimum temperature, site latitude and sowing date. However the plant parameters do vary with genotype, and we have considered only two cultivars, Hustler and Avalon. The determination of parameters for many different cultivars is a major task unless varietal differences can be grouped or follow predictable patterns.

The whole model is large (2000 lines of FORTRAN) and could be twice as large if responses to nitrogen and water were also included. Such large models have particular demands on their management. Few people will understand the purpose of all the lines of code, and particular care is necessary if modifications are not to lead to the introduction of errors; potential users need adequate documentation, explaining both how to run the program for their purposes, and the premises on which the model is built. Papers such as this can help by indicating the limitations and strengths of the model.

ACKNOWLEDGEMENTS

Dr. J. R. Porter and Dr. J. H. Rayner have been fully involved in the development and use of the model. We thank Dr. Anne Willington, Broom's Barn Experimental Station, and Dr. P. V. Biscoe, of Broom's Barn and now of ICI p.l.c., for making data from their experiments available to us.

REFERENCES

Baker, C. K., Gallagher, J. N., and Monteith, J. L., 1980, Daylength change and leaf appearance in winter wheat, Pl. Cell Environ., 3:285.

Fischer, R. A., 1980, Wheat, in: "Potential productivity of wheat crops", International Rice Research Institute, Manila.

Gales, K., 1983, Yield variation of wheat and barley in Britain in relation to crop growth and soil conditions - a review, J. Sci. Fd Agric., 34:1085.

Gallagher, J. N., 1979, Field studies of cereal leaf growth. I. Initiation and expansion in relation to temperature and ontogeny, J. exp. Bot., 30:625.

Porter, J. R., 1983, Modelling stage development in winter wheat, in: "Aspects of Applied Biology, 4, Influence of environmental factors on herbicide performance and crop and weed biology", Association of Applied Biologists, Wellesbourne.

Porter, J. R., 1984, A model of canopy development in winter wheat, J. agric. Sci., Camb., 102:383.

Prew, R. D., Church, B. M., Dewar, A. M., Lacey, J., Penny, A., Plumb, R. T., Thorne, G. N., Todd, A. D., and Williams, T. D., 1983, Effects of eight factors on the growth and nutrient uptake of winter wheat and on the incidence of pest and diseases, J. agric. Sci., Camb., 100:363.

Thorne, G. N., and Wood, D. W., 1983, Factors limiting yield of wheat: growth and development, Rothamsted Experimental Station Report for 1982, Part 1:20.

Tottman, D. R., 1977, The identification of growth stages in winter wheat with reference to the application of growth-regulator herbicides, Ann. appl. Biol., 87:213.

Weir, A. H., Bragg, P. L., Porter, J. R., and Rayner, J. H., 1984, A winter wheat crop simulation model without water or nutrient limitations, J. agric. Sci., Camb., 102:371.

Willington, V. B. A., and Biscoe, P. B., 1983, Growth and development of winter wheat, I.C.I. Agricultural Division Financed Research Programme Annual Report 2, Broom's Barn Experimental Station.

APPLICATION OF REMOTELY SENSED DATA IN WHEAT GROWTH MODELLING

E. T. Kanemasu, G. Asrar, and M. Fuchs

Kansas State University
Manhattan, Kansas

INTRODUCTION

Non-contact measurements play a major role in agricultural production. For example, the observant farmer makes maximum use of his visual senses to schedule irrigation, fertilizer, insecticide and herbicide applications. In research, the scientist requires a quantitative interpretation of many of these same casual observations e.g. leaf colour, leaf area index, plant development, leaf orientation. Direct measurements of these plant and canopy attributes are laborious and require a sampling strategy to account for the high spatial variability within plots and fields. Remote sensing offers an opportunity to provide a rapid means of estimating these canopy attributes.

Simulation models of crop growth are popular among the scientific community. The development, validation, testing and application of these models require quantitative estimates of plant characteristics (leaf area, leaf orientation and plant development) for assessing transpiration and photosynthetic rates. Because of the tedious nature of these measurements, complete data sets for model development and testing are not readily available.

The operational use of simulation models to predict crop production over large regions is aided by within-season measurements of crop parameters, such as leaf area and soil moisture. The inaccuracy of these predictions is not so much the result of the failure of the simulation model to reflect the plant response adequately as of the unavailability of necessary information from the field. Field information, such as rainfall and its effectiveness in increasing soil water, spatial (vertical and horizontal) non-

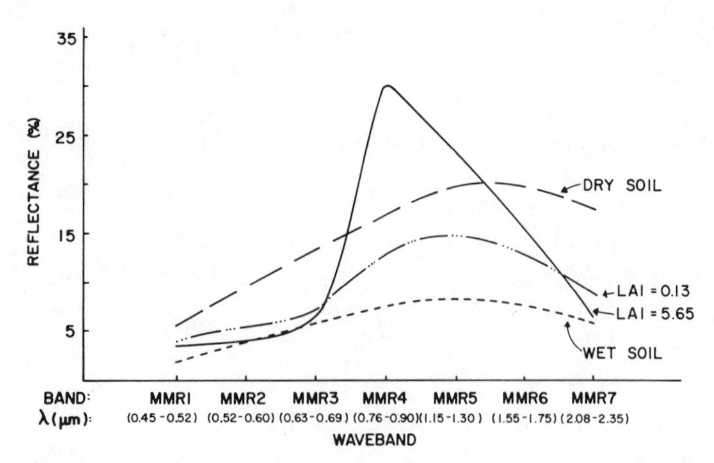

Fig. 1. Typical spectral reflectances of soil and vegetation. The wavebands are for the modular multispectral radiometer (MMR).

uniformity of soil texture, cultural practices (planting date, seeding rate, cultivar, fertility status, irrigation amounts, etc.), disease, insects and weed infestations, are largely unavailable over most of the agricultural regions of the world.

Remote sensing offers a new capability for evaluating crop condition. Because the reflectance spectra of soil and green plant leaves differ (Fig. 1), especially in the visible and near-infrared wavelengths, the amount of soil covered by vegetation is well correlated with the ratio of the reflectance in near-infrared and visible wavelengths. Leaf pigments strongly absorb in the blue and red wavelengths; therefore, the leaf reflectance in the visible waveband is low. The cellular structure of vegetation enhances the reflectance and transmission of the near-infrared radiation. The sharp increase in near-infrared reflectance of leaves with increasing wavelength contrasts with the nearly linear increase in visible and near-infrared reflectances for soil.

Pollock and Kanemasu (1979) have used data from the LANDSAT multispectral scanner (MSS) to assess leaf area indices (LAI) of wheat. The MSS has four sensors - two in the visible (500-600 nm and 600-700 nm) and two in the near-infrared (700-800 nm and 800-1100 nm) wavelength bands. Pollock and Kanemasu measured LAI in several commercial wheat fields and correlated the MSS digital counts from those fields with LAI. The major operational problem with LANDSAT data is the 18 day repeat cycle of the polar-orbiting satellite. Because of cloud conditions, there can be a gap of several weeks or months between observations of an area. The recent development of portable multispectral radiometers have made it possible to re-examine small plots and fields several times (weather permitting) using aircraft, land vehicles or ground-mounted platforms.

Using data taken by a hand-held radiometer, Hatfield et al. (1984a) found that the ratio of near-infrared to red reflectances could be used to estimate leaf area index of wheat. The regression equation obtained (with r = 0.91) was

$$LAI = -0.422 + 0.279\ (\rho_{nr}/\rho_r) \tag{1}$$

where ρ_{nr} and ρ_r are the canopy reflectances in the near-infrared and red wavelengths respectively. Kumar and Monteith (1981) and Hatfield et al. (1984b) have found that combinations of the near-infrared and red reflectances could be used to estimate light interception. Asrar et al. (1983) and Fuchs et al. (1984) have developed the methodology to estimate LAI from interception calculations.

The proportion of light intercepted, p, can be represented (Monsi and Saeki, 1953) as:

$$p = 1 - e^{-k'.LAI} \tag{2}$$

where k' is the leaf angular shape coefficient, dependent upon the leaf angle distribution of the canopy. Asrar et al. (1984) found a linear relationship between the normalised difference (ND) and light interception in the 400-700 nm wavelength, where ND is given by:

$$ND = \frac{\rho_{nr} - \rho_r}{\rho_{nr} + \rho_r}. \tag{3}$$

In the early part of the growing season, when leaf area index is low, ND is nearly equivalent to the ratio of the reflectances in the near-infrared and red wavelengths. ND increases with increasing LAI to a nearly constant value at LAI values above 5.0. Asrar et al. (1984) found the following relationship between ND and light interception in wheat (with r = 0.97):

$$ND = 0.087 + 0.798p \tag{4}$$

ND can be estimated from spectral reflectance measurements, and p can then be calculated using Equation (4). Equation (2) can be inverted to estimate LAI assuming a suitable leaf angle distribution. Asrar et al. (1984) and Fuchs et al. (1984) found that a spherical leaf angle distribution was a reasonable approximation for the wheat canopies studied; therefore, the angular shape coefficient is given as:

$$k' = 0.5/\cos\eta \tag{5}$$

where η is the solar zenith angle.

Thus we have two methods for estimating the LAI from spectral measurements: a regression equation given in Equation (1) and an inversion technique (Equation (2)).

It is the purpose of this paper to illustrate how spectral data can be used in a wheat growth and yield model. For this report, we have selected a wheat growth model developed at Kansas State University (Hodges and Kanemasu, 1977; Brakke and Kanemasu, 1979) to illustrate the use of spectral data as a model input. Simulated estimates of grain yield will be compared using observed LAI, simulated LAI, and spectrally-derived LAI.

WINTER WHEAT GROWTH MODEL

The flow diagram in Fig. 2 shows how daily data input influences the various plant processes in the model. The growth and evapotranspiration sub-model was developed for use over large areas; hence, a minimum of daily weather data were required - solar radiation, maximum and minimum air temperature and precipitation. It is envisaged that each of these meteorological parameters can or will be able to be assessed from meteorological satellites (e.g. GOES, NOAA, etc.). The yield model has been applied to 67 commercial fields across the USA with reasonable success (r = 0.81) using ground-based meteorological observations.

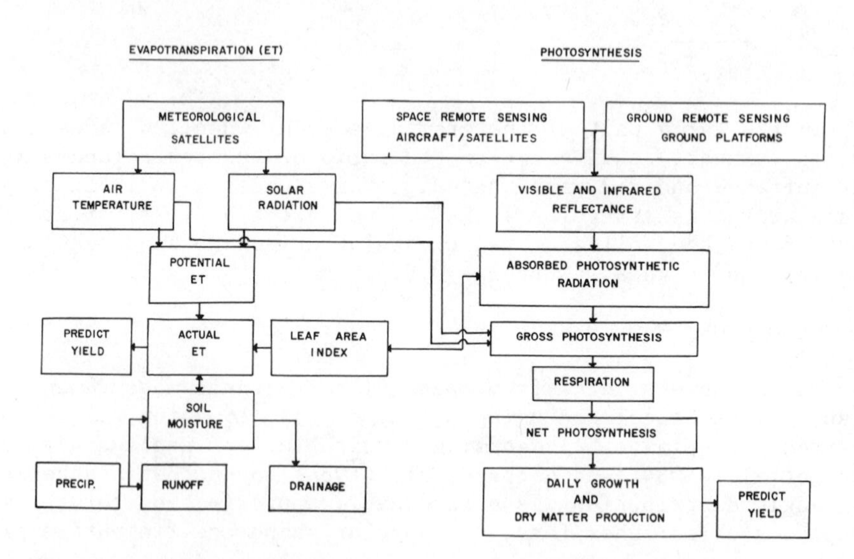

Fig. 2. Conceptual diagram illustrating the use of remote sensing information as direct inputs into evapotranspiration and plant growth models.

The wheat yield model begins by estimating evaporation and transpiration. The transpiration is estimated using LAI, the potential evapotranspiration, and soil water content in the root zone. Daily potential transpiration (T_p) is computed from the potential evapotranspiration (PET) and leaf area

$$T_p = (1 - \tau)\ PET$$

where $\tau = \exp(-0.737\ LAI)$. The evaporation from the soil surface is separated into an energy-limited and a soil-limited stage (Ritchie, 1972). The energy-limited stage is dependent upon the LAI since more energy reaches the soil under conditions of low LAI than high LAI.

A water balance is computed from the precipitation, evapotranspiration, and soil water profiles to allow estimates of soil water content, deep percolation and runoff. The soil water contents are then used to modify daily transpiration and photosynthetic rates. It is assumed that, if less than 35% of the available soil water content is present in the root zone, both processes are adversely affected and therefore the daily potential photosynthetic and transpiration rates are multiplied by a stress coefficient (between 1.0 and 0). If the available soil water is greater than 35% the stress coefficient is 1.0, and below 35% it declines linearly to zero when there is no available water.

For plant development in the model, observed dates of specified growth stages (emergence, double ridge, stem extension, boot, heading, anthesis, soft dough, and physiological maturity) can be entered. If observations are not available, a biological time clock is used to predict growth stages (Feyerherm and Paulsen, 1976). This time scale is dependent upon both daylength and temperature.

The LAI model consists of a tillering and a leaf area submodel. The components of LAI (tillers, leaves and leaf area) are estimated as a function of air temperature, growth stage and soil water status. During vegetative growth, increases in the number of tillers per plant are driven by accumulated temperature (thermal time, Tt, °C d) with a base temperature of 0°C. Tillers are formed in the axils of leaves and tiller prophylls. The tillering pattern follows a Fibonacci series (Friend, 1965) at 100 °C d intervals; this means that, the number of tillers in any interval will be the sum of the number of tillers in the two preceding intervals. Tillering can be reduced by multiplying Tt by a competition factor based on the previous day's estimate of LAI (i.e. competition for light) and/or a water stress factor based on soil water content in the root zone. On the day when the double ridge stage occurs, tillering is assumed to stop and the final number of surviving or spike-bearing tillers is calculated. Excess tillers are then lost at an exponential rate over a period of 640 °C d.

Throughout most of the growing season, the number of leaves per plant is determined from the number of tillers per plant assuming about three leaves per tiller. Prior to the double ridge stage, LAI and leaf area per leaf are calculated from an exponential relationship between leaf area per plant and the number of leaves per plant. During the winter, leaf area per leaf is reduced by temperatures below -3°C. From the double ridge until the boot stage, increases in leaf area per leaf are dependent on Tt. After boot stage, LAI is reduced each day reaching a value of zero at soft dough stage. Details of the model are given by Baker (1982).

The gross photosynthesis is dependent upon light absorption, temperature and soil water content. The amount of absorbed light is a function of the extinction coefficient and LAI. Both high and low temperatures adversely affect photosynthesis. Respiration is dependent upon temperature and gross photosynthesis. Net photosynthetic rates are used to estimate total dry matter accumulation.

The grain numbers and grain weights are estimated separately and multiplied to predict final grain yield (Brakke and Kanemasu, 1979). Grain numbers are estimated from the accumulated net photosynthate production between emergence and anthesis; grain weight is obtained by the accumulation of net photosynthate between anthesis and soft dough stage.

MATERIALS AND METHODS

The data used in this study were obtained from experiments conducted at two separate geographical locations. The first experiment was at the US Water Conservation Laboratory in Phoenix, Arizona (112° 01' W, 32° 26' N) in 1979-1980. The treatments were five planting dates and typically three irrigation levels on spring wheat (*Triticum aestivum* Desf., cv. Produra). Planting, emergence and irrigation dates, and the quantity of water applied to each treatment are presented in Table 1. The nominal irrigation amount at each date was 8 cm. On two occasions each week, six plants were selected randomly from each treatment to determine leaf area and total dry matter production. The second experiment was conducted during the 1982-83 season near Manhattan, Kansas (96° 37' W, 39° 09' N). Winter wheat (*Triticum aestivum* L., cv. Newton) was planted in east-west and north-south orientations at a row spacing of 18 cm. Twenty-five plants were harvested at random from each plot at least once a week, and used for leaf area and total dry matter determination.

Canopy spectral reflectance was measured using a boom-truck assembly equipped with an Exotech Model 100-A radiometer (Asrar et al., 1984). The Exotech radiometer detects in four wavebands corresponding to the Multispectral Scanner (MSS) aboard LANDSAT

Table 1. Planting, emergence, irrigation dates (day of year) and total applied water for the 1979-80 experiment on Produra wheat at Phoenix, Arizona.

Treatment	Planting date	Emergence date	Irrigation dates	Irrigation and rain (cm)
A1	271	275	271,278,289,334	51.7
B1	271	275	271,278,290,313,345	67.6
C1	271	275	272,278,290,324	52.9
A2	295	302	295,302,334	47.4
B2	295	302	296,302,324	46.1
C2	295	302	296,302,317,345	52.7
A3	318	330	319,351	34.7
B3	318	330	319,079	35.7
C3	318	330	320,351	34.3
A4	352	363	353	24.1
B4	352	363	353,079,099	43.7
C4	352	363	354,098	36.4
A5	036	047	039,100	32.1
B5	036	047	039,079,079,106,123,134	59.8
C5	036	047	039,093,114	41.0

satellites. In the following analysis only data from the red (600-700 nm) and one of the near infrared (800-1100 nm) wavebands were used for both Manhattan and Phoenix experiments. Reflectance was measured several times between mid-morning and mid-afternoon during days with clear sky conditions. Reflected radiation from a white barium sulphate panel was measured concurrently. Canopy reflectance factors were determined as the ratio of the reflectance of the canopy and the white panel.

RESULTS AND DISCUSSION

Figures 3-6 illustrate the LAI values obtained by direct observations (measured), model simulation, or spectrally derived by regression (Equation (1)) and inversion (Equation (2)) methods. Figures 3 and 4 are for winter wheat at Manhattan, Kansas sown at two row spacings (18 and 36 cm). At the normal row spacing (18 cm), the LAI estimates from all the techniques agree reasonably well, especially considering the standard deviation of the measured LAI

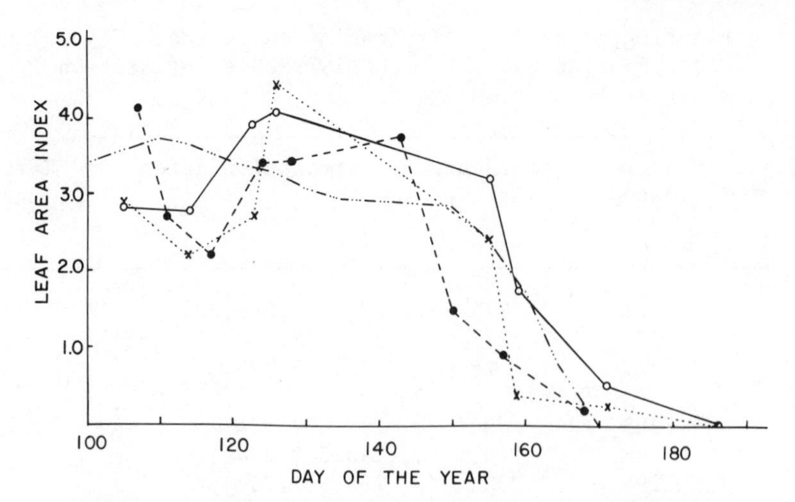

Fig. 3. Seasonal trends in leaf area index of winter wheat cv. Newton (18 cm row spacing, N-S rows) at Manhattan, Kansas in 1983 as estimated from direct measurements (●---●), regression (Equation (1); ○——○), inversion (Equation (2); X---X) and as simulated (-----).

(normally between 0.5 and 1.0). Figure 4 shows the LAI values (observed and estimated) for the wide row spacing (36 cm). The simulated values of LAI underestimate the observed values; the LAI model does not perform as well at the higher spacings probably because it was developed for 18 cm row spacings. When the yield

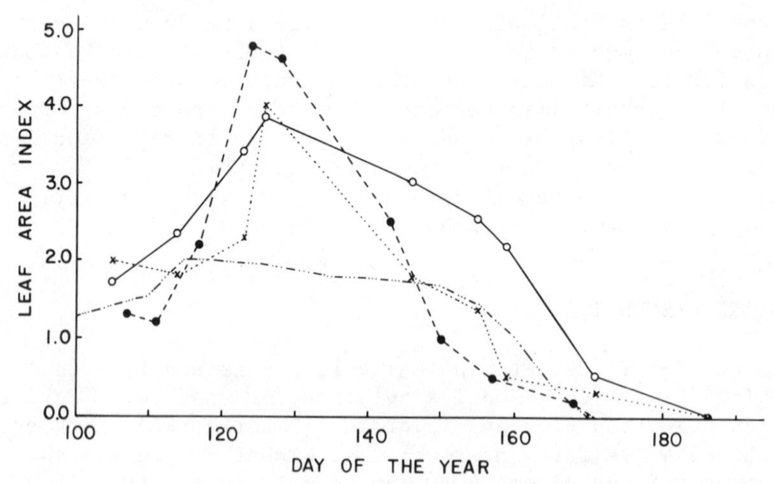

Fig. 4. Seasonal trends in leaf area index of cv. Newton (36 cm row spacing) at Manhattan, Kansas (see Fig. 3 for details).

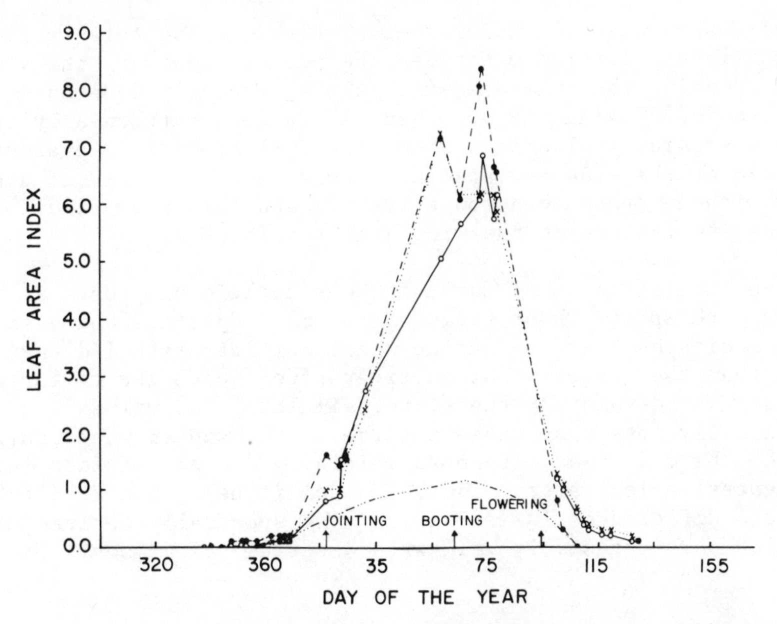

Fig. 5. Seasonal trends in leaf area index of spring wheat cv. Produra, at Phoenix, Arizona in 1979-1980 (see Fig. 3 for definition of symbols).

model was applied to these data, the yield estimated using simulated LAI and that using spectrally-derived LAI agreed with modelled yields

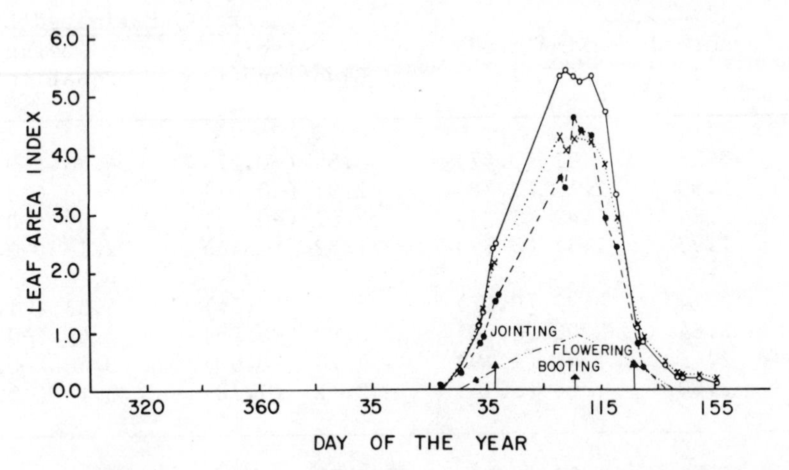

Fig. 6. Seasonal trends in leaf area index of cv. Produra at Phoenix, Arizona in 1979-1980 (see Fig. 3 for definition of symbols).

obtained using observed LAI values (Table 2). The t-tests indicate no significant difference between the means. However, the yields estimated using the simulated LAI values generally underestimate those predicted using the observed LAI values, particularly at the wide row spacings. The inversion procedure appears to overestimate the yield at the wide row spacings; this technique assumes a uniform canopy without pronounced row structure and this assumption probably explains its failure at the wider row spacings.

The simulation model for LAI shows additional problems when dealing with spring wheat (Figs. 5 and 6). Detailed analysis (not shown) indicates that the spring wheat cultivars studied have larger leaves than the winter wheat cultivars from which the tillering sub-model was developed; therefore, simulated LAI values are significantly less than those observed. The winter wheat cultivars on which the model was developed initially did not possess large differences in leaf area. The LAI model is being modified to include genotypic differences in leaf area. The spectrally-derived LAI values compare reasonably well with observed LAI values. The yields

Table 2. Comparison between simulated grain yields of Newton winter wheat for different methods of obtaining leaf area index (LAI) in 1982-83 season at Manhattan, Kansas. Differences between yield obtained from observed LAI and other techniques are given in brackets. Treatment numbers refer to the row spacings in cm.

	Simulated yield (t ha^{-1})			
			Spectrally Estimated LAI	
Plot No.	Observed LAI	Simulated LAI	Using Equation (1)	Using Equation (2)
EW-18	3.28	2.81 (0.47)	3.55 (-0.27)	4.03 (-0.75)
EW-36	2.90	2.12 (0.78)	2.91 (-0.01)	3.84 (-0.94)
EW-53	2.84	1.49 (1.35)	3.07 (-0.23)	3.59 (-0.75)
EW-72	1.93	1.31 (0.62)	1.52 (0.41)	2.13 (-0.20)
NS-18	3.24	3.07 (0.17)	3.20 (0.04)	4.33 (-1.09)
NS-36	2.64	2.06 (0.58)	2.92 (-0.28)	3.60 (-0.96)
NS-53	2.63	1.44 (1.19)	2.70 (-0.07)	3.48 (-0.85)
NS-72	2.60	1.66 (0.94)	2.32 (0.28)	3.10 (-0.50)
Mean	2.76	2.00	2.77	3.51
	SD	0.28	0.29	0.29
t(0.01)=3.45	t	2.77	0.06	2.65

estimated from the simulated LAI values are significantly lower than those estimated from observed LAI values as anticipated from LAI comparison (Table 3). The yields estimated from regression and inversion procedure are similar, though the inversion procedure results in slightly greater yields than the regression. We would anticipate that for uniform canopies the two estimates from spectral data would provide comparable yield results.

Table 3. Comparison between simulated grain yields of Produra spring wheat for different methods of obtaining leaf area index (LAI) in 1979-80 season at Phoenix, Arizona. Differences between yield obtained from observed LAI and other techniques are given in parenthesis.

	Simulated yield (t ha^{-1})				
				Spectrally Estimated LAI	
Plot No.	Observed LAI		Simulated LAI	Using Equation (1)	Using Equation (2)
A-1	2.08		1.59 (0.49)	1.71 (0.37)	1.96 (0.12)
B-1	2.55		1.68 (0.87)	1.81 (0.74)	2.05 (0.50)
C-1	2.66		1.79 (0.87)	1.94 (0.72)	2.17 (0.49)
A-2	5.00		2.91 (2.09)	4.72 (0.28)	5.15 (-0.15)
B-2	4.94		2.83 (1.11)	4.64 (0.30)	5.18 (-0.24)
C-2	4.74		2.87 (1.87)	7.18 (-2.44)	7.18 (-2.44)
A-3	4.63		2.88 (1.75)	4.48 (0.15)	4.56 (0.07)
B-3	4.20		3.05 (1.15)	6.00 (-1.80)	6.27 (-2.07)
C-3	4.20		2.92 (1.28)	5.12 (-0.92)	5.44 (-1.24)
A-4	4.30		2.56 (1.74)	4.20 (0.10)	4.28 (0.02)
B-4	5.35		2.59 (2.76)	6.44 (-1.09)	6.67 (-1.32)
C-4	5.26		2.61 (2.65)	5.97 (-0.71)	6.18 (-0.92)
A-5	2.61		1.88 (0.73)	3.13 (-0.52)	3.22 (-0.61)
B-5	4.08		1.86 (2.22)	4.92 (-0.84)	4.83 (-0.75)
C-5	3.27		1.82 (1.45)	4.27 (-1.00)	4.34 (-1.07)
Mean	3.99		2.39	4.44	4.63
		SD	0.31	0.51	0.51
t(0.01)=2.98		t	5.26	0.88	1.25

CONCLUSIONS

Spectral data have been used to assess LAI as an input into a growth and yield model and the results show that spectrally-derived LAI is an improvement over our leaf area simulation model. The leaf area simulation model was less accurate at wide row spacings and with plants that had large leaves. These problems occurred because the leaf area model was used to estimate LAI under conditions outside those included in the data base from which the model was developed. This emphasises the need for collecting data bases over a wide range of cultural and environmental conditions. In addition, these examples illustrate the merit of being able to assess LAI from remotely sensed data (e.g. aircraft or spacecraft) during the current growing season.

At present, the following problems exist in using satellite data: 1) during cloud cover, optical data are not available; 2) the LANDSAT repeat cycle is 16-18 days; and 3) satellite data are expensive and often not immediately available.

There is optimism about the future of earth observations. Satellite payloads are being designed to carry microwave sensors, that can "see through" clouds, and pointable sensors. Further, the high temporal resolution (short repeat cycle) required for agricultural or vegetal observations is being recognised as a critical element. The availability of satellite data to the user community is being reviewed critically before the satellite is launched instead of after. These elements of an earth observing system will play an important role in assessing crop production over agricultural regions of the globe.

REFERENCES

Asrar, G., Fuchs, M., Kanemasu, E. T., and Hatfield, J. L., 1984, Estimating absorbed photosynthetic radiation and leaf area index from spectral reflectance in wheat, Agron. J., 76:300.

Baker, J., 1982, Modeling tiller production and components of leaf area in winter wheat as affected by temperature, water and plant population, M.S. Thesis, Kansas State University, Manhattan, Kansas.

Brakke, T. W., and Kanemasu, E. T., 1979, Estimated winter wheat yields from LANDSAT MSS using spectral techniques, 13th International Symposium on Remote Sensing of the Environment, Ann Arbor, Michigan.

Feyerherm, A. M., and Paulsen, G. M., 1976, A biometeorological time scale applied to winter wheat, in: "Agronomy Abstracts of the 68th Annual Meeting", American Society of Agronomy, Madison, Wisconsin.

Friend, D. J. C., 1965, Tillering and leaf production in wheat as affected by temperature and light intensity, Can. J. Bot., 43:1063.

Fuchs, M., Asrar, G., Kanemasu, E. T., and Hipps, L. E., 1984, Leaf area estimates from measurements of photosynthetically active radiation in wheat canopies, Agric. Forest Meteorol., (in press).

Hatfield, J. L., Asrar, G., and Kanemasu, E. T., 1984, Intercepted photosynthetically active radiation estimated by spectral reflectance, Rem. Sens. Environ., 14:65.

Hatfield, J. L., Kanemasu, E. T., Asrar, G., Jackson, R. D., Pinter, P. J., Jr., Reginato, R. J., and Idso, S. B., 1984, Leaf area estimates from spectral measurements over various planting dates of wheat, Int. J. Rem. Sens., (in press).

Hodges, T., and Kanemasu, E. T., 1977, Modelling daily dry matter production of winter wheat, Agron. J., 69:974.

Kumar, M., and Monteith, J. L., 1981, Remote sensing of crop growth, in: "Plants and the Daylight Spectrum", H. Smith, ed., Academic Press, London.

Monsi, M., and Saeki, T., 1953, Uber den lichtfaktor in den pfanzengesellschaften und seine bedeutung fur die stoffproduktion, Jap. J. Bot., 14:22.

Pollock, R. B., and Kanemasu, E. T., 1979, Estimating leaf area index of wheat with LANDSAT data, Rem. Sens. Environ., 8:307.

Ritchie, J. T., 1972, Model for predicting evaporation from a row crop with incomplete cover, Water Resour. Res., 8:1204.

STATISTICAL MODELLING OF WINTER YIELD AT A REGIONAL SCALE

M. Guerif,* O. Philipe,‡ and R. Delécolle*

*INRA Station de Bioclimatologie
Montfavet, France

‡INRA Laboratoire de Biométrie
Montfavet, France

INTRODUCTION

In the late sixties and early seventies many statistical crop-weather models appeared, especially for wheat. For various reasons (non-reliability, lack of biological sense, etc.) they were not satisfactory and so, in the late seventies, mechanistic integrated models of wheat production were developed. These are much more satisfactory in terms of biological sense but cannot yet be extrapolated to all genotypes and field conditions and therefore to the "regional scale", where crop conditions are contrasted over large regions.

Wheat yield prediction at a regional scale requires data representative of the whole region and continues to use statistical models. These models, especially regression models, have of course the potential to be improved by integrating more analytical concepts. In this paper we discuss the major improvements so far, which have been:

i) reduction of spatial unreliability by limiting geographical areas for model application ("regionalization");

ii) introduction of agronomic factors;

iii) use of information on the formation of the yield components where possible.

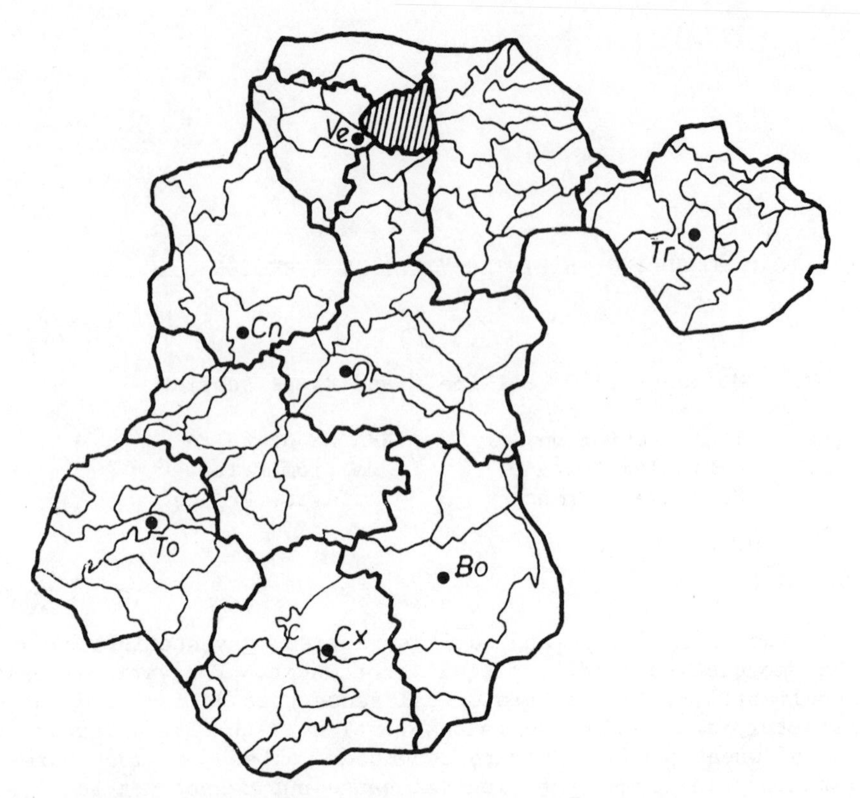

Fig. 1. Eleven Department and seven climatic stations (●) of the "Centre" and "Ile de France" French regions.

THE DATA

There must be numerous spatial and annual replicates in the data in order to cover a lot of agronomic and climatic situations, and thus to ensure that the model is representative of the region. Good information is also required on cultivars, cultural practices and yield components.

The data used were from surveys by the French Statistical Reporting Service (S.C.E.E.S.),carried out from 1971 to 1978 in 11 Departments of the "Centre" and "Ile de France" regions (Fig. 1) in order to determine departmental and national wheat yield (data from experimental plots were excluded). Each year, for each department a sample of 80 fields was available. For each field, information on cultivar, main cultural practices (preceding crop, approximate sowing date, soil tillage technique, fertilization) and measurements of yield and one of its components, the number of ears m^{-2} measured on two 1 m^2 areas in the field, were available.

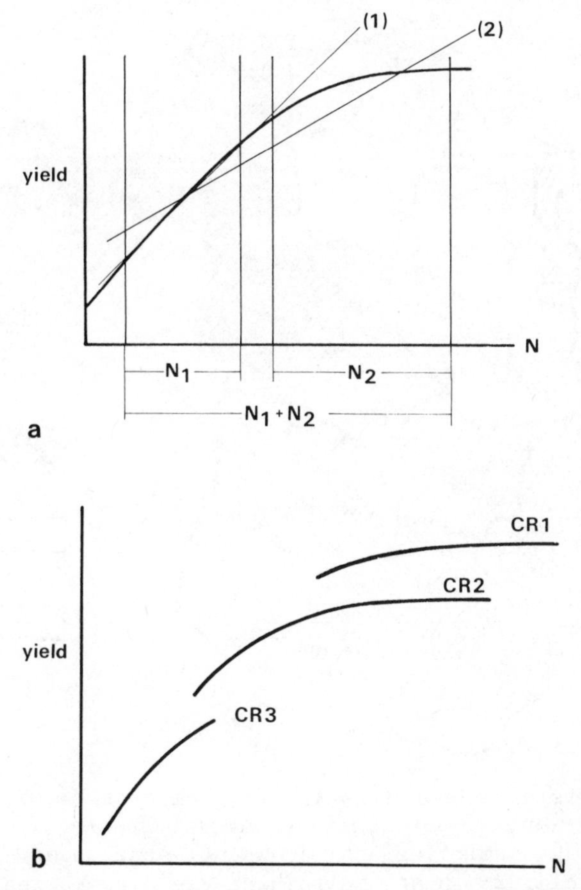

Fig. 2. Problems in statistical fitting of yield/predictor relationships, as illustrated by the response of yield to applied N (a) and crop rotation (b).

REGIONALIZATION

Statistical adjustments rely greatly on the data from which they are developed. This problem may easily be illustrated from linear regression. First, the slope, intercept and correlation coefficient of the linear relationship between yield and a predictor X, e.g. applied nitrogen or any climatic parameter, vary with the range of X (Fig. 2a). A straight line fitted in a situation where N lies in the range N_1 (line 1) will be a very poor predictor for a situation where N lies in the range N_2. The same for a line fitted in the range N_{1+2} (line 2) and applied to situations N_1 or N_2. Secondly, the predictors are generally numerous and they interact so that the response of yield to one predictor depends on the range of another,

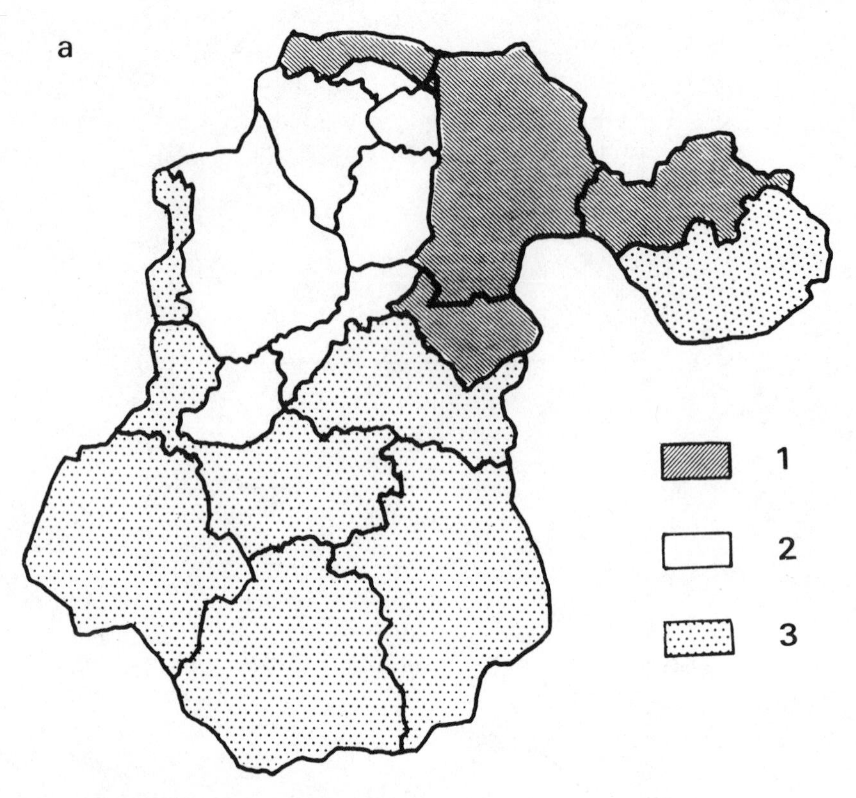

Fig. 3. Results of regionalization (T, temperature; R, solar radiation; E, potential evapotranspiration).
a) 3 agronomic regions: 1- wheat, sugar beet, maize/high yield; 2- wheat, maize/high yield; 3- cereals, oleaginous crops, forage/low yield.

e.g. the interaction between applied nitrogen and crop rotation (Fig. 2b). Here again a general fit for all crop rotations will not work for one specific situation.

These two cases of the instability of fitting show the interest in limiting areas so that predictors lie within reduced ranges and some factors are fixed. This is what we call regionalization. Such a regionalization was carried out for the cultivation system, which integrates a lot of soil, technical, climatic, social and economic factors. The regionalization can allow for implicit and hard-to-measure variables, which define regional potential. In our data, the cultivation system was characterized by the combined frequencies of previous crop and yield level; we chose homogeneous areas by a multivariate analysis of these frequencies. The results are mapped in Fig. 3a.

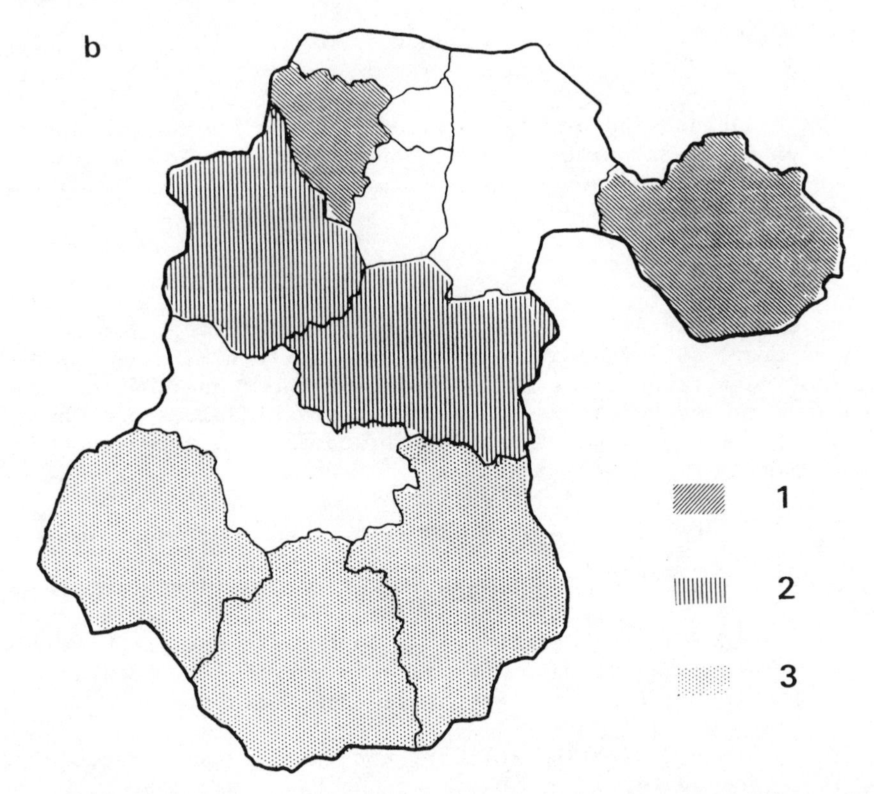

Fig. 3 (cont.) Results of regionalization (T, temperature; R, solar radiation; E, potential evapotranspiration).
b) 3 climatic regions: 1- low T, R, E_T/high soil moisture; 2- mean T, R, E_T, mean soil moisture; 3- high T, R, E_T/low soil moisture.

For the climatic aspect, such an analysis was carried on data from seven climatic stations. Assuming that a climatic station is representative of the department it belongs to, the results can be mapped (Fig. 3b); they are in good agreement with agronomic regions, proving - if that were necessary - the adaptation of agronomic technique to climate.

YIELD MODELLING ATTEMPTS INVOLVING YIELD COMPONENTS AND AGRONOMIC DATA

Number of ears per m^2, NE, is the only separately determined yield component in our data. One possible way of modelling yield, Y, is to use NE as a predictor. It can be determined at heading by field surveys and therefore provides a possible prediction of yield at that time. In the future, ear number may itself by predicted by other models.

The relationship for the whole data set between yield and number of ears observed is quite genotype-dependent, and for one genotype it is year- and location-dependent (Fig. 4). So, for a single genotype, the following model was used:

$$Y_{ij} = a_i \, NE_{ij} + b_i \, NE_{ij}^2 + E_{ij}$$

where i describes the 15 levels of an agronomic factor combining previous crop, soil tillage type and organic fertilization or 11 levels of departmental factor, j describes the number of replicates, a_i and b_i are constants and E_{ij} represents residual error. The observed collinearity between a_i and b_i coefficients led to the following model:

$$Y_{ij} = a_i \, NE_{ij} + (u - va_i) \, NE_{ij}^2 + e_{ij}$$

This model explains between 50 and 70% of the yield variability, depending on year, without using any climatic data. It could be improved by introducing climatic data to explain either the variation in the a_i coefficients, which are clearly affected by interannual (i.e. climatic) variability, or the residual error terms e_{ij}.

This form of model is, however, not entirely satisfactory, in that it is not evolutive - i.e. it cannot be improved by introduction of more analytical sub-models. Another way in which we can seek to model yield from the number of ears is to come back to the relationship:

Yield = Number of Ears x Grain Weight per Ear

For our purposes, the number of ears per unit ground area is known though it could also be estimated using other analytical models; the grain weight per ear is the complementary unknown yield component to be estimated.

From the first trials it seems that number of ears especially is influenced by agronomic factors. Therefore it could be expressed as a function of these factors and climatic data from sowing to heading (the heading date is not very variable and can be determined regionally from experimental plots) and grain weight per ear could be expressed as a function of climatic data from heading to harvest.

→

Fig. 4. Yield plotted against number of ears per m^2 (NE) for a) cv. Capitole and b) cv. Hardi data from 1972 for the agronomic regions 1 and 2 (see caption to Fig. 3a for definitions of the agronomic regions.

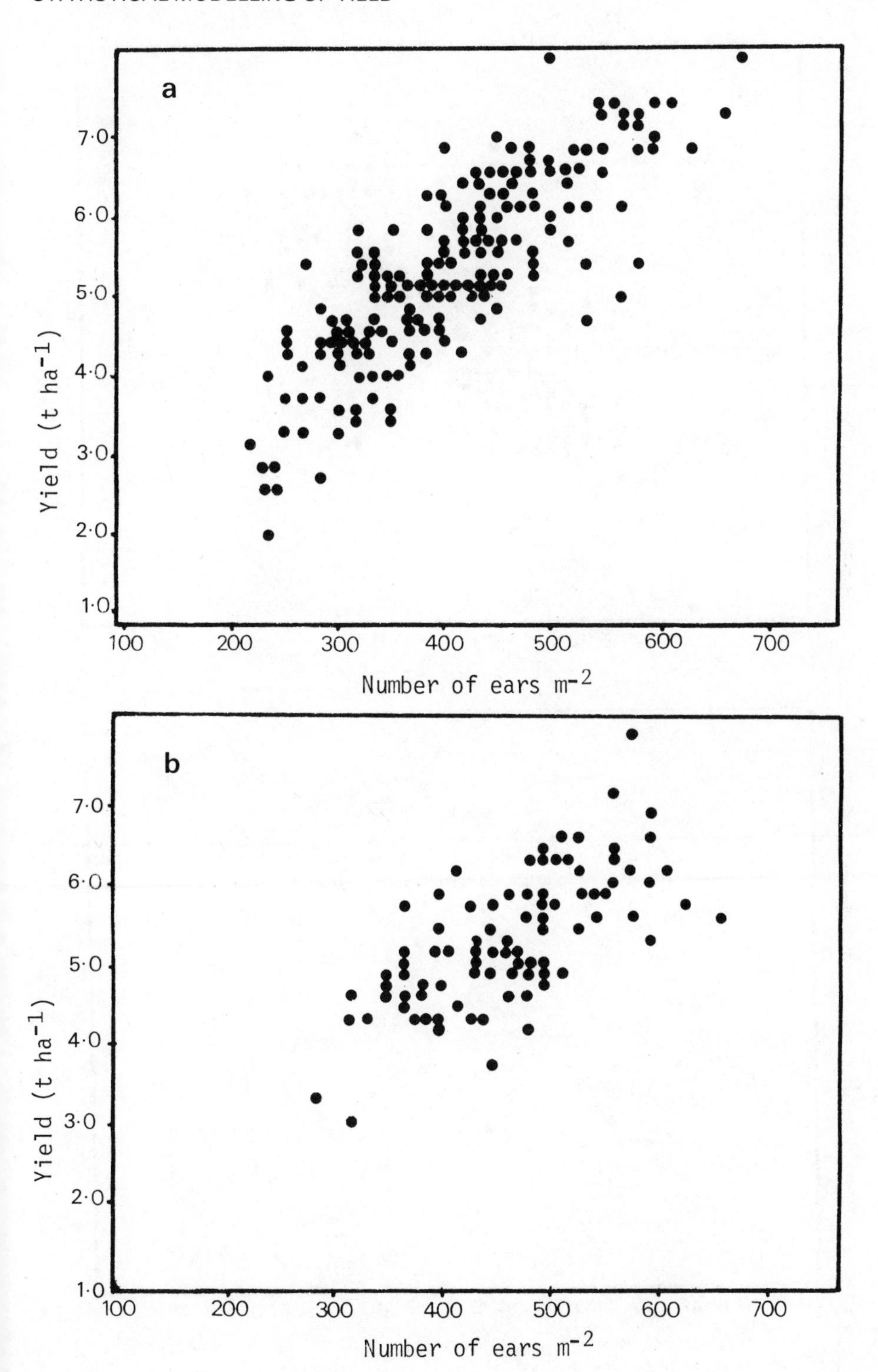
a
b
Yield (t ha^{-1})
1·0
2·0
3·0
4·0
5·0
6·0
7·0
100
200
300
400
500
600
700
Number of ears m^{-2}

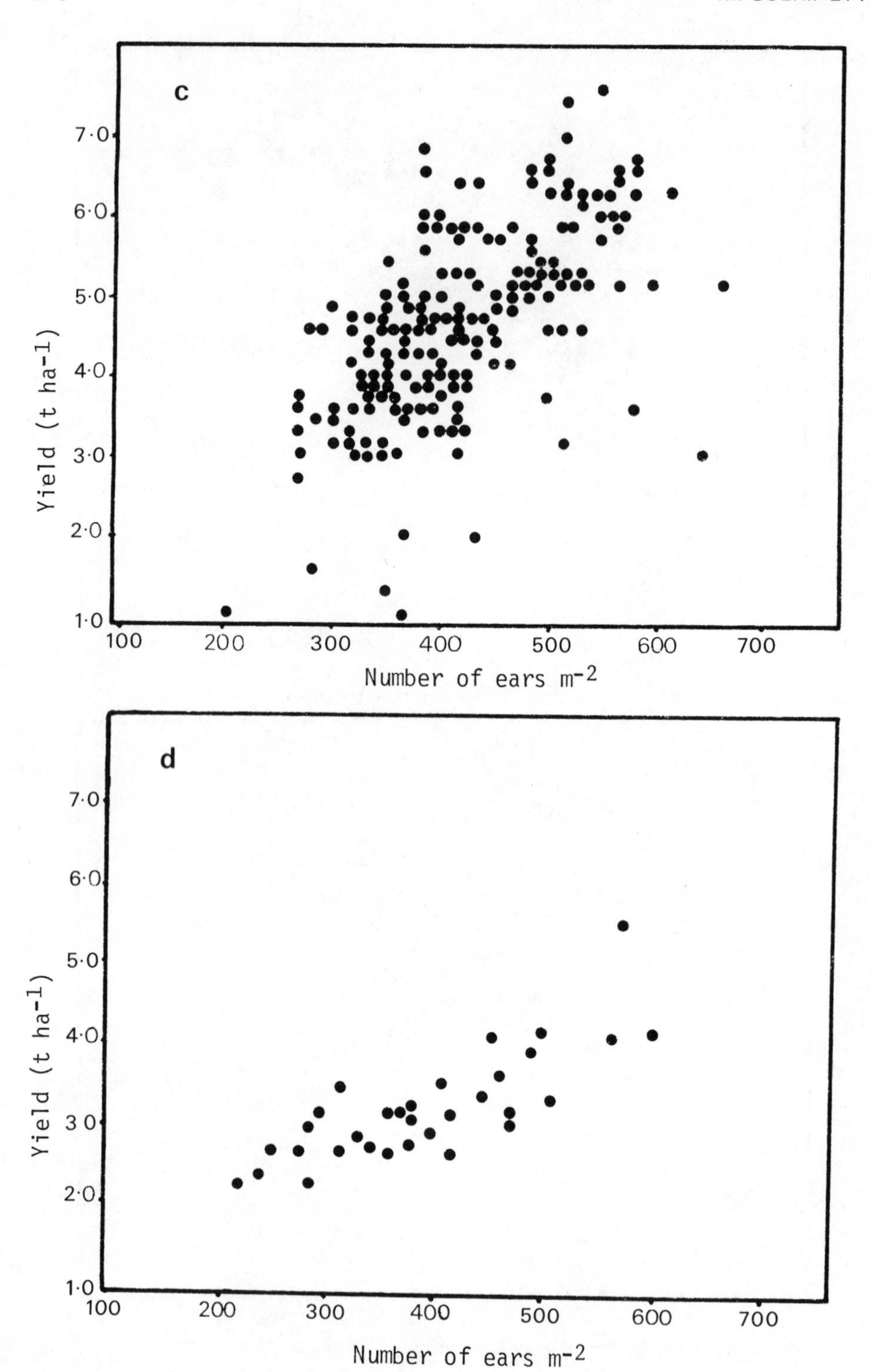
c
7·0
6·0
5·0
4·0
3·0
2·0
1·0
Yield (t ha-1)
100
200
300
400
500
600
700
Number of ears m-2
d
7·0
6·0
5·0
4·0
3 0
2·0
1·0
Yield (t ha-1)
100
200
300
400
500
600
700
Number of ears m-2

Fig. 4. (cont.) Yield plotted against number of ears per m^2 (NE) ← for cv. Capitole in 1977 for c) agronomic regions 1 and 2, d) agronomic region 3 (see caption to Fig. 3a).

These two components are of course dependent: they can interact through compensation or competition which will vary from one year to another. Further investigations are necessary to quantify this interaction and make the model applicable.

WHEAT GROWTH AND MODELLING: CONCLUSIONS

J. Moorby

AFRC Headquarters
London, UK

The strongest impression produced in the meeting, and reinforced by reading the papers, is of the large differences between the models described. This illustrates that there is no one perfect model of wheat, or any other crop. Models should be produced to serve a particular purpose, which can be quite limited. They may be totally unsuitable for any other purpose.

Various members of the meeting produced definitions of models and modelling. Two were:-

"a small imitation of reality" and
"an explicit statement of one's prejudices or imagination".

With regard to the latter, it is as well to remember the advice of the Canadian novelist, Robertson Davies - "I don't mean fantasy and poetry and moonshine; imagination is a good horse to carry you over the ground, not a flying carpet to set you free from probability". What this means in modelling terms is that models have to be realistic. Elegant mathematical speculations can be intriguing, but may be little more than that.

The modeller has to decide, therefore, the type of model required. It is obviously possible to construct models of varying degrees of complexity; from those which are extremely complex and physiologically realistic such as CERES (Chapters 27 and 28) and the AFRC model (Chapters 7 and 31), to simple empirical models which lead to no physiological insight but which can produce adequate descriptions or predictions. Examples of the latter are the use of a Beer's Law analogue for radiation interception by crop canopies or use of a logarithmic function to describe the decrease in root density through the soil profile. In addition, the various parts of a complex model have to be complementary. It is a waste of time to formulate one aspect of a model so that it can lead to overall predictions with an accuracy of say 5%, if another part of the model is so vague or imprecise that it can affect the output of the model by an order of magnitude. The realization that the levels of understanding of the parts of the system are so different can be very valuable as a means of directing future research, and is often cited as one of the benefits of the modelling approach. It is more likely, however, that the model merely confirms what is already known, or at least suspected, by the modeller.

A trend which can be seen in the literature, and was apparent in the workshop, is the progression with time from very simple models to very complex and back to simple. An example would be the development of early models of dry matter accumulation by crops in terms of net assimilation rate and leaf area, followed in the late 1960s and early 1970s by complex descriptions of radiation penetration into crop canopies, photosynthesis and respiration. More recently, attention has turned to the amount of radiation intercepted by the crop canopy and the efficiency of conversion of radiation to dry matter.

These changes have taken place not because the complicated models of physiological and physical processes were unsuccessful, but because they were often too complicated for regular use in whole crop models, and because simple models have proved to be adequate for many purposes. There is still interest in more complex models, but often as a means of studying individual processes or for validating simpler models. Considerable thought is still needed when applying models to agronomic problems: the agronomist may not need sophistication or accuracy (Chapter 23) but he does need flexibility and an appropriate model structure. It was apparent during the meeting that the current link between wheat modelling and its application to agronomy is often forced and unrealistic. At present clear distinctions are rarely drawn between descriptive and predictive models.

If due consideration is given to the limitations of the various approaches now available, and if they are carefully validated for the particular situation, it now seems possible to produce a variety of quite simple models of what can be termed the supply functions necessary for crop growth - radiation interception, photosynthesis, respiration, water use and nutrient uptake. Most of these can be treated as relatively simple chemical or physical processes. There were several examples of this type of approach in the workshop. Thus there are very many models of photosynthesis, of varying degrees of complexity, but the simple rectangular hyperbolic relationship between photosynthesis and intercepted radiation proposed by Rabinowitch is still found to be adequate by many workers. There seems, however, to be a greater awareness of the importance of reserve materials, both organic and inorganic, which can buffer the growing points against fluctuations in supply. As shown by Spiertz and Vos (Chapter 12) this effect is particularly marked towards the end of crop growth where the large post-anthesis transfer of nitrogen from leaves leads to a decline in the rate of photosynthesis (Chapter 14). The existence of genetic variation in this decline is beyond the scope of existing models, in which genetic characteristics have only been included in relation to plant development.

The treatment of respiration has tended to become more complex. The concepts of growth and maintenance respiration are well established. The work of Wilson and Robson (Robson, 1980) has shown that there is genetic variation in the efficiency of respiration, but the biochemical basis of these differences is still unclear. A possible explanation is now emerging from the work of Lambers (Chapter 11). He has shown that part of the total respiratory loss, which can be distinguished because of its sensitivity to cyanide or salicylhydroxamic acid (SHAM), is inefficient because there is little production of ATP. Despite these advances there is still no generally accepted way to model respiration for use in whole crop models. There is no general agreement, for example, on the temperature coefficients of growth and maintenance respiration. Because of this uncertainty, several members of the workshop suggested that it was safer to treat respiration as a single loss, usually about 50%, of the carbon fixed by photosynthesis, rather than give a spurious precision to a model by using a more complex formulation. Further, this general uncertainty could be one reason why the omission of root exudation from crop models has so little effect even though considerable amounts of material are lost from root systems (see below). In the post anthesis period, the large crop mass, the active transport processes and the declining net CO_2 uptake processes and the declining net CO_2 uptake by the canopy will all tend to highlight inadequacies in the definition of respiration in crop models.

The loss of water from crops as a function of the amount of radiation intercepted by the crop is now well understood and was not considered explicitly in the workshop. However, Passioura (Chapter 18), Hanks (Chapter 20) and Wilson (Chapter 21) discussed the concept of the water use efficiency of crops, i.e. the linear relationship between dry matter production and amount of the water transpired. Passioura reminded the workshop of the high root densities which are often achieved by crops. Even with a relatively low density of 0.1 cm cm^{-3}, the time constant for removal of the water, calculated using a simple root model, is only of the order of 3 to 6 days, so that all available water should be removed in a week or two. However, it is unusual for plants to absorb more than about 80% of the nominally available soil water. Growth of real roots in real soil also provides many difficulties for accurate modelling. The roots may preferentially grow in cracks or wormholes (Chapter 18), and certainly in many soils physical conditions can be seen to impede growth. As yet crop models do not treat this aspect mechanistically, e.g. some use an empirical root preference factor (Chapter 27) to account for the "friendliness" of the soil. Many conditions, for example oxygen shortage resulting from water logging (Chapter 19), can also restrict root growth and function: modellers will need to be cautious in their approach to the complex, heterogeneous system that is the soil.

There are several published methods of modelling nutrient uptake and its effect on the yield of a variety of crops. There are, however, still problems with several aspects of mineral nutrition. These were discussed by Vlassak (Chapter 22). It is disappointing, but maybe not too surprising in view of the complexity of the problem, that there is still no fully acceptable method for assessing nitrogen availability from the soil. Nor are there good rapid methods to measure tissue nitrogen as an indication of the current nitrogen status of the crop. Even if the latter were available they would be difficult to use because of the phloem mobility of many nutrients and our lack of understanding of the factors which regulate the direction of this movement.

There was not much disagreement about the simulation of these supply functions, but there were difficulties when the more biological inputs to crop models were considered. Nevertheless, it is in these areas of plant development, and the processes which underlie development, where most progress seems to have been made and where much of the discussion at the workshop was concentrated.

Various approaches to modelling the growth of leaves, tillers, ears and roots were described by Delécolle, Masle, Kirby and Klepper (Chapters 3, 4, 2 and 8 respectively). What is particularly interesting is the way in which common themes emerge from these four papers and others that were concerned with various aspects of development. For example, many workers now use thermal time (day degrees above a critical base temperature, which in most instances seems to be in the range 0 to 4°C, but can vary depending on the date of sowing or emergence) as a means of measuring developmental progression. This use carries with it the suggestion that the plant might be integrating over time something which is temperature sensitive. However, developmental processes are too complicated for there to be only one limiting process and it is probably safer to accept these relationships with thermal time simply as useful empirical correlations.

More fundamental correlations are those based on physiological and anatomical associations such as the relationships between the growth of roots, leaves and tillers described by Klepper (Chapter 8). Her observations that the application of fertilizer affected the number and length of root laterals, but not the time of their initiation, is a useful confirmation in the field of laboratory-based work. Topics not mentioned, however, were root senescence and death, or the significance of the root exudates. The continued production and senescence of roots is well documented in plants grown in nutrient solutions and grasses grown in the field. There is no reason to suppose that this turnover of roots does not occur in cereals also, with the production of new roots and the growth of existing roots sometimes continuing after anthesis even though by then the major sinks on the plant are the developing ears.

The loss of material from root systems - so-called root exudation - is now well established. What is more puzzling is its extent. No doubt there is a loss of cellular material by abrasion, and predation and grazing by the soil fauna. Soluble material is also lost and roots respire. In some of the experiments with wheat and barley the total loss has amounted to 20-40% of the carbon fixed by the plants. No models seem to take account explicitly of this major loss of dry matter and yet many can produce good simulations of the total amount of dry matter made by the crop. This suggests that the other aspects of the carbon balance are not as accurate as might have been hoped.

Several models now seem to be reasonably successful in predicting the timing of various developmental stages, but there is less success in the simulation of the subsequent growth. The reason for this is the almost complete absence of any understanding of the controls which determine the allocation of mobile nutrients between competing meristems. Modellers have tended to rely therefore on simple empirical relationships, rather than a more mechanistic approach. Two attempts at the latter approach were described in the workshop. The first, by Hansen and Svendsen (Chapter 17), is a development of a model which described partitioning into structural and storage components within a single organ. The second, by Thorpe (Chapter 16), is a completely new approach which describes movement into and out of a series of pools in the plant in terms of the "potentials" and "permeances" of the pools. Both of these approaches will need further independent testing to determine their more general utility. A point raised in discussion which is important in relation to partitioning is why do some organs stop growing, for example cereal grains and potato tubers. Is it simply because they are formed from a specific number of cells which have a limited capability for growth, or is there a limitation imposed by the supply of assimilates (Charles-Edwards, Chapter 26). The concept of a strong coupling between nutrient supply and the rate of differentiation and the survival of cells or organs will influence the future formulation of many models.

These and other problems were unresolved. They would form the basis of another workshop, but real progress towards their resolution is probably dependent on more experimentation rather than an increased effort on modelling *per se*.

REFERENCE

Robson, M. J., 1980, A physiologist's approach to raising the potential yield of the grass crop through breeding, in: "Opportunities for Increasing Crop Yields", R. G. Hurd, P. V. Biscoe and C. Dennis, eds., Pitman, London.

PARTICIPANTS

Austin, R. B.	Plant Breeding Institute, Maris Lane, Trumpington, Cambridge CB2 2LQ, U.K.
Belford, R.K.	Plant Protection Division, Imperial Chemical Industries plc., Fernhurst, Haslemere, Surrey GU27 3JE, U.K.
Belmans, C.	Katholieke Universiteit Leuven, Kardinaal Mercierlaan 92, B 3030 Leuven (Heverlee), Belgium
Biscoe, P.V.	Imperial Chemical Industries plc., Agricultural Division, Fertilizer Sales Department, P.O. Box 1, Billingham, Cleveland PS23 1LB, U.K.
Cavazza, L.	Istituto di Agronomia Generale e Coltivazioni Erbacee, University of Bologna, Via Filippo Re 6-8, 40126 Bologna, Italy
Charles-Edwards, D. A.	The Cunningham Laboratory, CSIRO, 306 Carmody Road, St. Lucia, Queensland 4067, Australia
Day, W.	Physiology and Environmental Physics Department, Rothamsted Experimental Station, Harpenden, Herts. AL5 2JQ, U.K.
Delécolle, R.	I.N.R.A., Station de Bioclimatologie, Centre de Recherches Agronomiques d'Avignon, Domaine Saint-Paul, 84140 Montfavet, France
Fischer, R. A.	CSIRO Division of Plant Industry, P.O. Box 1600, Canberra City, ACT 2601, Australia
Gallagher, J. N.	Department of Plant Sciences, Lincoln College, Canterbury, New Zealand
Godwin, D. C.	International Fertilizer Development Center, P.O. Box 2040, Muscle Shoals, Alabama 35660, U.S.A.
Goudriaan, J.	Department of Theoretical Production Ecology, Agricultural University, P.O. Box 430, Wageningen, The Netherlands

Guerif, M.	I.N.R.A., Station de Bioclimatologie, Centre de Recherches Agronomiques d'Avignon, Domaine Saint-Paul, 84140 Montfavet, France
Hanks, R. J.	Department of Soil Science and Biometeorology, (College of Agriculture), Utah State University, Logan, Utah 84322, U.S.A.
Hansen, G. K.	Department of Crop Husbandry and Plant Breeding, Royal Veterinary and Agricultural University, Forsogsgarden Hojbakkegard, Agrovej 10, DK-2630 Taastrup, Denmark
Hunt, L. A.	Department of Crop Science, University of Guelph, Guelph, Ontario, Canada
Kanemasu, E. T.	Evapotranspiration Laboratory, Department of Agronomy, Kansas State University, Waters Annex, Manhattan, Kansas 66506, U.S.A.
Kirby, E. J. M.	Plant Breeding Institute, Maris Lane, Trumpington, Cambridge CB2 2LQ, U.K.
Klepper, B.	USDA-ARS-WR, Columbia Plateau Conservation Research Center, P.O. Box 370, Pendleton, Oregon 97801, U.S.A.
Lambers, H.	Lab. voor Plantenfysiologie, Universiteit Groningen, Haren (Gr), Postbus 14, 9750 AA, Kerklaan 30, 9751 NN, The Netherlands
Masle, J.	I.N.R.A-AGRONOMIE, 16 Rue Claude-Bernard, 75231 Paris Cedex 05, France
Meyer, W. S.	CSIRO Centre for Irrigation Research, Private Mail Bag, Griffith, New South Wales, 2680, Australia
Moorby, J.	Agricultural and Food Research Council, 160 Great Portland Street, London W1N 6DT, U.K.
Olsen, C. C.	Government Research Station, Ronhave, Hestehave 20, DK 6400 Sonderborg, Denmark
Otter, S.	USDA-ARS-SR, Grassland, Soil and Water Research Laboratory, P.O. Box 748, Temple, Texas 76503, U.S.A.
Passiours, J. B.	CSIRO Division of Plant Industry, P.O. Box 1600, Canberra City, ACT 2601, Australia
Porter, J. R.	Long Ashton Research Station, Long Ashton, Bristol BS18 9AF, U.K.
Rayner, J. H.	Soils and Plant Nutrition Department, Rothamsted Experimental Station, Harpenden, Herts. AL5 2JQ, U.K.

Ritchie, J. T.	USDA-ARS-SR, Grassland, Soil and Water Research Laboratory, P.O. Box 748, Temple, Texas 76503, U.S.A.
Spiertz, J. H. J.	Research Station for Arable Farming and Field Production of Vegetables, P.O. Box 430, 8200 AK Lelystad, The Netherlands
Thorne, G. N.	Physiology and Environmental Physics Department, Rothamsted Experimental Station, Harpenden, Herts. AL5 2JQ, U.K.
Thorpe, M. R.	Physics and Engineering Laboratory, DSIR, Private Bag, Lower Hutt, New Zealand
Uehara, G.	Department of Agronomy and Soil Science, 2500 Dole Street, Krauss Hall 22, Honolulu, Hawaii 96822, U.S.A.
Verstreaten, L. M. J.	Laboratory of Soil Fertility and Soil Biology, Katholieke Universitat Leuven, Kardinaal Mercierlaan 92, 3030 Leuven (Heverlee), Belgium
Vlassak, K.	Laboratory of Soil Fertility and Soil Biology, Katholieke Universitat Leuven, Kardinaal Mercierlaan 92, 3030 Leuven (Heverlee), Belgium
Vos, J.	Centrum voor Agrobiologisch Onderzoek, Bornsesteeg 65, P.O. Box 14, 6700 AA, Wageningen, The Netherlands
Watts, D. G.	USAID (MIAC-MAROC), Casablanca, Department of State, Washington DC 20523, U.S.A.
Weir, A. H.	Soils and Plant Nutrition Department, Rothamsted Experimental Station, Harpenden, Herts. AL5 2JQ, U.K.
Wilson, D. R.	Crop Research Division, DSIR, Private Bag, Christchurch, New Zealand

AUTHOR INDEX

Numbers underlined indicate the page on which references are given in full.

SUBJECT INDEX